Data Structures and Algorithms in Graphics and Geometry

Data Structures and Algorithms in Graphics and Geometry

by

Veer Chanda | Ramandeep Bhavsar

Wordpen Academics
Publishing Science That Matter

ISBN: 978-93-6084-272-7

Published by : **WORDPEN ACADEMICS**
4736/23, Ansari Road, Darya Ganj,
New Delhi – 110 002
Phone: +91-011-35004336
E-mail: info@wordpenacademics.com

Printed at **:** **Replika Press Pvt. Ltd.**

Exclusive Distributor: ***Bio-Green Books, New Delhi***

Preface

"Data Structures and Algorithms in Graphics and Geometry" represents a pioneering effort to bridge the realms of computer science, mathematics, and visual arts, offering readers a comprehensive exploration of the foundational principles, techniques, and applications that underpin the synthesis of data structures and algorithms with graphics and geometry.

In an era marked by the proliferation of digital technologies and the increasing demand for immersive visual experiences, the integration of data structures and algorithms with graphics and geometry holds immense promise for innovation and creativity across diverse domains. From computer-aided design and animation to virtual reality, gaming, and computational geometry, this convergence of disciplines empowers practitioners to push the boundaries of visual storytelling, simulation, and interactive design.

As editors of this volume, we have endeavored to assemble a diverse array of perspectives from researchers, educators, and practitioners at the forefront of this interdisciplinary frontier. Through a blend of theoretical insights, practical methodologies, and illustrative examples, this book offers readers a roadmap for navigating the intricate landscapes of data-driven graphics and geometric computation.

Our overarching goal in presenting this book is to inspire curiosity, spark innovation, and foster collaboration among scholars and practitioners across disciplines. By showcasing the latest advancements in data-driven graphics and geometric algorithms, we aim to empower readers to harness the transformative potential of these technologies for creative expression, problem-solving, and visual communication.

We extend our sincere gratitude to all the contributors who have generously shared their expertise, experiences, and passion for the intersection of data structures, algorithms, graphics, and geometry. It is our hope that this volume will serve as a catalyst for interdisciplinary dialogue, interdisciplinary research, and interdisciplinary action, driving forward the frontier of knowledge and creativity in the digital age.

Veer Chanda | Ramandeep Bhavsar

Contents

Part-I

Programming environments for motion, graphics, and geometry

The purpose of an artificial programming environment

A *program* can be *designed* with the barest of tools, paper and pencil, or in the programmer's head. in the realm of such informal environments, a program design may contain vague concepts expressed in an informal notation. Before he or she can execute this program, the programmer needs a *programming environment*, typically a complex system with many distinct components: a computer and its operating system, utilities, and program libraries; text and program editors; various programming languages and their processors. Such real programming environments force programmers to express themselves in formal notations.

Programming is the realization of a solution to a problem, expressed in terms of those operations provided by a given programming environment. Most programmers Work in environments that provide very powerful operations and tools.

The more powerful a programming environment, the simpler the programming task, at least to the expert who has achieved mastery of this environment. Even an experienced programmer may need several months to master a new programming environment, and a novice may give up in frustration at the multitude of concepts and details he or she must understand before writing the simplest program.

The simpler a programming environment, the easier it is to write and run small programs, and the more Work it is to write substantial, useful programs. in the early days of computing, before the proliferation of programming languages during the 1960s, most programmers Worked in environments that were exceedingly simple by modern standards: Acquaintance with an assembler, a loader, and a small program library sufficed. The programs they wrote were small compared to what a professional programmer writes today. The simpler a programming environment is, the better suited it is for learning to program. Alas, today simple environments are hard to find! Even a home computer is equipped with complex software that is not easily ignored or bypassed. For the sake of education it is useful to invent artificial programming environments. Their only purpose is to illustrate some important concepts in the simplest possible setting and to facilitate insight. Part I of this book introduces such a toy programming environment suitable for programming graphics and motion, and illustrates how it can gradually be enriched to approach a simple but useful graphics environment.

Textbooks on computer graphics. The computer-driven graphics screen is a powerful new medium for communication. Visualization often makes it possible to present the results of a computation in intuitively appealing ways that convey insights not easily gained in any other manner. To exploit this medium, every programmer must master basic visualization techniques. We refer the reader interested in a systematic introduction to computer graphics to such excellent textbooks as [BG 89], [FDFH 90], [NS 79], [Rog 85], [Wat 89], and [Wol 89].

1

Reducing a Task to Given Primitives: Programming Motion

A robot car, its capabilities, and the task to be performed

Some aspects of programming can be learned without a computer, by inventing an artificial programming environment as a purely mental exercise. The example of a vehicle that moves under program control in a fictitious landscape is a microcosmos of programming lore. In this section we introduce important concepts that will reappear later in more elaborate settings.

The environment. Consider a two-dimensional square grid, a portion of which is enclosed by a wall made up of horizontal and vertical line segments that run halfway between the grid points (Exhibit 1.1). A robot car enclosed within the wall moves along this grid under computer control, one step at a time, from grid point to adjacent grid point. Before and after each step, the robot's state is described by a location (grid point) and a direction (north, east, south, or west).

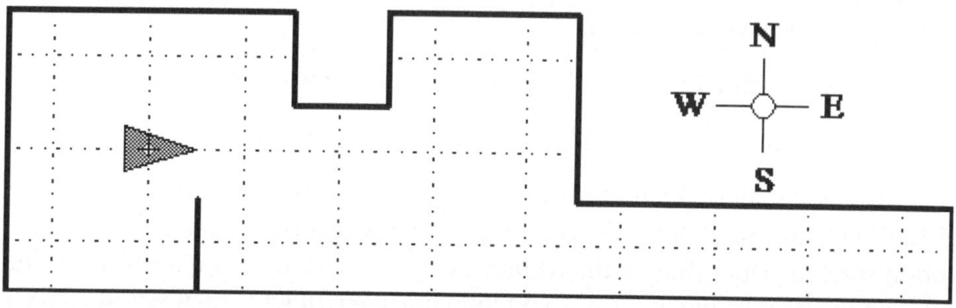

Exhibit 1.1: The robot's crosshairs show its current location on the grid.

The robot is controlled by a program that uses the following commands:

```
   left               Turn 90 degrees counterclockwise.
   right                 Turn 90 degrees clockwise.
   forward            Move one step, to the next grid point in front of
you
   goto #             Send program control to the label #.
   if touch goto #    If· you are touching a wall to your front, send
                      program control to the label #.
```

A program for the robot is a sequence of commands with distinct labels. The labels serve merely to identify the commands and need not be arranged either consecutively or in increasing order. Execution begins with the first command and proceeds to successive commands in the order in which they appear, except when flow of control is redirected by either of the goto commands.

Example

The following program moves the robot forward until it bumpsin to a wall:

```
1 if touch goto 4
2 forward
3 goto 1
4 { there is no command here; just a label }
```

In developing programs for the robot, we feel free to use any high-level language we prefer, and embed robot commands in it. Thus we might have expressed our wall-finding program by the simpler statement

```
while not touch do forward;
```

and then translated it into the robot's language.

A program for this robot car to patrol the walls of a city consists of two parts: First, find a wall, the problem we just solved. Second, move along the wall forever while maintaining two conditions:

- ❖ Never lose touch with the wall; at all times, keep within one step of it.
- ❖ Visit every spot along the wall in a monotonic progression.

The mental image of walking around a room with eyes closed, left arm extended, and the left hand touching the wall at all times will prove useful. To mirror this solution we start the robot so that it has a wall on its immediate left rather than in front. As the robot has no sensor on its left side, we will let it turn left at every step to sense the wall with its front bumper, then turn right to resume its position with the wall to its left.

Wall-following algorithm described informally

Idea of solution: Touch the wall with your left hand; move forward, turning left or right as required to keep touching the wall.

Wall-following algorithm described in *English:* Clockwise, starting at left, look for the first direction not blocked by a wall, and if found, take a step in that direction.

Let us test this algorithm on some critical configurations. The robot inside a unit square turns forever, never finding a direction to take a step (Exhibit 1.2). in Exhibit 1.3 the robot negotiates a left-hand spi$_k$e. After each step, there is a wall to its left-rear. in Exhibit 1.4 the robot enters a blind alley. At the end of the alley, it turns clockwise twice, then exits by the route it entered.

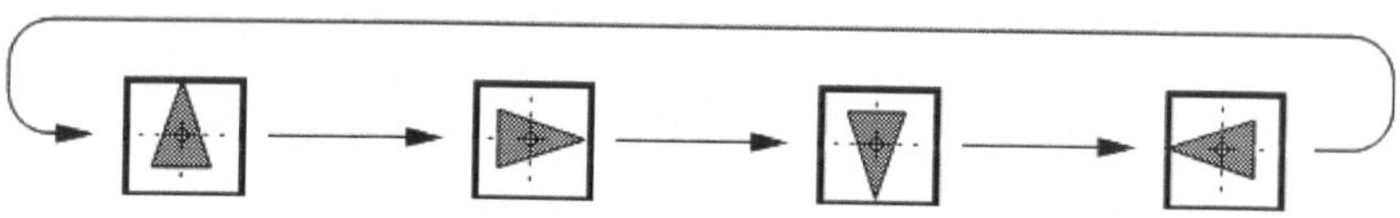

Exhibit 1.2: Robot in a box spins on its heels.

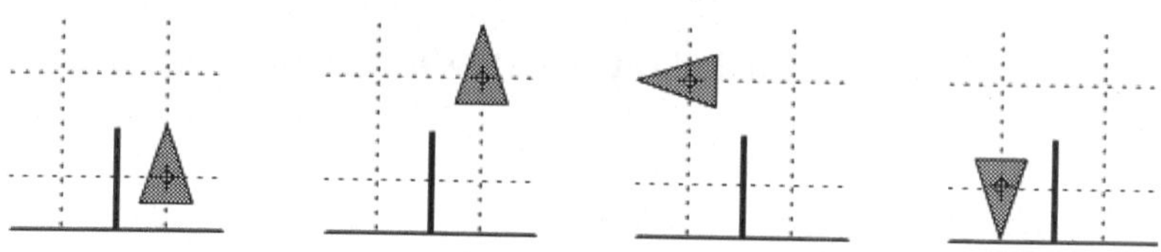

Exhibit 1.3: The robot turns around a spi$_k$e.

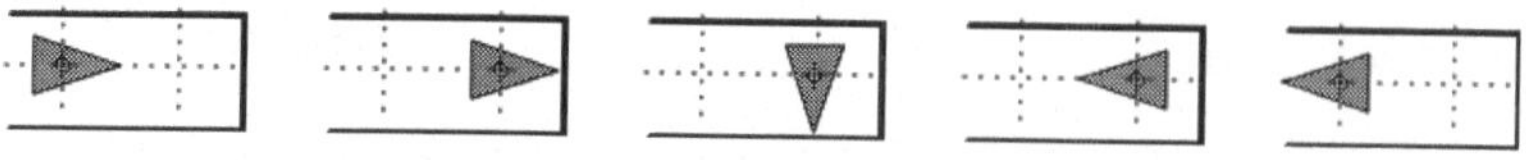

Exhibit 1.4: Backing up in a blind alley.

Algorithm specified in a high-level language

The ideas presented informally in above section are made precise in the following elegant, concise program:

```
{ wall to left-rear }
loop
   { wall to left-rear }
   left;
   { wall to left-front }
   while touch do
      { wall to right-front }
      right;
      { wall to left-front }
   endwhile;
   { wall to left-front }
   forward;
   { wall to left-rear }
forever;
{ wall to left-rear }
```

Program verification. The comments in braces are *program* inv*ariants*: Assertions about the state of the robot that are true every time the flow of control reaches the place in the program where they are written. We need three types of invariants to verify the wall-following program: “wall to left-rear”, “wall to left-front”, and “wall to right-front”. The relationships between the robot’s position and the presence of a nearby wall that must hold for each assertion to be true are illustrated in Exhibit 1.5. Shaded circles indicate points through which a wall must pass. Each robot command transforms its *precondition* (i.e. the assertion true before the command is executed) into its *postcondition* (i.e. the assertion true after its execution). Thus each of the commands ‘left’, ‘right’, and ‘forward’ is a *predicate transformer*, as suggested in Exhibit 1.6.

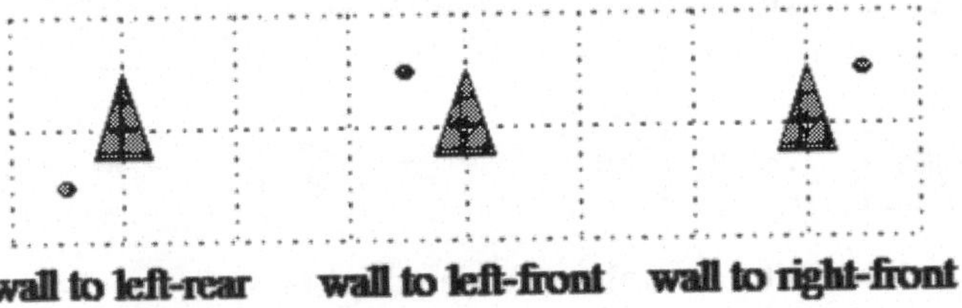

Exhibit 1.5: Three types of invariants relate the positions of robot and wall.

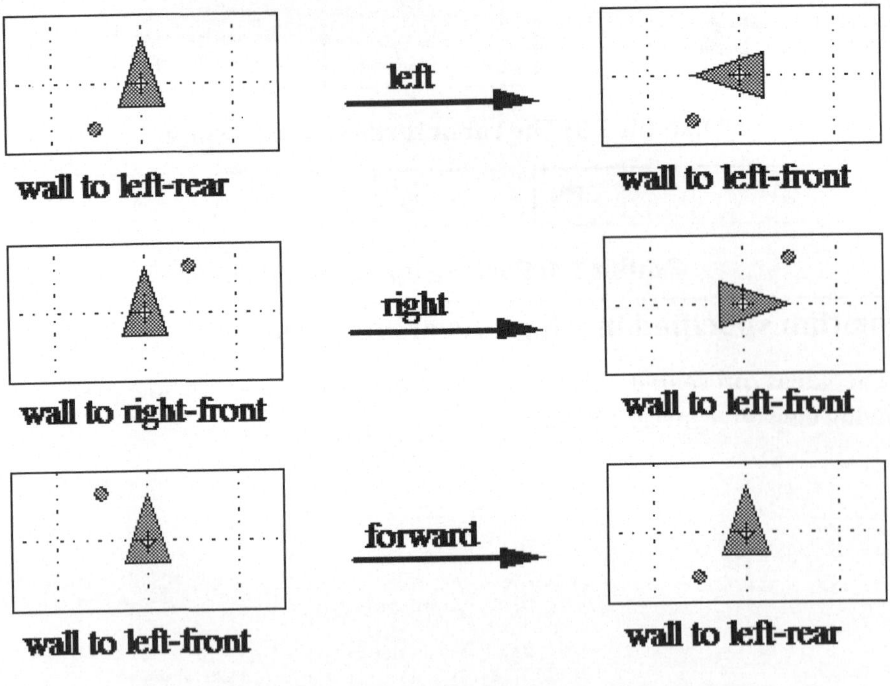

Exhibit 1.6: Robot motions as predicate transformers.

Algorithm programmed in the robot's language

A straightforward translation from the high-level program into the robot's low-level language yields the following seven-line wall-following program:

```
loop
   left;
   while touch do

      right;
   endwhile;
   forward;
forever;
```

```
1 left
2 if touch goto 4
3 goto 6
4 right
5 goto 2
6 forward
7 goto 1
```

The robot's program optimized

In *designing* a program it is best to follow simple, general ideas, and to decide on detailsin the most straightforward manner, without regard for the many alternative ways that are always available for handling details. Once a program is proven correct, and runs, then we may try to improve its efficiency, measured by time and memory requirements. This process of *program transformation* can often be done syntactically, that is merely by considering the definition of individual statements, not the algorithm as a whole. As an example, we derive a five-line version of the wall-following program by transforming the seven-line program in two steps.

If we have the complementary primitive 'if not touch goto #', we can simplify the flow of the program at the left as shown on the right side.

```
  { wall to left-rear }
1 left
2 if touch goto 4
3 goto 6
  { wall to right-front }
4 right
5 goto 2
6 forward
7 goto 1
```

```
  { wall to left-rear }
1 left
2 if not touch goto 6

  { wall to right-front }
4 right
5 goto 2
6 forward
7 goto 1
```

An optimization technique called *loop rotation* allows us to shorten this program by yet another instruction. it changes the structure of the program significantly, as we see from the way the labels have been permuted. The assertion "wall to right-front" attached to line 4 serves as an in*variant of the loop* "keep turning right while you can't advance".

```
  { wall to right-front }
4 right
2 if touch goto 4
6 forward
1 left
7 goto 2
```

Programming projects

Design a data structure suitable for storing a wall made up of horizontal and vertical line segmentsin a square grid of bounded size. Write a "wall-editor", i.e. an interactive program that lets the user define and modify an instance of such a wall.

Program the wall-following algorithm and animate its execution when tracking a wall entered with the wall- editor. Specifically, show the robot's position and orientation after each change of state.

2

Graphics Primitives and Environments

Turtle graphics: a basic environment

Seymour Papert [Pap80] introduced the term *turtle graphics* to denote a set of primitives for line drawing. Originally implemented in the programming language Logo, turtle graphics primitives are now available for several computer systems and languages. They come in different versions, but the essential point is the same as that introduced in the example of the robot car: The pen (or "turtle") is a device that has a state (position, direction) and is driven by incremental operations "move" and "turn" that transform the turtle to a new state depending on its current state:

```
move(s)        { take s unit steps in the direction you are facing }
turn(d)        { turn counterclockwise d degrees }
```

The turtle's initial state is set by the following operations:

```
moveto(x,y)    { move to the position (x,y) in absolute coordinates }
turnto(d)      { face d degrees from due east }
```

In addition, we can specify the color of the trail drawn by the moving pen:

```
pencolor(c) { where c = white, black, none, etc. }
```

Example

The following program fragment approximates a circle tangential to the x-axis at the origin by drawing a 36- sided polygon:

```
moveto(0, 0);  { position pen at origin }
turnto(0);     { face east }
step := 7;     { arbitrarily chosen step length }
do 36 times    { 36 sides · 10° = 360° }
  { move(step);  turn(10) } { 10 degrees counterclockwise }
```

In graphics programming we are likely to use basic figures, such as circles, over and over again, each time with a different size and position. Thus we wish to turn a program fragment such as the circle approximation above into a reusable procedure.

Procedures as building blocks

A program is built from components at many different levels of complexity. At the lowest level we have the constructs provided by the language we use: constants, variables, operators, expressions, and simple (unstructured) statements. At the next higher level we have *procedures*: they let us refer to a program fragment of arbitrary size and complexity as a single entity, and build hierarchically nested structures. Modern programming languages provide yet another level of packaging: *modules*, or *packages*, useful for grouping related data and procedures. We limit our discussion to the use of procedures.

Programmers accumulate their own collection of useful program fragments. Programming languages provide the concept of a *procedure* as the major tool for turning fragmentsinto *reusable* building blocks. A procedure consists of two parts with distinct purposes:

- The *heading* specifies an important part of the procedure's external behavior through the list of *formal parameters*: namely, what type of data movesin and out of the procedure.
- The *body* implements the action performed by the procedure, processing the input data and generating the output data.

A program fragment that embodies a single coherent concept is best written as a procedure. This is particularly true if we expect to use this fragment again in a different context. The question of how general we want a procedure to be deserves careful thought. If the procedure is too specific, it will rarely be useful. If it is too general, it may be unwieldy: too large, too slow, or just too difficult to understand. The generality of a procedure depends primarily on the choice of formal parameters.

Example: the long road toward a procedure "circle"

Let us illustrate these issues by discussing design considerations for a procedure that draws a circle on the screen. The program fragment above for drawing a regular polygon is easily turned into

```
procedure ngon(n,s: integer);   { n = number of sides, s = step
                                  size }
var  i,j: integer;
begin
   j := 360 div n;
   for i := 1 to n do  { move(s);  turn(j) }
end;
```

But, a useful procedure to draw a circle requires additional arguments. Let us start with the following:

```
procedure circle(x, y, r, n: integer);
  { centered at (x, y);  r = radius;  n = number of sides }
var  a, s, i: integer;  { angle, step, counter }
begin
  moveto(x, y - r);  { bottom of circle }
  turnto(0);  {. east }
  a := 360 div n;
  s := r · sin(a);  { between inscribed and circumscribed polygons }
  for  i := 1  to  n  do  { move(s);  turn(a) }
end;
```

This procedure places the burden of choosing n on the programmer. A more sophisticated, "adaptive" version might choose the number of sides on its own as a function of the radius of the circle to be drawn. We assume that lengths are measured in terms of pixels (picture elements) on the screen. We observe that a circle of radius r is of length 2πr. We approximate it by drawing short-line segments, about 3 pixels long, thus needing about 2·r line segments.

```
  procedure circle(x, y, r: integer);  { centered at (x, y);  radius
r}
  var  a, s, i: integer;  { angle, step, counter }
  begin
    moveto(x, y - r);  { bottom of circle }
    turnto(0);  { east }
    a := 180 div r;  { 360 / (# of line segments) }
    s := r · sin(a);  { between inscribed and circumscribed polygons }
    for  i := 1  to  2 · r  do  { move(s);  turn(a) }
  end;
```

This circle procedure still suffers from severe shortcomings:

- If we discretize a circle by a set of pixels, it is an unnecessary detour to do thisin two steps as done above: first, discretize the circle by a polygon; second, discretize the polygon by pixels. This two-step process is a source of unnecessary Work and errors.
- The approximation of the circle by a polygon computed from vertex to vertex leads to *rounding errors* that accumulate. Thus the polygon may fail to close, in particular when using integer computation with itsinherent large rounding error.
- The procedure attempts to draw its circle on an infinite screen. Computer screens are finite, and attempted drawing beyond the screen boundary may or may not cause an error. Thus the circle ought to be clipped at the boundaries of an arbitrarily specified rectangle.

Writing a good circle procedure is a demanding task for professionals. We started this discussion of desiderata and difficulties of a simple library procedure so that the reader may appreciate the thought and effort that go into building a useful programming environment. in chapter 14 we return to this problem and present one possible goal of "the long road toward a procedure 'circle'". We now make a huge jump from the artificially small environments discussed so far to one of today's realistic programming environments for graphics

Quick Draw: a graphics toolbox

For the sake of concreteness, the next few sections show programs written for a specific programming environment: MacPascal using the QuickDraw library of graphics routines [App 85]. it is not our purpose to duplicate a manual, but only to convey the flavor of a realistic graphics package and to explain enough about QuickDraw for the reader to understand the few programs that follow. So our treatment is highly selective and biased.

Concerning the circle that we attempted to program above, QuickDraw offers five procedures for drawing circles and related figures:

```
procedure FrameOval(r: Rect);
procedure PaintOval(r: Rect);
procedure EraseOval(r: Rect);
procedure InvertOval(r: Rect);
procedure FillOval(r: Rect; pat: Pattern);
```

Each one inscribes an oval in an aligned rectangle r (sides parallel to the axes) so as to touch the four sides of r.

If r is a square, the oval becomes a circle. We quote from [App 85]:

FrameOval draws an outline just inside the oval that fitsinside the specified rectangle, using the current grafPort's pen pattern, mode, and size. The outline is as wide as the pen width and as tall as the pen height. it's drawn with the pnPat, according to the pattern transfer mode specified by pnMode. The pen location is not changed by this procedure.

Right away we notice a trade-off when comparing QuickDraw to the simple turtle graphics environment we introduced earlier. At one stroke, "FrameOval" appears to be able to produce many different pictures, but before we can exploit this power, we have to learn about grafPorts, pen width, pen height, pen patterns, and pattern transfer modes. 'FrameOval' draws the perimeter of an oval, 'PaintOval' paints the interior as well, 'EraseOval' paints an oval with the current grafPort's background pattern, 'invertOval' complements the pixels: 'white' becomes 'black', and vice versa. 'FillOval' has an additional argument that specifies a pen pattern used for painting the interior.

We may not need to know all of thisin order to use one of these procedures, but we do need to know how to specify a rectangle. QuickDraw has predefined a type 'Rect' that, somewhat ambiguously at the programmer's choice, has either of the following two interpretations:

```
   type Rect  = record  top, left, bottom, right: integer  end;
   type Rect  = record  topLeft, botRight: Point  end;
with one of the interpretations of type 'Point' being
   type Point = record  v, h: integer  end;
```

Exhibit 2.1 illustrates and provides more information about these concepts. it shows a plane with first coordinate v that runs from top to bottom, and a second coordinate h that runs from left to right. (The reason for v running from top to bottom, rather than vice versa as used in math books, is compatibility with text coordinates where lines are naturally numbered from top to bottom.) The domain of v and h are the integers from –215= –32768 to 215– 1 = 32767. The points thus addressed on the screen are shown asintersections of grid lines. These lines and grid points are infinitely thin - they have no extension. The pixels are the unit squares between them. Each pixel is paired with its top left grid point. This may be enough information to let us draw a slightly fat point of radius 3 pixels at the grid point with integer coordinates (v, h) by calling

```
PaintOval(v - 3, h - 3, v + 3, h + 3);
```

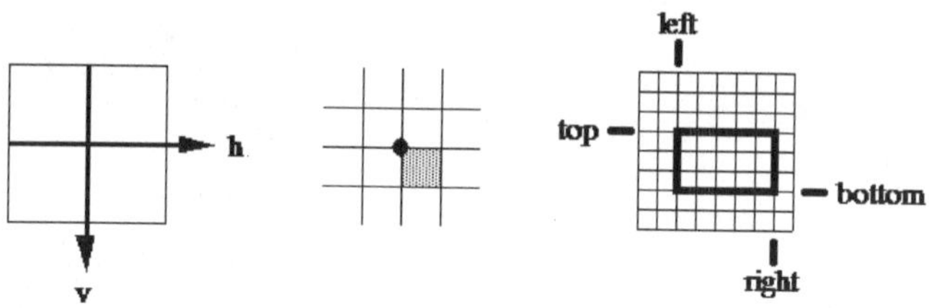

Exhibit 2.1: Screen coordinates define the location of pixels.

To understand the procedures of this section, the reader has to understand a few details about two key aspects of interactive graphics:

- timing and synchronization of devices and program execution
- how screen pictures are controlled at the pixel level

Synchronization

In interactive applications we often wish to specify a grid point by letting the user point the mouse-driven cursor to some spot on the screen. The 'procedure GetMouse(v, h)' returns the coordinates of the grid point where the cursor is located at the moment 'GetMouse' is executed. Thus we can track and paint the path of the mouse by a loop such as

```
  repeat  GetMouse(v, h);  PaintOval(v - 3, h - 3, v + 3, h + 3)
until stop;
```

This does not give the user any timing control over when he or she wants the computer to read the coordinates of the mouse cursor. Clicking the mouse button is the usual way to tell the computer "Now!". A predefined boolean function 'Button' returns 'true' when the mouse button is depressed, 'false' when not. We often synchronize program execution with the user's clicks by programming *busy waiting loops*:

```
    repeat until Button;  { waits for the button to be pressed }
    while Button do; { waits for the button to be released }
  The following procedure waits for the next click:
    procedure waitForClick;
    begin  repeat until Button;  while Button do  end;
```

Pixel acrobatics

The QuickDraw pen has four parameters that can be set to draw lines or paint textures of great visual variety: pen location 'pnLoc', pen size 'pnSize' (a rectangle of given height and width), a pen pattern 'pnPat', and a drawing mode 'pnMode'. The pixels affected by a motion of the pen are shown in Exhibit 2.2.

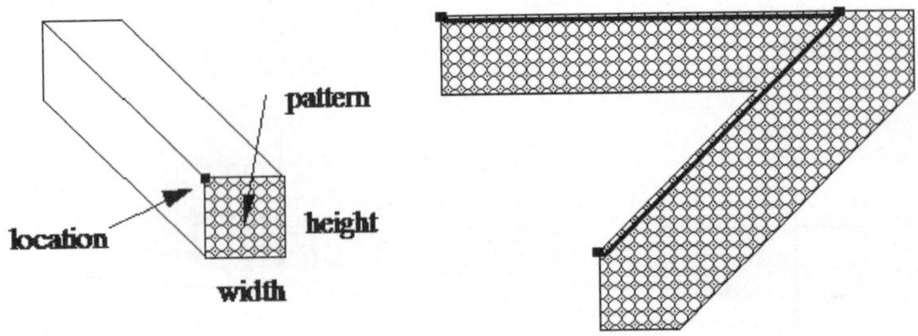

Exhibit 2.2: Footprint of the pen.

Predefined values of 'pnPat' include 'black', 'gray', and 'white'. 'pnPat' is set by calling the predefined 'procedure PenPat(pat: Pattern)' [e.g. 'PenPat(gray)']. As 'white' is the default background, drawing in 'white' usually serves for erasing.

The result of drawing also depends critically on the transfer mode 'pnMode', whose valuesinclude 'patCopy', 'patOr', and 'patXor'. A transfer mode is a boolean operation executed in parallel on each pair of pixelsin corresponding positions, one on the screen and one in the pen pattern.

- 'patCopy' uses the pattern pixel to overwrite the screen pixel, ignoring the latter's previous value; it is the default and most frequently used transfer mode.
- 'patOr' paints a black pixel if either or both the screen pixel or the pattern pixel were black; it progressively blackens the screen.
- 'patXor' (*exclusive-or*, also known as "odd parity") sets the result to black iff exactly one of (screen pixel, pattern pixel) is black. A white pixel in the pen leaves the underlying screen pixel unchanged; a black pixel complementsit. Thus a black pen inverts the screen.

'pnMode' is set by calling the predefined 'procedure PenMode(mode: integer)' [e.g. 'PenMode(patXor)'].

The meaning of the remaining predefined procedures our examples use, such as 'MoveTo' and 'LineTo', is easily guessed. So we terminate our peep into some key details of a powerful graphics package, and turn to examples of its use.

A graphics frame program

Reusable software is a time saving concept that can be practiced profitably in the small. We keep a program that contains nothing but a few of the most useful input/output procedures, displays samples of their results, and conducts a minimal dialog so that the user can step through its execution. We call this a *frame program* because its real purpose is to facilitate development and testing of new procedures by embedding them in a ready-made, tested environment. A simple frame program like the one below makesit very easy for a novice to write his first interactive graphics program.

This particular frame program contains procedures 'GetPoint', 'DrawPoint', 'ClickPoint', 'DrawLine', 'DragLine', 'DrawCircle', and 'DragCircle' for input and display of points, lines, and circles on a screen idealized as a part of a Euclidean plane, disregarding the discretization due to the raster screen. Some of these procedures are so short that one asks why they are introduced at all. 'GetPoint', for example, only convertsinteger mouse coordinates v, h into a point p with real coordinates. it enables us to refer to a point p without mentioning its coordinates explicitly. Thus, by bringing us closer to standard geometric notation, 'GetPoint' makes programs more readable.

The procedure 'DragLine', on the other hand, is a very useful routine for interactive input of line segments. it uses the *rubber-band technique*, which is familiar to users of graphics editors. The user presses the mouse button to fix the first endpoint of a line segment, and keepsit depressed while moving the mouse to the desired second endpoint. At all times during this motion the program keeps displaying the line segment asit would look if the button were released at that moment. This rubber band keeps getting drawn and erased asit moves across other objects on the screen. The user should study a key detail in the procedure 'DragLine' that prevents other objects from being erased or modified as they collide with the ever-refreshed rubber band: We temporarily set 'PenMode(patXor)'. We encourage you to experiment by modifying this procedure in two ways:

- Change the first call of the 'procedure DrawLine(L.p_1, L.p_2, black)' to 'DrawLine(L.p_1, L.p_2, white)'. You will have turned the procedure 'DragLine' into an artful, if somewhat random, painting brush.
- Remove the call 'PenMode(patXor)' (thus reestablishing the default 'pnMode = patCopy'), but leave the first 'DrawLine(L.p_1, L.p_2, white)', followed by the

second 'DrawLine(L.p_1, L.p_2, black)'. You now have a naive rubber-band routine: it alternates erasing (draw 'white') and drawing (draw 'black') the current rubber band, but in so doing it modifies other objects that share pixels with the rubber band. This,s our first example of the use of the versatile *exclusive-or*; others will follow later in the book.

```
program Frame;
  { provides mouse input and drawing of points, line segments,
circles }

type point = record  x, y: real  end;
  lineSegment = record  p1, p2: point  { endpoints }  end;

var  c, p: point;
  r: real;  { radius of a circle }
  L: lineSegment;

procedure WaitForClick;
begin  repeat until Button;  while Button do  end;

procedure GetPoint (var p: point);
var  v, h: integer;
begin
  GetMouse(v, h);
  p.x := v;  p.y := h  { convert integer to real }
end;

procedure DrawPoint(p: point; pat: Pattern);
const  t = 3;  { radius of a point }
begin
  PenPat(pat);
  PaintOval(round(p.y) - t, round(p.x) - t, round(p.y) + t,
round(p.x) + t)
end;

procedure DrawPoint(p: point; pat: Pattern);
const  t = 3;  { radius of a point }
begin
  PenPat(pat);
  PaintOval(round(p.y) - t, round(p.x) - t, round(p.y) + t,
round(p.x) + t)
end;

procedure ClickPoint(var p: point);
begin  WaitForClick;  GetPoint(p);  DrawPoint(p, Black)  end;

function Dist(p, q: point): real;
begin  Dist := sqrt(sqr(p.x - q.x) + sqr(p.y - q.y))  end;

procedure DrawLine(p1, p2: point; pat: Pattern);
begin
  PenPat(pat);
  MoveTo(round(p1.x), round(p1.y));
  LineTo(round(p2.x), round(p2.y))
end;
```

```
procedure DragLine(var L: lineSegment);
begin
  repeat until Button;  GetPoint(L.p1);  L.p2 := L.p1;
PenMode(patXor);
  while Button do  begin
    DrawLine(L.p1, L.p2, black);
    { replace 'black' by 'white' above to get an artistic drawing
tool }
    GetPoint(L.p2);
    DrawLine(L.p1, L.p2, black)
  end;
  PenMode(patCopy)
end;  { DragLine }

procedure DrawCircle(c: point; r: real; pat: Pattern);
begin
  PenPat(pat);
  FrameOval(round(c.y - r), round(c.x - r), round(c.y + r),
round(c.x + r))
end;

procedure DragCircle(var c: point; var r: real);
var  p: point;
begin
  repeat until Button;  GetPoint(c);  r := 0.0;  PenMode(patXor);
  while Button do  begin
    DrawCircle(c, r, black);
    GetPoint(p);
    r := Dist(c, p);
    DrawCircle(c, r, black);
  end;
  PenMode(patCopy)
end;  { DragCircle }

procedure Title;
begin
  ShowText;  { make sure the text window and … }
  ShowDrawing;  { … the graphics window show on the screen }
  WriteLn('Frame program');
  WriteLn('with simple graphics and interaction routines.');
  WriteLn('Click to proceed.');
  WaitForClick
end;  { Title }

procedure What;
begin
  WriteLn('Click a point in the drawing window.');
  ClickPoint(p);
  WriteLn('Drag mouse to enter a line segment.');
  DragLine(L);
  WriteLn('Click center of a circle and drag its radius');
  DragCircle(c, r)
end;  { What }

procedure Epilog;
begin  WriteLn('Bye.')  end;

begin  { Frame }
  Title;  What;  Epilog
end.  { Frame }
```

Example of a graphics routine: polyline input

Let us illustrate the use of the frame program above in developing a new graphics procedure. We choose interactive polyline input as an example. A *polyline* is a chain of directed straight-line segments—the starting point of the next segment coincides with the endpoint of the previous one. 'Polyline' is the most useful tool for interactive input of most drawings made up of straight lines. The user clicks a starting point, and each subsequent click extends the polyline by another line segment. A double click terminates the polyline.

We developed 'PolyLine' starting from the frame program above, in particular the procedure 'DragLine', modifying and adding a few procedures. Once 'Polyline' Worked, we simplified the frame program a bit. For example, the original frame program uses reals to represent coordinates of points, because most geometric computation is done that way. A polyline on a graphics screen only needsintegers, so we changed the type 'point' to integer coordinates. At the moment, the code for polyline input is partly in the procedure 'NextLineSegment' and in the procedure 'What'. in the next iteration, it would probably be combined into a single self-contained procedure, with all the subproceduresit needs, and the frame program would be tossed out—it has served its purpose as a development tool.

```
program PolyLine;
  { enter a chain of line segments and compute total length }
  { stop on double click }

type point = record  x, y: integer;  end;
var  stop: boolean;
  length: real;

  p, q: point;

function EqPoints (p, q: point): boolean;
begin  EqPoints := (p.x = q.x) and (p.y = q.y)  end;

function Dist (p, q: point): real;
begin  Dist := sqrt(sqr(p.x - q.x) + sqr(p.y - q.y))  end;

procedure DrawLine (p, q: point; c: Pattern);
begin  PenPat(c);  MoveTo(p.x, p.y);  LineTo(q.x, q.y)  end;

procedure WaitForClick;
begin  repeat until Button;  while Button do  end;

  procedure NextLineSegment (var stp, endp: point);
  begin
    endp := stp;
    repeat
      DrawLine(stp, endp, black);  { Try 'white' to generate artful
pictures! }
      GetMouse(endp.x, endp.y);
      DrawLine(stp, endp, black)
    until Button;
    while Button do
  end;  { NextLineSegment }

  procedure Title;
  begin
    ShowText;  ShowDrawing;
    WriteLn('Click to start a polyline.');
    WriteLn('Click to end each segment.');
    WriteLn('Double click to stop.')
  end;  { Title }
```

```
procedure What;
begin
  WaitForClick;  GetMouse(p.x, p.y);
  stop := false;  length := 0.0;
  PenMode(patXor);
  while  not stop  do  begin
    NextLineSegment(p, q);
    stop := EqPoints(p, q);  length := length + Dist(p, q);  p := q
  end
end;  { What }

procedure Epilog;
begin  WriteLn('Length of polyline = ', length);  WriteLn('Bye.')
end;

begin  { PolyLine }
  Title;  What;  Epilog
end.  { PolyLine }
```

Programming projects

- Implement a simple package of turtle graphics operations on top of the graphics environment available on your computer.
- Use this package to implement and test a procedure 'circle' that meets the requirements listed at the end of the section "Turtle graphics: a basic environment".
- Implement your personal graphics frame program as described in "A graphics frame program". Your effort will pay off in time saved later, as you will be using this program throughout the entire course.

3

Algorithm Animation

Computer-driven visualization: characteristics and techniques

The computer-driven graphics screen is a powerful new communications medium; indeed, it is the only two-way mass communications medium we know. Other mass communications media–the printed e.g. recorded audio and video—are one-way streets suitable for delivering a monolog. The unique strength of our new medium isinteractive presentation of information. Ideally, the viewer drives the presentation, not just by pushing a start button and turning a channel selector, but controls the presentation at every step. He controls the flow not only with commands such as "faster", "slower", "repeat", "skip", "play this backwards", but more important, with a barrage of "what if?" questions. What if the area of this triangle becomes zero? What if we double the load on this beam? What if world population grows a bit faster? This powerful new medium challenges us to use it well.

When using any medium, we must ask: What can it do well, and what doesit do poorly? The computer-driven screen is$_{i}$deally suited for rapid and accurate display of information that can be deduced from large amounts of data by means of straightforward algorithms and lengthy computation. it can do so in response to a variety of user inputs as long as this variety is contained in an algorithmically tractable, narrow domain of discourse. it is not adept at tasks that require judgment, experience, or insight. By comparison, a speaker at the blackboard is slow and inaccurate and can only call upon small amounts of data and tiny computations; we hope she makes up for this technical shortcoming by good judgment, teaching experience, and insight into the subject. By way of another comparison, books and films may accurately and rapidly present results based on much data and computation, but they lack the ability to react to a user'sinput.

Algorithm animation, the technique of displaying the state of programsin execution, is$_{i}$deally suited for presentation on a graphics screen. There is a need for this type of computation, and there are techniques for producing them. The reasons for animating programsin execution fall into two major categories, which we label *checking* and *exploring*.

Checking

To understand an algorithm well, it is useful to understand it from several distinct points of view. One of them is the static point of view on which correctness proofs are based: Formulate invariants on the data and show that these are preserved under the program's operations. This abstract approach appeals to our rational mind. A second, equally important point of view, is dynamic: Watch the algorithm go through its paces on a variety of input data. This concrete approach appeals to our intuition. Whereas the static approach relies mainly on "thinking", the dynamic approach calls mostly for "doing" and "perceiving", and thus$_{i}$s a prime candidate for visual human computer interaction. in this use of algorithm animation, the user may be checking his understanding of the algorithm, or may be checking the algorithm's correctness—in principle, he could reason this out, but in practice, it is faster and safer to have the computer animation as a double check.

Exploring

in a growing number of applications, computer visualization cannot be replaced by any other technique. This$_{i}$s the case, for example, in exploratory data analysis, where a scientist may not know a priori what she is looking for, and the only way to look at a mass of data is to generate pictures from it (see a special issue on scientific visualization [Nie 89]). At times static pictures will do, but in simulations (e.g. of the onset of turbulent flow) we prefer to see an animation over time.

Turning to the *techniques of animation*, computer technology isin the midst of extremely rapid evolution toward ever-higher-quality interactive image generation on powerful graphics Workstations (see [RN 91] for a survey of the state of the art). Fortunately, animating algorithms such as those presented in this book can be done adequately with the graphics tools available on low-cost Workstations. These algorithms operate on discrete data configurations (such as matrices, trees, graphs), and use standard data structures, such as arrays and lists. For such limited classes of algorithms, there are software packages that help produce animations based on specifications, with a minimum of extra programming required. An example of an algorithm animation environment is the BALSA system [Bro 88, BS 85]. A more recent example is the XYZ GeoBench, which animates geometric algorithms [NSDAB 91].

in our experience, the bottleneck of algorithm animation is not the extra code required, but graphic design. *What do you want to show, and how do you display* it, *keeping* in *mind the limitations of the system you have to* Work *with?* The key point to consider is that data does not look li$_{k}$e anything until we have defined a mapping from the data space into visual space. Defining such a mapping ranges from trivial to practically impossible.

- For some kinds of data, such as geometric data in two- and three-dimensional space, or real-valued functions of one or two real variables, there are natural mappings that we learned in school. These help us greatly in getting a feel for the data.
- Multidimensional data (dimension ≥ 3) can be displayed on a two-dimensional screen using a number of straight forward techniques, such as projectionsinto a subspace, or using color or gray level as a fourth dimension. But our power of perception diminishes rapidly with increasing dimensionality.
- For discrete combinatorial data there is often no natural or accepted visual representation. As an example, we often draw a graph by mapping nodesinto points and edgesinto lines. This representation is natural for graphs that are embedded in Euclidean space, such as a road netWork, and we can readily make sense of a map with thousands of cities and road links. When we extend it to arbitrary graphs by placing a node anywhere on the screen, on the other hand, we get a random crisscrossing of lines of little intuitive value.

In addition to such inherent problems of visual representation, practical difficulties of the most varied type abound. *Examples:*

- Some screens are awfully small, and some data sets are awfully large for display even on the largest screens.
- An animation has to run within a narrow speed range. If it is too fast, we fail to follow, or the screen may flicker disturbingly; if too slow, we may lack the time to observe it.

In conclusion, we hold that it is not too difficult to animate simple algorithms as discussed here by interspersing drawing statementsinto the normal code. independent of the algorithm to be animated, you can call on your own collection of display and interaction procedures that you have built up in your frame program (in the section "A graphics frame program). But designing an adequate graphic representation is hard and requires a creative effort for each algorithm—that is where animators/programmers will spend the bulk of their effort. More on this topic in [NVH 86].

Example: the convex hull of pointsin the plane

The following program is an illustrative example for algorithm animation. 'ConvexHull' animates an *on-line* algorithm that constructs half the convex hull (say, the upper half) of a set of points presented incrementally. it accepts one point at a time, which must lie to the right of all preceding ones, and immediately extends the convex hull. The algorithm is explained in detail in "sample problems and algorithms".

```
program ConvexHull;  { of n ≤ 20 points in two dimensions }

const  nmax = 19;  { max number of points }
       r = 3; { radius of point plot }
var  x, y, dx, dy: array[0 .. nmax] of integer;
       b: array[0 .. nmax] of integer;  { backpointer }
       n: integer;  { number of points entered so far }
       px, py: integer;  { new point }

procedure PointZero;
begin
  n := 0;
  x[0] := 5;  y[0] := 20;  { the first point at fixed location }
  dx[0] := 0;  dy[0] := 1;  { assume vertical tangent }
  b[0] := 0;  { points back to itself }
  PaintOval(y[0] - r, x[0] - r, y[0] + r, x[0] + r)
end;

function NextRight: boolean;
begin
  if  n ≥ nmax  then
    NextRight := false
  else  begin
    repeat until Button;
    while Button do  GetMouse(px, py);
    if  px ≤ x[n]  then
      NextRight := false
    else  begin
      PaintOval(py - r, px - r, py + r, px + r);
      n := n + 1;  x[n] := px;  y[n] := py;
      dx[n] := x[n] - x[n - 1]; { dx > 0 } dy[n] := y[n] - y[n -1];
      b[n] := n - 1;
      MoveTo(px, py);  Line(-dx[n], -dy[n]);  NextRight := true
    end
  end
end;
  procedure ComputeTangent;
  var i: integer;
  begin
    i := b[n];
    while  dy[n] · dx[i] > dy[i] · dx[n]  do  begin  { dy[n]/dx[n] >
dy[i]/dx[i] }

    i := b[i];
    dx[n] := x[n] - x[i];  dy[n] := y[n] - y[i];
    MoveTo(px, py);  Line(-dx[n], -dy[n]);
    b[n] := i
  end;
  MoveTo(px, py);  PenSize(2, 2);  Line(-dx[n], -dy[n]);  PenNormal
end;
procedure Title;
begin
  ShowText;  ShowDrawing;  { make sure windows lie on top }
  WriteLn('The convex hull');
  WriteLn('of n points in the plane sorted by x-coordinate');
  WriteLn('is computed in linear time.');
  Write('Click next point to the right, or Click left to quit.')
end;

begin  { ConvexHull }
  Title;  PointZero;
  while  NextRight  do  ComputeTangent;
  Write('That's it!')
  end.
```

A gallery of algorithm snapshots

The screen dumps shown in Exhibit 3.1 were taken from demonstration programs that we use to illustrate topics discussed in class. Although snapshots cannot convey the information and the impact of animations, they may give the reader ideas to try out. We select two standard algorithm animation topics (sorting and random number generation), and an example showing the effect of cumulative rounding errors.

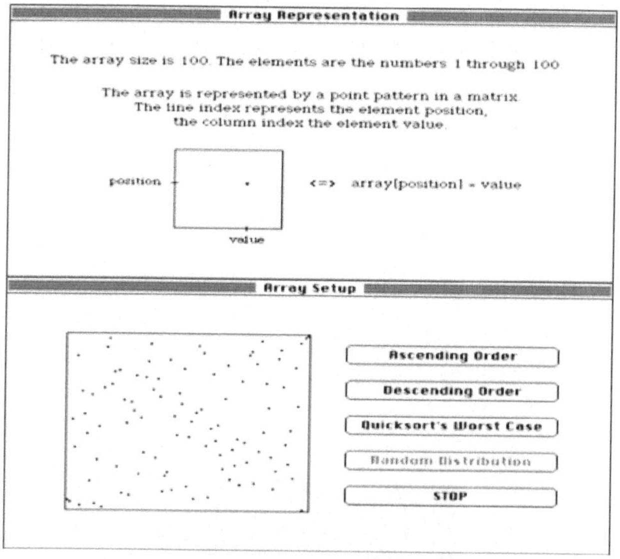

Exhibit 3.1: initial configuration of data, ...

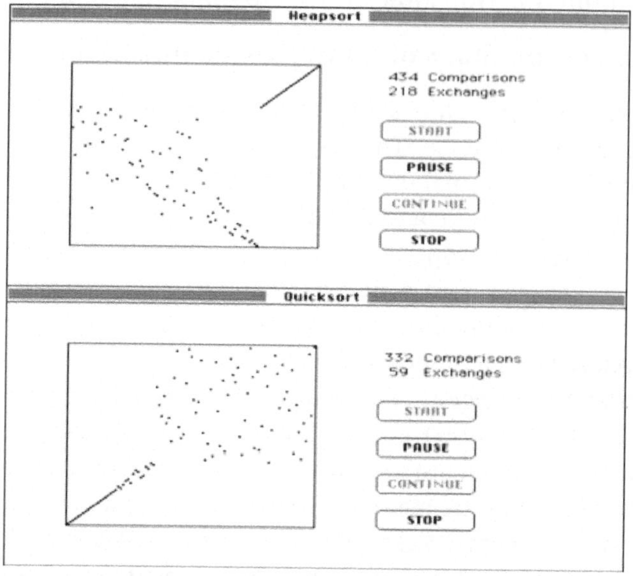

Exhibit 3.1: ... and snapshots from two sorting algorithms.

Visual test for randomness

Our visual system is amazingly powerful at detecting patterns of certain kindsin the midst of noise. Random number generators (RNGs) are intended to simulate "noise" by means of simple formulas. When patterns appear in the visual representation of supposedly random numbers, chances are that this RNG will also fail more rigorous statistical tests. The eyes' pattern detection ability serves well to disqualify a faulty RNG but cannot certify one as adequate. Exhibit 3.2 shows a simulation of the Galton board. in theory, the resulting density diagram should approximate a bellshaped Gaussian distribution. Obviously, the RNG used falls short of expectations.

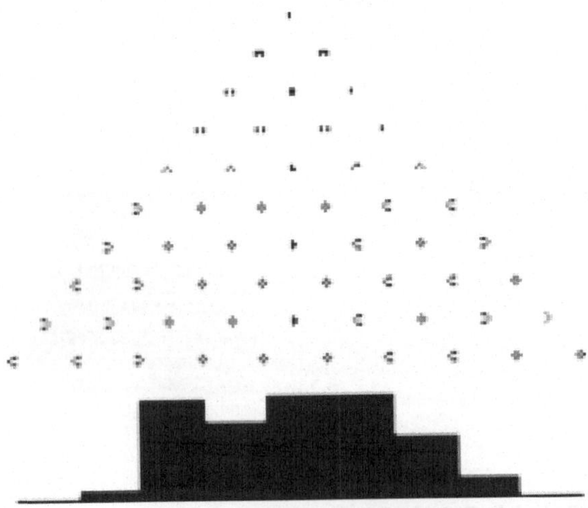

Exhibit 3.2: One look suffices to unmask a bad RNG.

Numerics of chaos, or chaos of numerical computation?

The following example shows the effect of rounding errors and precision in linear recurrence relations. The d- step linear recurrence with constant coefficientsin the domain of real or complex numbers,. is one of the most frequent formulas evaluated in scientific and technical computation (e.g. for the solution of differential equations). By proper choice of the constants ci and of initial values $z_0, z_1, \ldots, z_{d-1}$ we can generate sequences z_k that when plotted in the plane of complex numbers form many different figures. With d= 1 and $|\chi_1|= 1$, for example, we generate circles. The picturesin Exhibit 3.3 were all generated with d = 3 and conditions that determine a curve that is most easily described as a circle 3 running around the perimeter of another circle 2 that runs around a stationary circle 1. We performed this computation with a floating-point package that lets us pick precision P (i.e. the number of bitsin the mantissa). The resulting pictures look a bit chaotic, with a behavior we have come to associate with fractals—even if the mathematics of generating them is completely different, and linear recurrences computed without error would look much more regular. Notice that the first two images are generated by the same formula, with a single bit of difference in the precision used. The whim of this 1-bit difference in precision changes the image entirely.

$$z_k = \sum_{i=1}^{d} c_i z_{k-i}, \quad c_i \in C$$

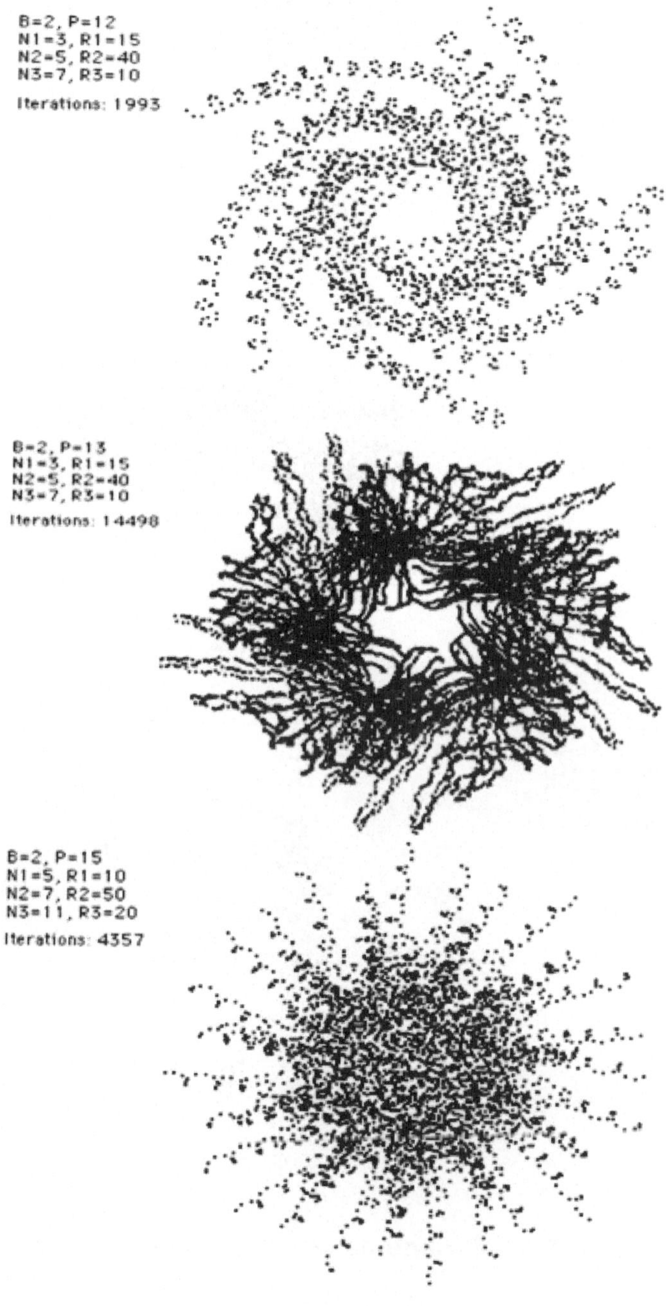

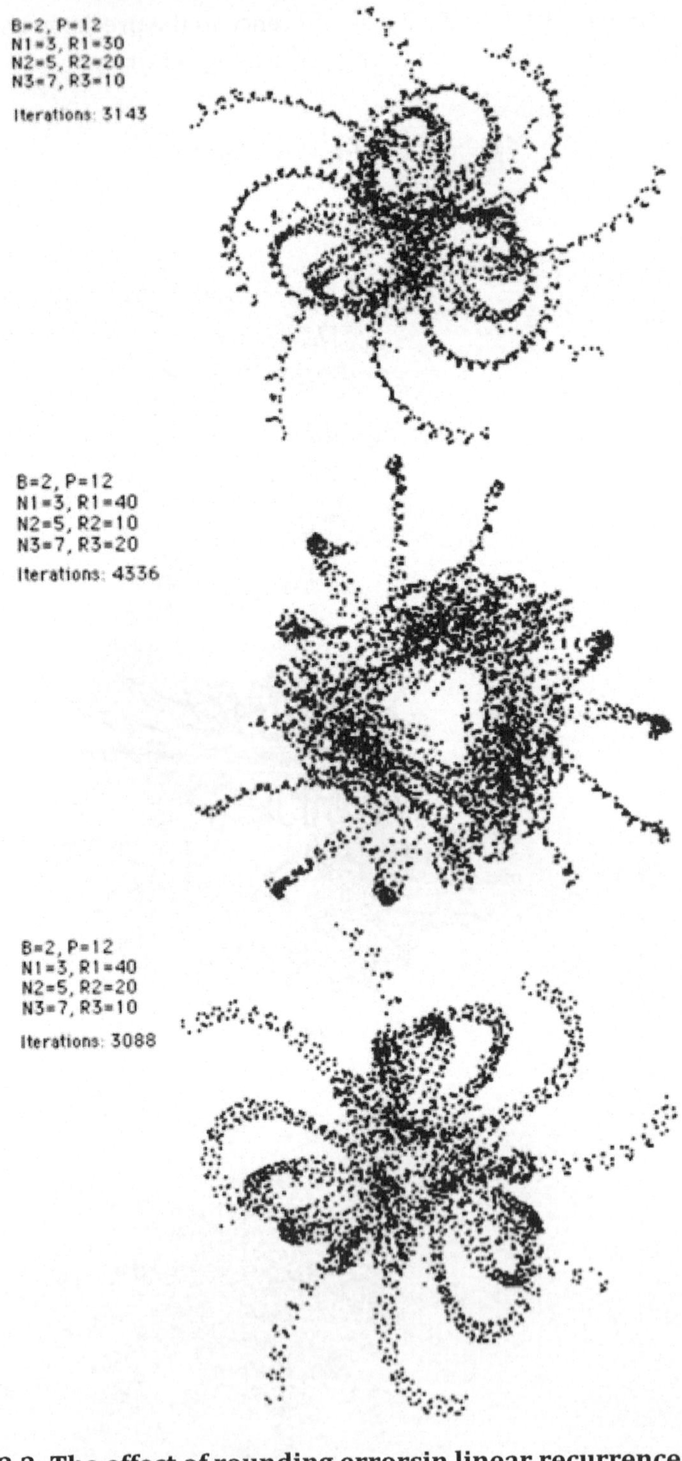

Exhibit 3.3: The effect of rounding errorsin linear recurrence relations.

Programming projects

- Use your personal graphics frame program (the programming project of "graphics primitives and environments") to implement and animate the convex hull algorithm example.
- Use your graphics frame program to implement and animate the behavior of recurrence relations as discussed in the section "A gallery of algorithm snapshots".
- Extend your graphics frame program with a set of dialog control operations sufficient to guide the user through the various steps of the animation of recurrence relations: in particular, to give him the options, at any time, to enter a new set of parameters, then execute the algorithm and animate it in either 'movie mode' (it runs at a predetermined speed until stopped by the user), or 'step mode' [the display changes only when the user enters a logical command 'next' (e.g. by clicking the mouse or hitting a specific key)].

Part-II

Programming Concepts: beyond Notation

Thoughts on the role of programming notations

A programming language is the main interface between a programmer and the physical machine, and a novice programmer will tend to identify "programming" with "programming in the particular language she has learned". The realization that there is much to programming "beyond notation" (i.e. principles that transcend any one language) is a big step forward in a programmer's development.

Part II aims to help the reader take this step forward. We present examples that are best understood by focusing on abstract principles of algorithm design, and only later do we grope for suitable notations to turn this principle into an algorithm expressed in sufficient detail to become executable. in keeping with our predilection for graphic communication, the first informal expression of an algorithmic idea is often pictorial. We show by example how such representations, although they may be incomplete, can be turned into programsin a formal notation.

The literature on programming and languages. There are many books that present principles of programming and of programming languages from a higher level of abstraction. The principles highlighted differ from author to author, ranging from intuitive understanding to complete formality. The following textbooks provide an excellent sample from the broad spectrum of approaches: [ASS 84], [ASU 86], [Ben 82], [Ben 85], [Ben 88], [Dij 76], [DF 88], [Gri 81], and [Mey 90].

4

Algorithms and Programs as Literature: Substance and Form

Programming in the large versus programming in the small

in studying and discussing the art of programming it is useful to distinguish between large programs and small programs, since these two types impose fundamentally different demands on the programmer.

Programming in the large

Large programs (e.g. operating systems, database systems, compilers, application packages) tax our *organizational ability*. The most important issues to be dealt with include requirements analysis, functional specification, compatibility with other systems, how to break a large program into modules of manageable size, documentation, adaptability to new systems and new requirements, how to organize the team of programmers, and how to test the software. These issues are the staple of software engineering. When compared to the daunting managerial and design challenges, the task of actual coding is relatively simple. Large programs are often *flat:* Most of the listing consists of comments, interface specifications, definitions, declarations, initializations, and a lot of code that is executed only rarely. Although the function of any single page of source code may be rather trivial when considered by itself, it is difficult to understand the entire program, as you need a lot of information to understand how this page relates to the whole. The classic book on programming in the large is [Bro 75].

Programming in the small

Small programs, of the kind discussed in this book, challenge our technical know-how and inventiveness. *Algorithmic issues* dominate the programmer's thinking: Among several algorithms that all solve the same problem, which is the most efficient under the given circumstances? How much time and space doesit take? What data structures do we use? in contrast to large programs, small programs are usually *deep*, consisting of short, compact code many of whose statements are executed very often. Understanding a small program may also be difficult, at least initially, since the chain of thought is often subtle. Once you understand it thoroughly, you can reproduce it at any time with much less effort than was first required. Mastery of interesting small programs is the best way to get started in computer science. We encourage the reader to Work out all the details of the examples we present.

This book is concerned only with programming in **the small.** This decision determines our choice of topics to be presented, our style of presentation, and the notation we use to express programs, explanations, and proofs, and heavily influences our comments on techniques of programming. Our style of presentation appeals to the reader'sintuition more than to formal rigor. We aim at highlighting the key idea of any argument that we make rather than belaboring the details. We take the liberty of using a free notation that suits the purpose of any specific argument we wish to make, trusting that the reader understands our small programs so well that he can translate them into the programming language of his choice. in a nut shell, we emphasize substance over form.

The purpose of Part II is to help engender a fluency in using different notations. We provide yet other examples of unconventional notations that match the nature of the problem they are intended to describe, and we show how to translate them into Pascal-like programs. Since much of the difference between programming languages is merely syntactic, we include two chapters that cover the basics of syntax and syntax analysis. These topics are important in their own right; we present them early in the hope that they will help the student see through differences of notation that are merely "syntactic sugar".

Documentation versus literature: isit meant to be read?

it isinstructive to distinguish two types of written materials, and two corresponding types of writing tasks: documents and literature. *Documents* are constrained by requirements of many kinds, are read when a specific need arises (rarely for pleasure), and their quality is judged by criteria such as formality, conformity to a standard, completeness, accuracy, and consistency. *Literature* is a form of art free from conventions, read for education or entertainment, and its quality is judged by aesthetic criteria much harder to enumerate than the ones above. The touchstone is the question: Isit meant to be read? If the answer is "only if necessary", then it's a document, not literature.

As the name implies, the documentation of large programs is a typical document-writing chore. Much has been written in software engineering about documentation, a topic whose importance grows with the size and complexity of the system to be documented. We hold that small programs are not documented, they are explained. As such, they are literature, or ought to be. The idea of programs as literature is widely held (see, e.g. [Knu 84]). The key idea is that an algorithm or program is part of the text and meltsinto the text in the same way as a paragraph, a formula, or a picture does. There are also formal notations and systems designed to support a style of programming that integrates text and code to form a package that is both readable for humans and executable by machines [Knu 83].

Whatever notation is used for literate programming, it has to describe all phases of a program's evolution, from idea to specification to algorithm to program. Details of a good program cannot be understood, or at least not appreciated, without an awareness of the grand design that guided the programmer. Whereas details are usually well expressed in some formal notation, grand designs are not. For this reason we renounce formality and attempt to convey ideasin whatever notation suits our purpose of insightful explanation. Let us illustrate this philosophy with some examples.

A snowflake

Fractal pictures are intuitively characterized by the requirement that any part of the picture, of any size, when sufficiently magnified, looks like the whole picture. Two pieces of information are required to define a specific fractal:

- A picture primitive that serves as a building-block: Many copies of this primitive, scaled to many different sizes, are composed to generate the picture.
- A recursive rule that defines the relative position of the primitives of different size.

A picture primitive is surely best defined by a drawing, and the manner of composing primitivesin space again calls for a pictorial representation, perhaps augmented by a verbal explanation. in this style we define the fractal 'Snowflake' by the following *production* rule, which we read as follows: A line segment, as shown on the left-hand side, must be replaced by a polyline, a chain of four shorter segments, as shown at the right-hand side (Exhibit 4.1). We start with an initial configuration (the zero-generation) consisting of a single segment (Exhibit 4.2). If we apply the production rule just once to every segment of the current generation, we obtain successively a first, second, and third generation, as shown in Exhibit 4.3. Further generations quickly exhaust the resolution of a graphics screen or the printed page, so we stop drawing them. The curve obtained as the limit when this process is continued indefinitely is a *fractal*. Although we cannot draw it exactly, one can study it as a mathematical object and prove theorems about it.

Exhibit 4.1: Production for replacing a straight-line segment by a polyline

Exhibit 4.2: The simplest initial configuration

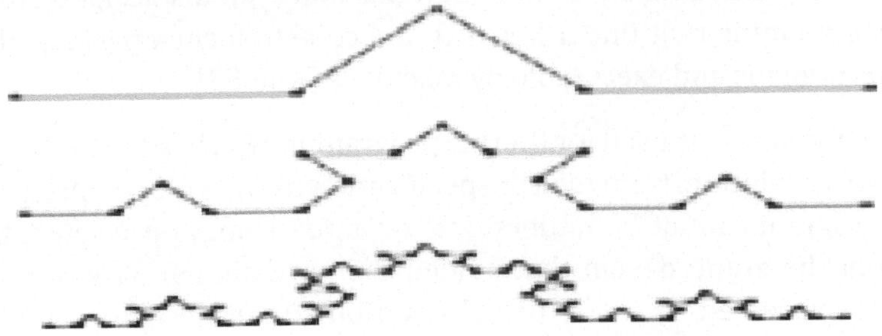

Exhibit 4.3: The first three generations

The production rule drawn above is the essence of this fractal, and of the sequence of pictures that lead up to it. The initial configuration, on the other hand, is quite arbitrary: If we had started with a regular hexagon, rather than a single line segment, the pictures obtained would really have lived up to their name, snowflake. Any other initial configuration still generates curves with the unmistakable pattern of snowflakes, as the reader is encouraged to verify.

After having familiarized ourselves with the objects described, let us turn our attention to the method of description and raise three questions about the formality and executability of such notations.

- Is our notation sufficiently formal to serve as a program for a computer to draw the family of generations of snowflakes? Certainly not, as we stated certain rulesin colloquial language and left others completely unsaid, implying them only by sample drawings. As an example of the latter, consider the question: If a segment is to be replaced by a "plain with a mountain in the center", on which side of the segment should the peak point? The drawings above suggest that all peaks stick out on the same side of the curve, the outside.
- Could our method of description be extended and formalized to serve as a programming language for fractals? Of course. As an example, the production shown in Exhibit 4.4 specifies the side on which the peak is to point. Every segment now has a + side and a – side. The production above states that the new peak is to grow over the + side of the original segment and specifies the + sides and – sides of each of the four new segments. For every other aspect that our description may have left unspecified, such as placement on the screen, some notation could readily be designed to specify every detail with complete rigor. in "Syntax" and "Syntax analysis" we introduce some of the basic techniques for designing and using formal notations.
- Should we formalize this method of description and turn it into a machine-executable notation? it depends on the purpose for which we plan to use it. Often

in this book we present just one or a few examples that share a common design. Our goal is for the reader to understand these few examples, not to practice the design of artificial programming languages. To avoid being sidetracked by a pedantic insistence on rigorous notation, with itsinevitable overhead of introducing formalisms needed to define all details, we prefer to stop when we have given enough information for an attentive reader to grasp the main idea of each example.

Exhibit 4.4: Refining the description to specify a "left-right" orientation.

Hilbert's space-filling curve

Space-filling curves have been an object of mathematical curiosity since the nineteenth century, as they can be used to prove that the cardinality of an interval, considered as a set of points, equals the cardinality of a square (or any other finite two-dimensional region). The term *space-filling* describes the surprising fact that such a curve visits every point within a square. in mathematics, space-filling curves are constructed as the limit to which an infinite sequence of curves Ci converges. On a discretized plane, such as a raster-scanned screen, no limiting process,s needed, and typically one of the first dozen curvesin the sequence already paints every pixel, so the term *space-filling* is quickly seen to be appropriate.

Let us,llustrate this phenomenon using Hilbert's space-filling curve (David Hilbert, 1862-1943), whose first six approximations are shown in Exhibit 4.5. As the pictures suggest, Hilbert curves are best-described recursively, but the composition rule is more complicated than the one for snowflakes. We propose the two productions shown in Exhibit 4.6 to capture the essence of Hilbert (and similar) curves. This pictorial program requires explanation, but we hope the reader who has once understood it will find this notation useful for inventing fractals of her own. As always, a production is read: "To obtain an instance of the left-handside, get instances of all the things listed on the right-handside", or equivalently, "to do the task specified by the left-hand side, do all the tasks listed on the right- hand side".

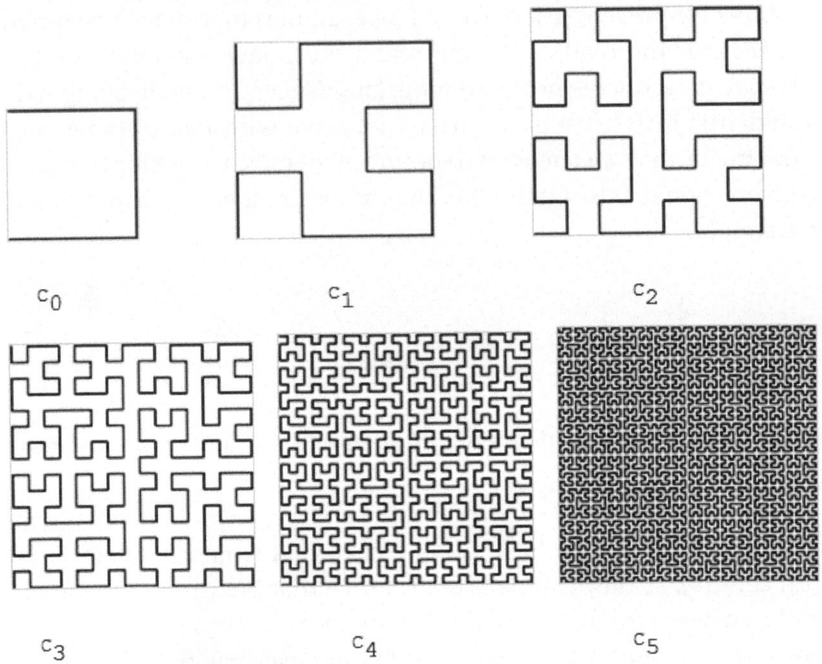

Exhibit 4.5: Six generations of the family of Hilbert curves

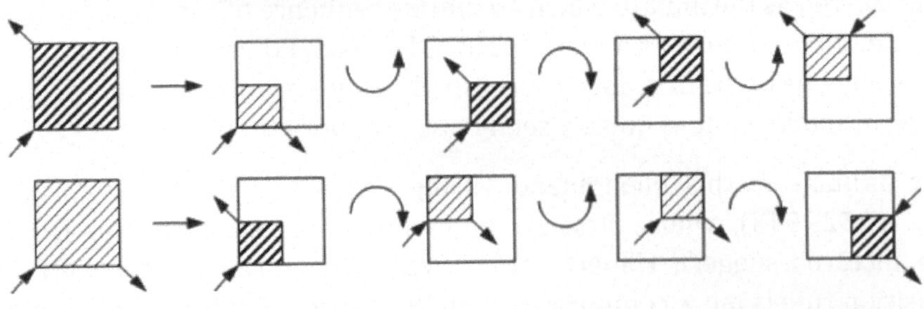

Exhibit 4.6: Productions for painting a square in terms of its quadrants

The left-hand side of the first production stands for the task: paint a square of given size, assuming that you enter at the lower left corner facing in the direction indicated by the arrow and must leave in the upper left corner, again facing in the direction indicated by that arrow. We assume turtle graphics primitives, where the state of the brush is given by a position and a direction. The hatching indicates the area to be painted. it lies to the right of the line that connects entry and exit corners, which we read as "paint with your right hand", and the hatching isin thick strokes. The left-hand side of the second production is similar: Paint a square "with your left hand" (hatching isin thin strokes), entering and exiting asindicated by the arrows.

The right-hand sides of the productions are now easily explained. They say that in order to paint a square you must paint each of its quadrants, in the order indicated. They give explicit instructions on where to enter and exit, what direction to face, and whether you are painting with your right or left hand. The last detail is to make sure that when the brush exits from one quadrant it getsinto the correct state for entering the next. This requires the brush to turn by 90°, either left or right, as the curved arrowsin the picturesindicate. in the continuous plane we imagine the brush to "turn on its heels", whereas on a discrete grid it also moves to the first grid point of the adjacent quadrant.

These productions omit any rule for termination, thus simulating the limiting process of true space-filling curves. To draw anything on the screen we need to add some termination rules that specify two things: (1) when to invoke the termination rule (e.g. at some fixed depth of recursion), and (2) how to paint the square that invokes the termination rule (e.g. paint it all black). As was the case with snowflakes and with all fractals, the primitive pictures are much less$_{\text{i}}$mportant than the composition rule, so we omit it.

The following program implements a specific version of the two pictorial productions shown above. The procedure 'Walk' implements the curved arrowsin the productions: the brush turns by 'halfTurn', takes a step of length s, and turns again by 'halfTurn'. The parameter 'halfTurn' isintroduced to show the effect of cumulative small errorsin recursive procedures. 'halfTurn = 45' causes the brush to make right-angle turns and yields Hilbert curves. The reader is encouraged to experiment with 'halfTurn = 43, 44, 46, 47', and other values.

```
program PaintAndWalk;
const  pi = 3.14159;   s = 3;   { step size of walk }
var    turtleHeading: real;   { counterclockwise, radians }
       halfTurn, depth: integer;   { recursive depth of painting }

procedure TurtleTurn(angle: real);
{ turn the turtle angle degrees counterclockwise }
begin  { angle is converted to radian before adding }
  turtleHeading := turtleHeading + angle · pi / 180.0
end;  { TurtleTurn }

procedure TurtleLine(dist: real);
{ draws a straight line, dist units long }
begin
  Line(round(dist · cos(turtleHeading)), round(-dist·sin(turtle
Heading)))

end;  { TurtleLine }

procedure Walk (halfTurn: integer);
begin  TurtleTurn(halfTurn);  TurtleLine(s);  TurtleTurn(halfTurn)
end;
```

```
procedure Qpaint (level: integer; halfTurn: integer);
begin
  if  level = 0  then
    TurtleTurn(2 · halfTurn)
  else  begin
    Qpaint(level - 1, -halfTurn);
    Walk(halfTurn);
    Qpaint(level - 1, halfTurn);
    Walk(-halfTurn);
    Qpaint(level - 1, halfTurn);
    Walk(halfTurn);
    Qpaint(level - 1, -halfTurn)
  end
end;  { Qpaint }

begin  { PaintAndWalk }
    ShowText;  ShowDrawing;
    MoveTo(100, 100);  turtleHeading := 0;  { initialize turtle
state }
    WriteLn('Enter halfTurn 0 .. 359  (45 for Hilbert curves): ');
    ReadLn(halfTurn);
    TurtleTurn(-halfTurn);  { init turtle turning angle }
    Write('Enter depth 1 .. 6: ');  ReadLn(depth);
    Qpaint(depth, halfTurn)
  end.  { PaintAndWalk }
```

As a summary of this discourse on notation, we point to the fact that an executable program necessarily has to specify many details that are irrelevant from the point of view of human understanding. This book assumes that the reader has learned the basic steps of programming, of thinking up such details, and being able to express them formally in a programming language. Compare the verbosity of the one-page program above with the clarity and conciseness of the two pictorial productions above. The latter state the essentials of the recursive construction, and no more, in a manner that a human can understand "at a glance". We aim our notation to appeal to a human mind, not necessarily to a computer, and choose our notation accordingly.

Pascal and its dialects: lingua franca of computer science

Lingua franca (1619):

- A common language that consists of italian mixed with French, Spanish, Greek and Arabic and is spoken in Mediterranean ports
- Any of various languages used as common or commercial tongues among peoples of diverse speech
- Something resembling a common language

(From *Webster's Collegiate Dictionary*)

Pascal as representative of today's programming languages

The definition above fits Pascal well: in the mainstream of the development of programming languages for a couple of decades, Pascal embodies, in a simple design, some of the most important language features that became commonly accepted in the 1970s. This simplicity, combined with Pascal's preference for language features that are now well understood, makes Pascal a widely understood programming notation. A few highlightsin the development of programming languages may explain how Pascal got to be a lingua franca of computer science.

Fortran emerged in 1954 as the first high-level programming language to gain acceptance and became *the* programming language of the 1950s and early 1960s. its appearance generated great activity in language design, and suddenly, around$_1$960, dozens of programming languages emerged. Three among these, Algol 60, COBOL, and Lisp, became milestonesin the development of programming languages, each in its own way. Whereas COBOL became the most widely used language of the 1960s and$_1$970s, and Lisp perhaps the most innovative, Algol 60 became the most influential in several respects: it set new standards of rigor for the definition and description of a language, it pioneered hierarchical block structure as the major technique for organizing large programs, and through these major technical contributions became the first of a family of mainstream programming languages that includes PL/1, Algol 68, Pascal, Modula-2, and Ada.

The decade of the 1960s remained one of great ferment and productivity in the field of programming languages. PL/1 and Algol 68, two ambitious projects that attempted to integrate many recent advancesin programming language technology and theory, captured the lion's share of attention for several years. Pascal, a much smaller project and language designed by Ni$_k$laus Wirth during the 1960s, ended up eclipsing both of these major efforts. Pascal took the best of Algol 60, in streamlined form, and added just one major extension, the then novel type definitions [Hoa 72]. This lightweight edifice made it possible to implement efficient Pascal compilers on the microcomputers that mushroomed during the mid$_1$970s (e.g. UCSD Pascal), which opened the doors to universities and high schools. Thus Pascal became the programming language most widely used in introductory computer science education, and every computer science student must be fluent in it.

Because Pascal is so widely understood, we base our programming notation on it but do not adhere to it slavishly. Pascal is more than 20 years old, and many of its key ideas are 30 years old. With today'sinsights into programming languages, many details would probably be chosen differently. indeed, there are many "dialects" of Pascal, which typically extend the standard defined in 1969 [Wir 71] in different directions. One extension relevant for a publication language is that with today's hardware that supports large character sets and many different fonts and styles, a greater variety of symbols can be used to make the source more readable. The following examplesintroduce some of the conventions that we use often.

"Syntactic sugar": the look of programming notations

Pascal statements lack an explicit terminator. This makes the frequent use of begin-end brackets necessary, asin the following program fragment, which implements the insertion sort algorithm (see chapter 17 and the section "Simple sorting algorithms that Work in time"); –∞ denotes a constant ≤ any key value:

```
A[0] := -∞;
for i := 2 to n do  begin
  j := i;
  while  A[j] < A[j - 1]  do
    begin  t := A[j];  A[j] := A[j - 1];  A[j - 1] := t;  j := j - 1  end;
end;
```

We aim at brevity and readability but wish to retain the flavor of Pascal to the extent that any new notation we introduce can be translated routinely into standard Pascal. Thus we write the statements above as follows:

```
A[0] := -∞;
for i := 2 to n do  begin
  j := i;  { comments appear in italics }
  while  A[j] < A[j - 1]  do  { A[j] :=: A[j - 1];  j := j - 1 }
  { braces serve as general-purpose brackets, including begin-end }
  { :=: denotes the exchange operator }
end;
```

Borrowing heavily from standard mathematical notation, we use conventional mathematical signs to denote operators whose Pascal designation was constrained by the small character sets typical of the early days, such as:

```
    ≠  ≤  ≥  ≠  ¬  ∧ ∨ ∈ ∉ ∩ ∪ \  |x|  instead of
<> <= >= <> not and or in not in  ·  + -  abs(x) respectively
```

We also use signs that may have no direct counterpart in Pascal, such as:

⊃ ⊇ ⊄ ⊂ ⊆	Set-theoretic relations
∞	infinity, often used for a "sentinel" (i.e. a number larger than all numbers to be processed in a given application)
±	Plus-or-minus, used to define an interval [of uncertainty]
ΣΠ	Sum and product
⌈x⌉	Ceiling of a real number x (i.e. the smallest integer ≥ x)
⌊x⌋	Floor of a real number x (i.e. the largest integer ≤ x)
√	Square root
log	Logarithm to the base 2
ln	Natural logarithm, to the base e
iff	If and only if

Although we may take a cavalier attitude toward notational differences, and readily use concise notations such as ∧ ∨ for the more verbose 'and', 'or', we will try to remind readers explicitly about our assumptions when there is a question about semantics. As an example, we assume that the boolean operators ∧ and ∨ are *conditional*, also called 'cand' and 'cor': An expression containing these operators is evaluated from left to right, and the evaluation stops as soon as the result is known. in the expression x ∧ y, for example, x is evaluated first. If x evaluates to 'false', the entire expression is 'false' without y ever being evaluated. This convention makesit possible to leave y undefined when x is 'false'. Only if x evaluates to 'true' do we proceed to evaluate y. An analogous convention applies to x ∨ y.

Program structure

Whereas the concise notationsintroduced above to denote operators can be translated almost one-to-one into a single line of standard Pascal, we also introduce a few extensions that may affect the program structure. in our view these changes make programs more elegant and easier to understand. Borrowing from many modern languages, we introduce a 'return()' statement to exit from procedures and functions and to return the value computed by a function.

Example

```
function gcd(u, v: integer): integer;
{ computes the greatest common divisor (gcd) of u and v }
begin if v = 0 then return(u) else return(gcd(v, u mod v));
end;
```

in this example, 'return()' merely replaces the Pascal assignments 'gcd := u' and 'gcd := gcd(v, u mod v)'. The latter in particular illustrates how 'return()' avoids a notational blemish in Pascal: On the left of the second assignment, 'gcd' denotes a variable, on the right a function. 'Return()' also has the more drastic consequence that it causes control to exit from the surrounding procedure or function as soon asit is executed. Without entering into a controversy over the general advantages and disadvantages of this "flow of control" mechanism, let us present one example, typical of many search procedures, where 'return()' greatly simplifies coding. The point is that a search routine terminatesin one of (at least) two different ways: successfully, by having found the item in question, or unsuccessfully, because of a number of reasons (the item is not present, and some index is about to fall outside the range of a table; we cannot insert an item because the table is full, or we cannot pop a stack because it is empty, etc.). For the sake of efficiency as well as readability we prefer to exit from the routine as soon as a case has been identified and dealt with, as the following example from "Address computation:" illustrates:

```
function insert-into-hash-table(x: key): addr;
var  a: addr;
begin
  a := h(x);  { locate the home address of the item x to be
inserted }
  while  T[a] ≠ empty  do  begin
    { skipping over cells that are already occupied }
    if  T[a] = x  then  return(a);  { x is already present; return
its address }

    a := (a + 1) mod m    { keep searching at the next address }

  end;
  { we've found an empty cell; see if there is room for x to be
inserted }

  if  n < m - 1  then  { n := n + 1;  T[a] := x }  else  err-
msg('table is full');
  return(a)  { return the address where x was inserted }
end;
```

This code can only be appreciated by comparing it with alternatives that avoid the use of 'return()'. We encourage readers to try their hands at this challenge. Notice the three different ways this procedure can terminate: (1) no need to insert x because x is already in the table, (2) impossible to insert x because the table is full, and (3) the normal case when x isinserted. Standard Pascal incorporates no facilities for "exception handling" (e.g. to cover the first two cases that should occur only rarely) and forces all three outcomes to exit the procedure at its textual end.

Let us just mention a few other liberties that we may take. Whereas Pascal limits results of functions to certain simple types, we will let them be of *any* type: in particular, structured types, such as records and arrays. Rather than nesting if-then-else statementsin order to discriminate among more than two mutually exclusive cases, we use the "flat" and more legible control structure:

if B_1 then S_1 elsif B_2 then S_2 elsif ... else S_n ;

Our sample programs do not return dynamically allocated storage explicitly. They rely on a memory management system that retrieves free storage through "garbage collection". Many implementations of Pascal avoid garbage collection and instead provide a procedure 'dispose(...)' for the programmer to explicitly return unneeded cells. If you Work with such a version of Pascal and write list-processing programs that use significant amounts of memory, you must insert calls to 'dispose(...)' in appropriate placesin your programs.

The list above is not intended to be exhaustive, and neither do we argue that the constructs we use are necessarily superior to others commonly available. Our reason for extending the notation of Pascal (or any other programming language we might have chosen as a starting point) is the following: in addressing human readers, we believe an open-ended, somewhat informal notation is preferable to the straightjacket of any one programming language. The latter becomes necessary if and when we execute a program, but during the incubation period when our understanding slowly grows toward a firm grasp of an idea, supporting intuition is much more important than formality. Thus we describe data structures and algorithms with the help of figures, words, and programs as we see fit in any particular instance.

Programming project

- ❖ Use your graphics frame program of "Graphics primitives and environments" to implement an editor for simple graphics productions such as those used to define snowflakes (e.g. 'any line segment gets replaced by a specified sequence of line segments'), and an interpreter that draws successive generations of the fractals defined by these productions.

5

Divide-and-conquer and Recursion

An algorithmic principle

Let A(D) denote the application of an algorithm A to a set of data D, producing a result R. An important class of algorithms, of a type called divide-and-conquer, processes data in two distinct ways, according to whether the data is small or large:

- If the set D is small, and/or of simple structure, we invoke a simple algorithm A_0 whose application $A_0(D)$ yields R.
- If the set D is large, and/or of complex structure, we partition it into smaller subsets $d_1, \ldots, D_k$. For each i, apply $A(D_i)$ to yield a result R_i. Combine the results $r_1, \ldots, r_k$ to yield R.

This algorithmic principle of divide-and-conquer leads naturally to the notion of recursive procedures. The following example outlines the concept in a high-level notation, highlighting the role of parameters and local variables.

```
procedure A(D: data; var R: result);
var  D1, ... , Dk: data;  R1, ... , Rk: result;
begin
   if  simple(D) then R := A0(D)
     else { D1, ... , Dk := partition(D);
            R1 := A(D1); ... ; Rk := A(Dk);
            R := combine(R1, ... , Rk) }
end;
```

Notice how an initial data set D spawns sets $d_1, \ldots, D_k$ which, in turn, spawn children of their own. Thus the collection of all data sets generated by the partitioning scheme is a tree with root D. in order for the recursive procedure A(D) to terminate in all cases, the partitioning function must meet the following condition: Each branch of the partitioning tree, starting from the root D, eventually terminates with a data set D0 that satisfies the predicate 'simple(D0)', to which we can apply the algorithm.

Divide-and-conquer reduces a problem on data set D to k instances of the same problem on new sets d_1, ..., D_k that are "simpler" than the original set D. Simpler often means "has fewer elements", but any measure of "simplicity" that monotonically heads for the predicate 'simple' will do, when algorithm A_0 will finish the job. "D is simple" may mean "D has no elements", in which case A_0 may have to do nothing at all; or it may mean "D has exactly one element", and A_0 may just mar_k this element as having been visited.

The following sections show examples of divide-and-conquer algorithms. As we will see, the actual Workload is sometimes distributed unequally among different parts of the algorithm. in the sorting example, the step 'R:=combine(r_1, ... , r_k)' requires most of the Work; in the "Tower of Hanoi" problem, the application of algorithm A_0 takes the most effort.

Divide-and-conquer expressed as a diagram: merge sort

Suppose that we wish to sort a sequence of names alphabetically, as shown in Exhibit 5.1. We make use of the divide-and-conquer strategy by partitioning a "large" sequence D into two subsequences d_1 and d_2, sorting each subsequence, and then merging them back together into sorted order. $This_is$ our algorithm A(D). If D contains at most one element, we do nothing at all. A_0 is the identity algorithm, $A_0(D) = D$.

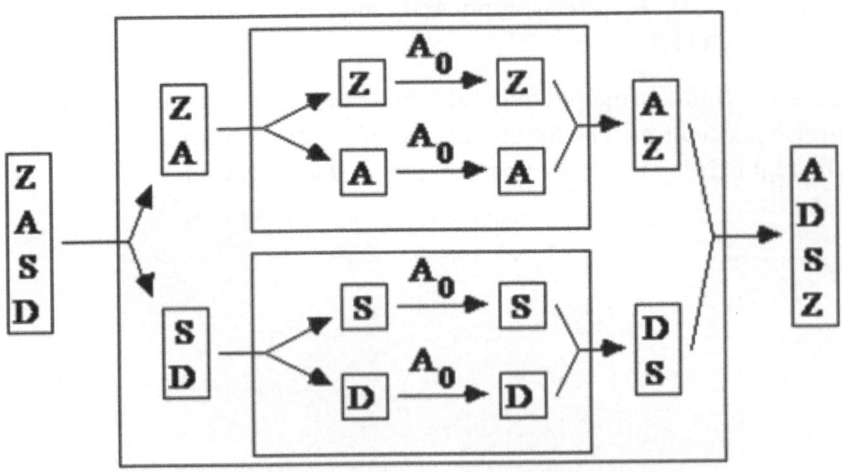

Exhibit 5.1: Sorting the sequence {Z, A, S, D} by using a divide-and-conquer scheme

```
procedure sort(var D: sequence);
var  D1, D2: sequence;

  function combine(D1, D2: sequence): sequence;
  begin  { combine }
    merge the two sorted sequences D1 and D2
    into a single sorted sequence D';
    return(D')
  end;  { combine }

begin  { sort}
  if  |D| > 1  then  { split D into two sequences D1 and D2 of
equal size;
        sort(D1);  sort(D2);  D := combine(D1, D2) }
  { if |D| ≤ 1, D is trivially sorted, do nothing }
end;  { sort }
```

In the chapter on "sorting and its complexity", under the section "merging and merge sorts" we turn this divide- and-conquer scheme into a program.

Recursively defined trees

A *tree*, more precisely, a rooted, ordered tree, is a data type used primarily to model any type of hierarchical organization. its primitive parts are *nodes* and *leaves.* it has a distinguished node called the *root*, which, in violation of nature, is typically drawn at the top of the page, with the tree growing downward. Each node has a certain number of children, either leaves or nodes; leaves have no children. The exact definition of such trees can differ slightly with respect to details and terminology. We may define a *binary tree*, for example, by the condition that each node has either exactly, or at most, two children.

The pictorial grammar shown in Exhibit 5.2 captures this recursive definition of 'binary tree' and fixes the details left unspecified by the verbal description above. it uses an *alphabet* of three symbols: the nonterminal 'tree symbol', which is also the start symbol; and two terminal symbols, for 'node' and for 'leaf'.

Exhibit 5.2: The three symbols of the alphabet of a tree grammar

There are two production or rewriting rules, p_1 and p_2 (Exhibit 5.3). The *derivation* shown in Exhibit 5.4 illustrates the application of the production rules to generate a tree from the nonterminal start symbol.

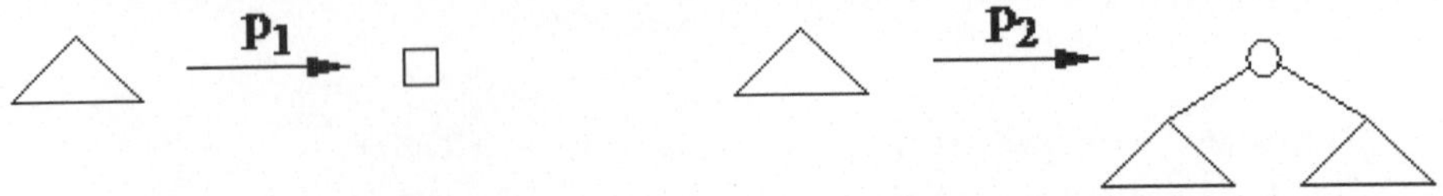

Exhibit 5.3: Rule p_1 generates a leaf, rule p_2 generates a node and two new trees

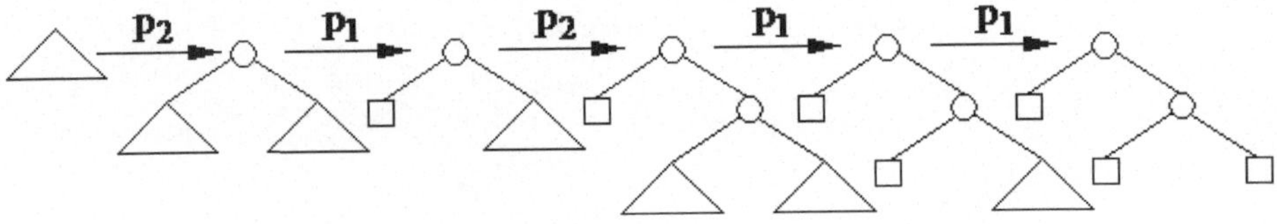

Exhibit 5.4: One way to derive the tree at right

We may make the production rules more detailed by explicitly naming the coordinates associated with each symbol. On a display device such as a computer screen, the x- and y-values of a point are typically Cartesian coordinates with the origin in the upper-left corner. The x-valuesincrease toward the bottom and the y-valuesincrease toward the right of the display. Let (x, y) denote the screen position associated with a particular symbol, and let d denote the depth of a node in the tree. The root has depth 0, and the children of a node with depth d have depth d^{+1}. The different levels of the tree are separated by some constant distance s. The separation between siblings$_i$s determined by a (rapidly decreasing) function t(d) which takes as argument the depth of the siblings and depends on the drawing size of the symbols and the resolution of the screen. These more detailed productions are shown in Exhibit 5.5.

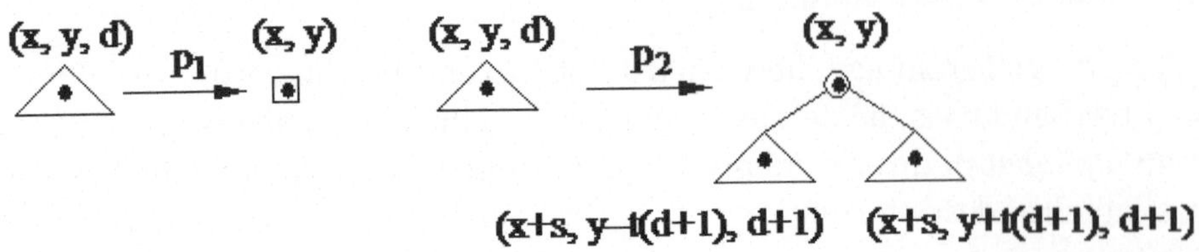

Exhibit 5.5: Adding coordinate information to productionsin order to control graphic layout The translation of these two rulesinto high-level code is now plain:

```
procedure p1(x, y: coordinate);
begin
   eraseTreeSymbol(x, y);
   drawLeafSymbol(x, y)
end;

procedure p2(x, y: coordinate; d: level);
begin
   eraseTreeSymbol(x, y);
   drawNodeSymbol(x, y);
   drawTreeSymbol(x + s, y - t(d + 1));
   drawTreeSymbol(x + s, y + t(d + 1))
end;
```

If we choose $t(d) = c \cdot 2^{-d}$, these two procedures produce the display shown in Exhibit 5.6 of the tree generated in Exhibit 5.4.

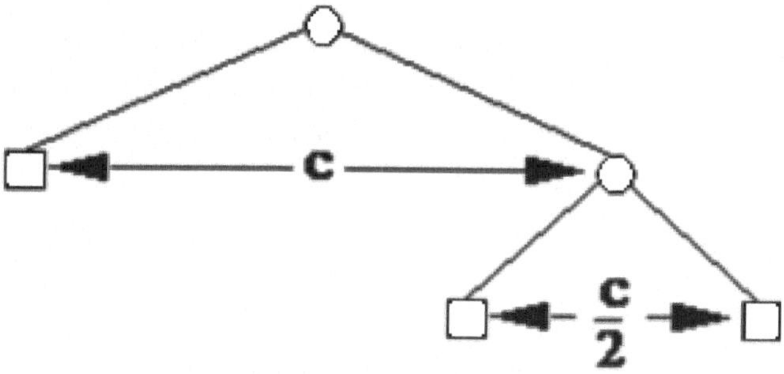

Exhibit 5.6: Sample layout obtained by halving horizontal displacement at each successive level

Technical remar$_k$ about the details of defining binary trees: Our grammar forces every node to have exactly two children: A child may be a node or a leaf. This lets us subsume two frequently occurring classes of binary trees under one common definition.

- 0-2 (binary) trees. We may identify leaves and nodes, making no distinction between them (replace the squares by circlesin Exhibit 5.3 and Exhibit 5.4). Every node in the new tree now has either zero or two children, but not one. The smallest tree has a single node, the root.
- (Arbitrary) Binary trees. Ignore the leaves (drop the squaresin Exhibit 5.3 and Exhibit 5.4 and the branches leading into a square). Every node in the new tree now has either zero, one, or two children. The smallest tree (which consisted of a single leaf) now has no node at all; it is empty.

For clarity's sake, the following examples use the terminology of nodes and leavesintroduced in the defining grammar. in some instances we point out what happens under the interpretation that leaves are dropped.

Recursive tree traversal

Recursion is a powerful tool for programming divide-and-conquer algorithmsin a straightforward manner. in particular, when the data to be processed is defined recursively, a recursive processing algorithm that mirrors the structure of the data is most natural. The recursive tree traversal procedure below illustrates this point.

Traversing a tree (in general: a graph, a data structure) means visiting every node and every leaf in an orderly sequence, beginning and ending at the root. What needs to be done at each node and each leaf is of no concern to the traversal algorithm, so we merely designate that by a call to a 'procedure visit( )'. You may think of inspecting the contents of all nodes and/or leaves, and writing them to a file.

Recursive tree traversals use divide-and-conquer to decompose a tree into its subtrees: At each node visited along the way, the two subtrees L and R to the left and right of this node must be traversed. There are three natural ways to sequence the node visit and the subtree traversals:

```
1. node; L; R   { preorder, or prefix }
2. L; node; R   { inorder or infix }
3. L; R; node   { postorder or suffix }
```

The following example translates this traversal algorithm into a recursive procedure:

```
procedure traverse(T: tree);
   { preorder, inorder, or postorder traversal of tree T with
leaves }
 begin
   if  leaf(T) then visitleaf(T)
     else { T is composite }
       { visit1(root(T));
         traverse(leftsubtree(T));
         visit2(root(T));
         traverse(rightsubtree(T);
         visit3(root(T)) }
 end;
```

When leaves are ignored (i.e. a tree consisting of a single leaf is considered to be empty), the procedure body becomes slightly simpler:

```
if  not empty(T)  then  { ... }
```

To accomplish the k-th traversal scheme (k = 1, 2, 3), 'visitk' performs the desired operation on the node, while the other two visits do nothing. If all three visits print out the name of the node, we obtain a sequence of node names called 'triple tree traversal', shown in Exhibit 5.7 along with the three traversal orders of which it is composed. During the traversal the nodes are visited in the following sequence:

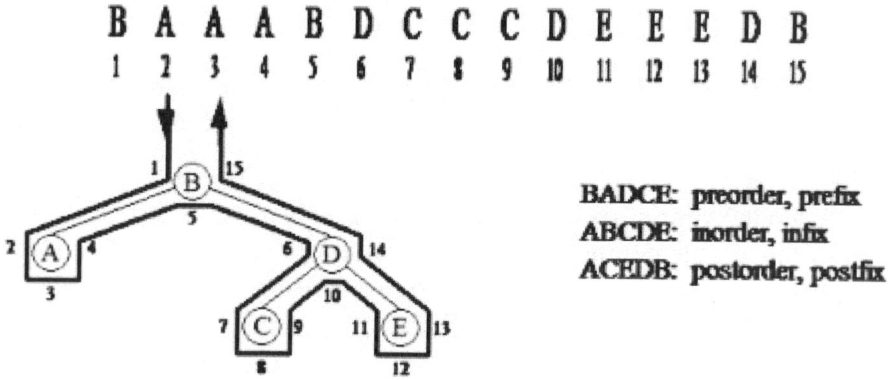

Exhibit 5.7: Three standard orders merged into a triple tree traversal

Recursion versusiteration: the Tower of Hanoi

The "Tower of Hanoi" is a stack of n disks of different sizes, held in place by a tall peg (Exhibit 5.8). The task is to transfer the tower from source peg S to a target peg T via an intermediate peg I, one disk at a time, without ever placing a larger disk on a smaller one. in this case the data set D is a tower of n disks, and the divide-and-conquer algorithm A partitions D asymmetrically into a small "tower" consisting of a single disk (the largest, at the bottom of the pile) and another tower D' (usually larger, but conceivably empty) consisting of the n – 1 topmost disks. The puzzle is solved recursively in three steps:

```
1. Transfer D' to the intermediate peg I.
2. Move the largest disk to the target peg T.
3. Transfer D' on top of the largest disk at the target peg T.
```

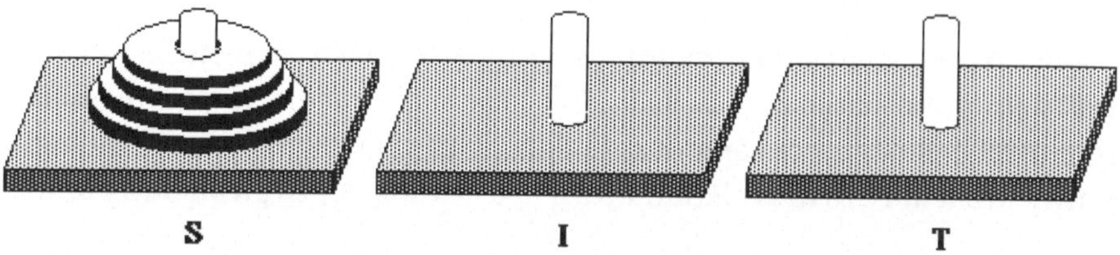

Exhibit 5.8: initial configuration of the Tower of Hanoi.

Step 1 deserves more explanation. How do we transfer the n – 1 topmost disks from one peg to another? Notice that they themselves constitute a tower, to which we may apply the same three-step algorithm. Thus we are presented with successively simpler problems to solve, namely, transferring the n – 1 topmost disks from one peg to another, for decreasing n, until finally, for n = 0, we do nothing.

```
procedure Hanoi(n: integer; x, y, z: peg);
  { transfer a tower with n disks from peg x, via y, to z }
begin
  if  n > 0  then  { Hanoi(n - 1, x, z, y);  move(x, z);  Hanoi(n -
1, y, x, z) }
end;
```

Recursion has the advantage of intuitive clarity. Elegant and efficient as this solution may be, there is some complexity hidden in the bookkeeping implied by recursion.

The following procedure is an equally elegant and more efficient it*erative* solution to this problem. it assumes that the pegs are cyclically ordered, and the target peg where the disks will first come to rest depends on this order and on the parity of n (Exhibit 5.9). For odd values of n, 'iterativeHanoi' moves the tower to peg I, for even values of n, to peg T.

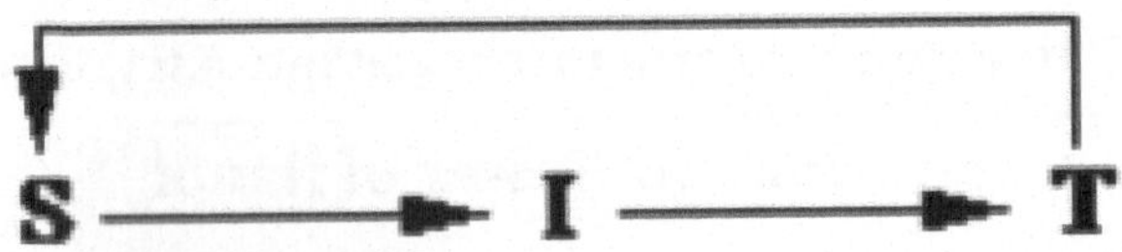

Exhibit 5.9: Cyclic order of the pegs.

```
procedure IterativeHanoi(n: integer);
var  odd: boolean;   { odd represents the parity of the move }
begin
  odd := true;
  repeat
    case  odd  of
      true:  transfer smallest disk cyclically to next peg;
      false: make the only legal move leaving the smallest in place
    end;
    odd := not odd
  until  entire tower is on target peg
end;
```

Exercise: recursive or iterative pictures?

Chapter 4 presented some beautiful examples of recursive pictures, which would be hard to program without recursion. But for simple recursive picturesiteration is just as natural. Specify a convenient set of graphics primitives and use them to write an iterative procedure to draw Exhibit 5.10 to a nesting depth given by a parameter d.

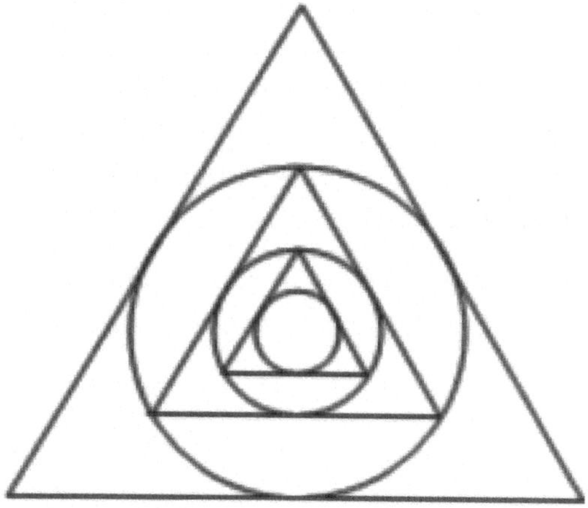

Exhibit 5.10: interleaved circles and equilateral triangles cause the radius to be exactly halved at each step.

Solution

There are many choices of suitable primitives and many ways to program these pictures. Specifying an equilateral triangle by its center and the radius of its circumscribed circle simplifies the notation. Assume that we may use the procedures:

```
  procedure circle(x, y, r: real); { coordinates of center and
radius }
  procedure equitr(x, y, r: real); { center and radius of
circumscribed circle}

 procedure citr(x, y, r: real; d: integer);
 var  vr: real;   { variable radius }
      i: integer;
 begin
   vr := r;
   for i := 1 to d do  { equitr(x, y, vr);  vr := vr/2;  circle(x, y,
vr) }
   { show that the radius of consecutively nested circles gets
     exactly halved at each step }
 end;
```

The flag of Alfanumerica: an algorithmic novel on iteration and recursion

In the process of automating its flag industry, the United States of Alfanumerica announced a competition for the most elegant program to print its flag:

```
                ****************  k blanks followed by k stars
        ********        ********  twice (k/2 blanks followed by k/2 stars)
    ****    ****    ****    ****  ...
  **  **  **  **  **  **  **  **  continue doubling and halving
 * * * * * * * * * * * * * * * *  down to runs length of 1.
```

All solutions submitted to the prize committee fell into one of two classes, the iterative and recursive programs. The proponents of these two algorithm design principles could not agree on a winner, and the selection process spar$_k$ed a civil war that split the nation into two: the iterative States of Alfanumerica (ISA) and the Recursive States of Alfanumerica (RSA). Both nations fly the same flag but use entirely different production algorithms.

- Write a to print the ISA flag, using an iterative algorithm, of course. Assume that k is a power of 2 and k ≤ (half the line length of the printer).

```
procedure ISA(k: integer);
```

- Explain why the printer industry in RSA is much more innovative than the one in ISA. All modern RSA printers include operations for positioning the writing head anywhere within a line, and line feed Works both forward and backward.
- Specify the precise operations for some RSA printer of your design. Using these operations, write a recursive to print the RSA flag.

```
procedure RSA(k: integer);
```

- Explain an unforeseen consequence of this drive to automate the flag industry of Alfanumerica: in both ISA and RSA, a growing number of flags can be seen fluttering in the breeze turned around by 90°.

Exercises

- Whereas divide-and-conquer algorithms usually attempt to divide the data in equal halves, the recursive Tower of Hanoi procedure presented in the section 'Recursion versusiteration: The Tower of Hanoi" divides the data in a very asymmetric manner: a single disk versus n – 1 disks. Why?
- Prove by induction on n that the iterative program 'iterativeHanoi' solves the problem in 2n–1 iterations.

6

Syntax

Syntax and semantics

Computer science has borrowed some important concepts from the study of natural languages (e.g. the notions of syntax and semantics). *Syntax* rules prescribe how the sentences of a language are formed, independently of their meaning. *Semantics* deals with their meaning. The two sentences "The child draws the horse" and "The horse draws the child" are both syntactically correct according to the accepted rules of grammar. The first sentence clearly makes sense, whereas the second sentence is baffling: perhaps senseless (if "draw" means "drawing a picture"), perhaps meaningful (if "draw" means "pull"). Semantic aspects—whether a sentence is meaningful or not, and if so, what it means—are much more difficult to formalize and decide than syntactic issues.

However, the analogy between natural languages and programming languages does not go very far. The choice of English words and phrases such as "begin", "end", "goto", "if-then-else" lends a programming language a superficial similarity to natural language, but no more. The possibility of verbal encoding of mathematical formulasinto pseudo-English has deliberately been built into COBOL; for example, "compute velocity times time giving distance" is nothing but *syntactic sugar* for "distance := velocity · time". Much more important is the distinction that natural languages are not rigorously defined (neither the vocabulary, nor the syntax, and certainly not the semantics), whereas programming languages should be defined according to a rigorous formalism. Programming languages are much closer to the *formal* notations of mathematics than to natural languages, and *programming notation* would be a more accurate term.

The *lexical part* of a modern programming language [the alphabet, the set of reserved words, the construction rules for the identifiers (i.e. the equivalent to the vocabulary of a natural language) and the *syntax* are usually defined formally. However, system-dependent differences are not always described precisely. The compiler often determinesin detail the syntactic correctness of a program with respect to a certain system (computer and operating system). The semantics of a programming language could also be defined formally, but this is rarely done, because formal semantic definitions are extensive and difficult to read.

The syntax of a programming language is not as important as the semantics, but good understanding of the syntax often helpsin understanding the language. With some practice one can often guess the semantics from the syntax, since the syntax of a well-designed programming language is the frame that supports the semantics.

Grammars and their representation: syntax diagrams and EBNF

The syntax of modern programming languages is defined by *grammars*. These are mostly of a type called *context-free grammars*, or close variants thereof, and can be given in different notations. *Backus-Naur form (BNF)*, a milestone in the development of programming languages, wasintroduced in 1960 to define the syntax of Algol. it is the basis for other notations used today, such as *EBNF (extended BNF)* and graphical representations such as *syntax diagrams*. EBNF and syntax diagrams are syntactic notations that describe exactly the *context-free* grammars of formal language theory.

Recursion is a central theme of all these notations: the syntactic correctness and structure of a large program text are reduced to the syntactic correctness and structure of its textual components. Other common notionsinclude: *terminal* symbol, *nonterminal* symbol, and *productions* or *rewriting rules* that describe how nonterminal symbols generate strings of symbols.

The set of terminal symbols forms the *alphabet* of a language, the symbols from which the sentences are built. in EBNF a terminal symbol is enclosed in single quotation marks; in syntax diagrams a terminal symbol is represented by writing it in an oval:

'+'

Nonterminal symbols represent syntactic entities: statements, declarations, or expressions. Each nonterminal symbol is given a name consisting of a sequence of letters and digits, where the first character must be a letter. in syntax diagrams a nonterminal symbol is represented by writing its name in a rectangular box:

T

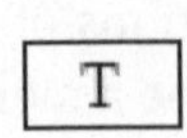

If a construct consists of the catenation of constructs A and B, this is expressed by

AB

If a construct consists of either A or B, this is denoted by

A | B

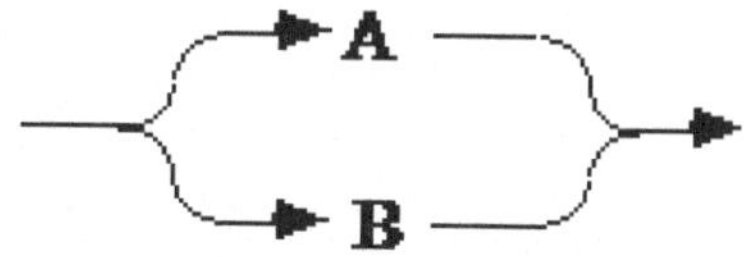

If a construct may be either construct A or nothing, this is expressed by

[A]

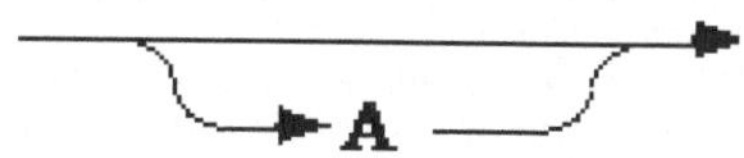

If a construct consists of the catenation of any number of A's (including none), this is denoted by

{ A }

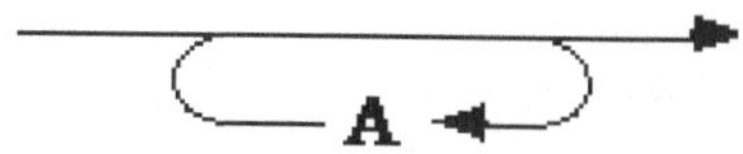

In EBNF parentheses may be used to group entities [e.g. (A | B)].

For each nonterminal symbol there must be at least one production that describes how this syntactic entity is formed from other terminal or nonterminal symbols using the composition constructs above:

A = ------ .

The following examples show productions and the constructs they generate. A, B, C, D may denote terminal or nonterminal symbols.

S = (A | B) (C | D) .

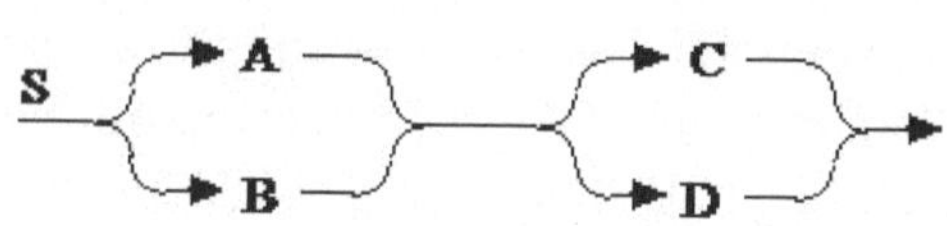

generates: AC AD BC BD

T = A [B] C .

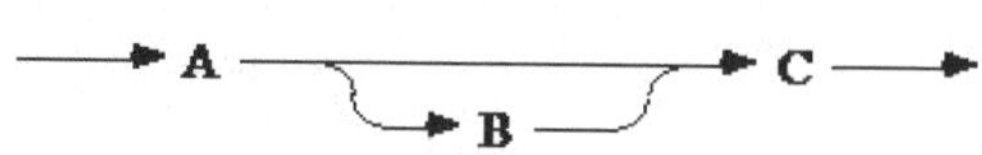

generates: AC ABC

U = A { B A } .

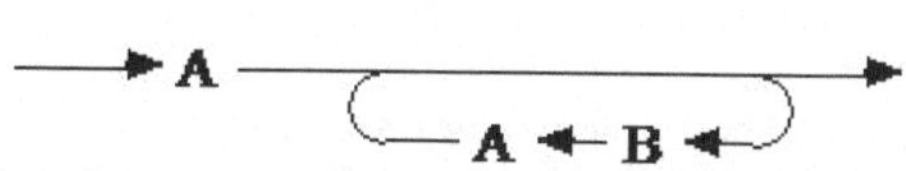

or

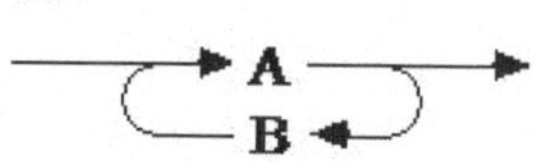

generates: A ABA ABABA ABABABA ...

EBNF is a formal language over a finite alphabet of symbolsintroduced above, built according to the rules explained above. Thusit is no great surprise that EBNF can be used to define itself. We use the following names for syntactic entities:

stmt	A syntactic equation.
expr	A list of alternative terms.
term	A concatenation of factors.
factor	A single syntactic entity or parenthesized expression.
nts	Nonterminal symbol that denotes a syntactic entity. it consists of a sequence of letters and digits where the first character must be a letter.
ts	Terminal symbol that belongs to the defined language's vocabulary. Since the vocabulary depends on the language to be defined there is no production for ts.

EBNF is now defined by the following productions:

EBNF = { stmt } .

stmt = nts '=' expr '.' .

expr = term { '|' term } .

term = factor { factor } .

factor = nts | ts | '(' expr ')' | '[' expr ']' | '{' expr '}' .

nts= letter { letter | digit } .

Example: syntax of simple expressions

The following productions for the three nonterminals E(xpression), T(erm), and F(actor) can be traced back to Algol 60. They form the core of all grammars for arithmetic expressions. We have simplified this grammar to define a class of expressions that lacks, for example, a unary minus operator and many other convenient notations. These details are but not important for our purpose: namely, understanding how this grammar assigns the correct structure to each expression. We have further simplified the grammar so that constants and variables are replaced by the single terminal symbol # (Exhibit 6.1):

```
E = T { ( '+' | '-' ) T } .
T = F { ( '·' | '/' ) F } .
F = '#' | '(' E ')' .
```

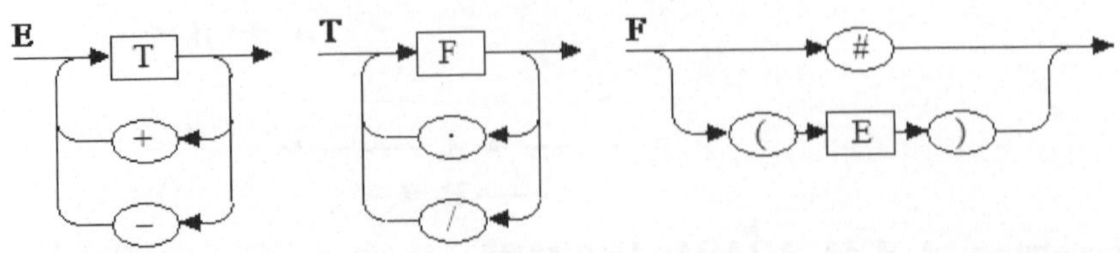

Exhibit 6.1: Syntax diagrams for simple arithmetic expressions.

From the nonterminal E we can *derive* different expressions. in the opposite direction we start with a sequence of terminal symbols and check by *syntactic analysis*, or *parsing*, whether a given sequence is a valid expression. If this is the case the grammar assigns to this expression a unique tree structure, the *parse tree* (Exhibit 6.2).

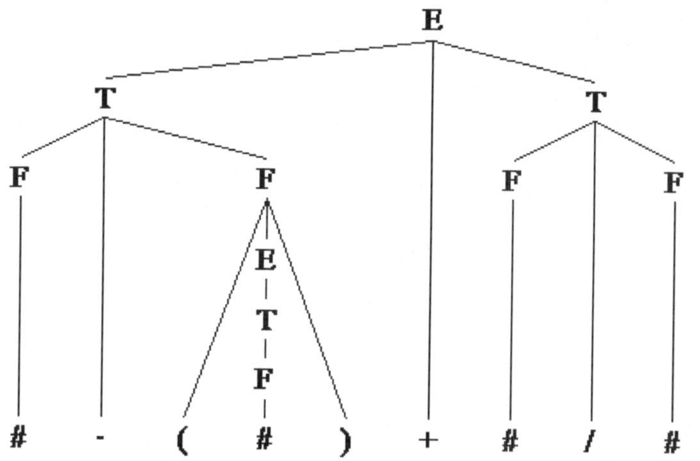

Exhibit 6.2: Parse tree for the expression # · (#) + # / # .

Exercise: syntax diagrams for palindromes

A palindrome is a string that reads the same when read forward or backward. *Examples:* 0110 and 01010. 01 is not a palindrome, asit differs from its reverse 10.

- ❖ What is the shortest palindrome?
- ❖ Specify the syntax of palindromes over the alphabet {0, 1} in EBNF-notation, and by drawing syntax diagrams.

Solution

- ❖ The shortest palindrome is the null or empty string.
- ❖ S = ['0' | '1'] | '0' S '0' | '1' S '1' (Exhibit 6.3).

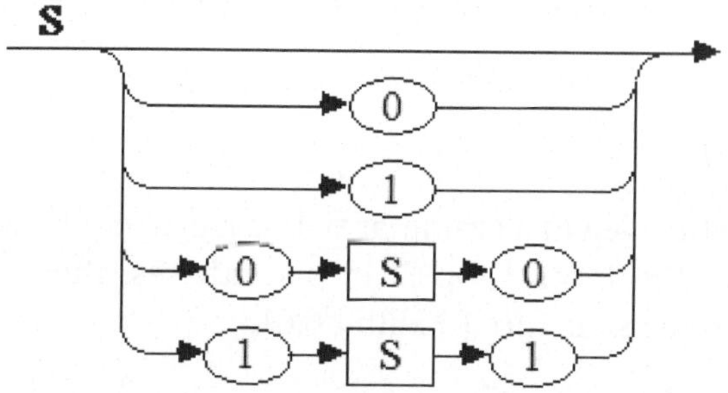

Exhibit 6.3: Syntax diagram for palindromes

An overly simple syntax for simple expressions

Why does the grammar given in previous section contain *term* and *factor*? An expression E that involves only binary operators (e.g. +, –, · and /) is either a primitive operand, abbreviated as #, or of the form 'E op E'. Consider a "simpler" grammar for simple, parenthesis-free expressions (Exhibit 6.4):

```
E = '#' | E ( '+' | '-' | '·' | '/' ) E .
```

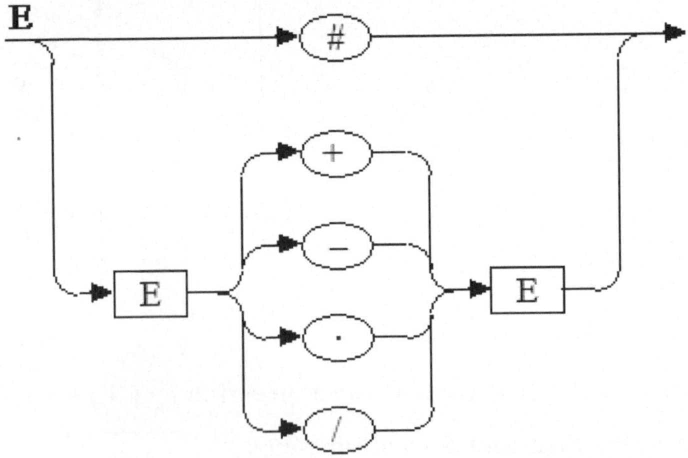

Exhibit 6.4: A syntax that generates parse trees of ambiguous structure

Now the expression # · # + # can be derived from E in two different ways (Exhibit 6.5). Such an *ambiguous* grammar is useless since we want to derive the semantic interpretation from the syntactic structure, and the tree at the left contradicts the conventional operator precedence of · over +.

Exhibit 6.5: Two incompatible structures for the expression # · # + # .

"Everything should be explained as simply as possible, but not simpler."

(Albert Einstein)

We can salvage the idea of a grammar with a single nonterminal E by enclosing every expression of the form 'E op E' in parentheses, thus ensuring that every expression has a unique structure (Exhibit 6.6):

```
E = '#' | '(' E ( '+' | '-' | '·' | '/' ) E ')' .
```

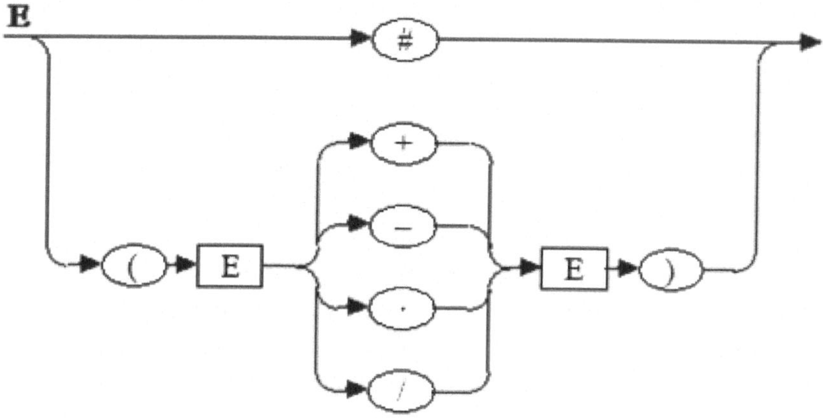

Exhibit 6.6: Parentheses serve to restore unique structure.

In doing so we change the language. The more complex grammar with three nonterminals E(xpression, T(erm), and F(actor) lets us write expressions that are only partially parenthesized and assigns to them a unique structure compatible with our priority conventions: · and / have higher priority than + and-.

Exercise: the ambiguity of the dangling "else"

The problem of the *dangl*ing *"else"* is an example of a syntax chosen to be "too simple" for the task it is supposed to handle. The syntax of several programming languages (e.g., Pascal) assigns to nested 'if-then[-else]' statements an ambiguous structure. it is left to the semantics of the language to disambiguate.

Let E, E_1, E_2, ... denote Boolean expressions, S, S_1, S_2, ... statements. Pascal syntax allows two types of if statements:

```
if E then S
```

and

```
if E then S else S
```

- Draw one syntax diagram that expresses both of these syntactic possibilities.
- Show all the possible syntactic structures of the statement

$$\texttt{if } E_1 \texttt{ then if } E_2 \texttt{ then } S_1 \texttt{ else } S_2$$

- Propose a small modification to the Pascal language that avoids the syntactic ambiguity of the dangling else. Show that in your modified Pascal any arbitrarily nested structure of 'if-then' and 'if-then-else' statements must have a unique syntactic structure.

Parenthesis-free notation for arithmetic expressions

In the usual in*fix* notation for arithmetic expressions a binary operator is written between its two operands. Even with operator precedence conventions, some parentheses are required to guarantee a unique syntactic structure. The selective use of parentheses complicates the syntax of infix expressions: Syntax analysis, interpretative evaluation, and code generation all become more complicated.

Parenthesis-free or Polish notation (named for the Polish logician Jan Lukasiewicz) is a simpler notation for arithmetic expressions. All operators are systematically written either before (*prefix* notation) or after (*postfix* or *suffix* notation) the operands to which they apply. We restrict our examples to the binary operators +, –, · and /. Operators with different *arities* (i.e. different numbers of arguments) are easily handled provided that the number of arguments used is uniquely determined by the operator symbol. To introduce the unary minus we simply need a different symbol than for the binary minus.

```
Infix    a+b  a+(b·c) (a+b)·c
Prefix   +ab  +a·bc   ·+abc
Postfix  ab+  abc·+   ab+c·
```

Postfix notation mirrors the sequence of operations performed during the evaluation of an expression. 'ab+' isinterpreted as: load a (find first operand); load b (find the second operand); add both. The syntax of arithmetic expressionsin postfix notation is determined by the following grammar:

```
S = '#' | S S ( '+' | '-' | '·' | '/' )
```

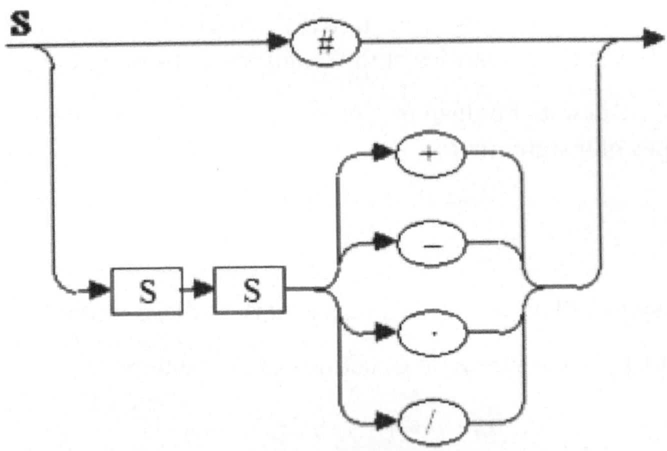

Exhibit 6.7: Suffix expressions have a unique structure even without the use of parentheses.

Exercises

- Consider the following syntax, given in EBNF: S = A.

S = A.

A = B | 'IF' A 'THEN' A 'ELSE' A.

B = C | B 'OR' C.

C = D | C 'AND' D.

D = 'x' | '(' A ')' | 'NOT' D.

- Determine the sets of terminal and nonterminal symbols.
- Give the syntax diagrams corresponding to the rules above.
- Which of the following expressions is correct corresponding to the given syntax? For the correct expressions show how they can be derived from the given rules:

x AND x

x NOT AND x

(x OR x) AND NOT x

IF x AND x THEN x OR x ELSE NOT x

x AND OR x

- Extend the grammar of Section 6.3 to include the 'unary minus' (i.e. an arithmetic operator that turns any expression into its negative, asin –x). Do this under two different assumptions:
- The unary minus is denoted by a different character than the binary minus, say ¬.
- The character – is 'overloaded' (i.e. it is used to denote both unary and binary minus). For any specific occurrence of –, only the context determines which operator it designates.
- Extended Backus-Naur form and syntax diagrams

Define each of the four languages described below using both EBNF and syntax diagrams. Use the following conventions and notations: Uppercase letters denote nonterminal symbols. Lowercase letters and the three separators ';' '(' and ')' denote terminal symbols. "" stands for the empty or null string. Notice that the blank character does not occur in these languages, so we use it to separate distinct sentences.

L ::= a | b | ... | z Letter

D ::= 0 | 1 | 2 | 3 | 4 | 5 | 6 | 7 | 8 | 9 Digit

S ::= D { D } Sequence of digits I ::= L { L | D } Identifier

- *Real numbers (constants)* in *Pascal*

  Examples: –3 + 3.14 10e–06 –10.0e6 but not 10e6

- *Nonnested lists of identifiers* (including the empty list)

  Examples: () (a) (year, month, day) but not (a,(b)) and not “”

- *Nested lists of identifiers* (including empty lists)

  Examples: in addition to the examplesin part (b), we have lists such as

  ((),()) (a, ()) (name, (first, middle, last)) but not (a)(b) and not “”

- *Parentheses expressions*

  Almost the same problem as part (c), except that we allow the null string, we omit identifiers and commas, and we allow multiple outermost pairs of parentheses.

  Examples: “” () ()() ()(()) ()(()())()()

- Use both syntax diagrams and EBNF to define the repeated if-then-else statement:

```
if B1 then S1 elsif B2 then S2 elsif ... else S
```

7

Syntax Analysis

The role of syntax analysis

The syntax of a language is the skeleton that carries the semantics. Therefore, we will try to get as much Work as possible done as a side effect of syntax analysis; for example, compiling a program (i.e. translating it from one language into another) is a mainly semantic task. However, a good language and compiler are designed in such a way that syntax analysis determines where to start with the translation process. Many processesin computer science are syntax-driven in this sense. Hence syntax analysis$_{i}$s$_{i}$mportant. in this section we derive algorithms for syntax analysis directly from syntax diagrams. These algorithms reflect the recursive nature of the underlying grammars. A program for syntax analysis$_{i}$s called a *parser*.

The composition of a sentence can be represented by a *syntax tree* or *parse tree*. The root of the tree is the start symbol; the leaves represent the sentence to be recognized. The tree describes how a syntactically correct sentence can be derived from the start symbol by applying the productions of the underlying grammar (Exhibit 7.1).

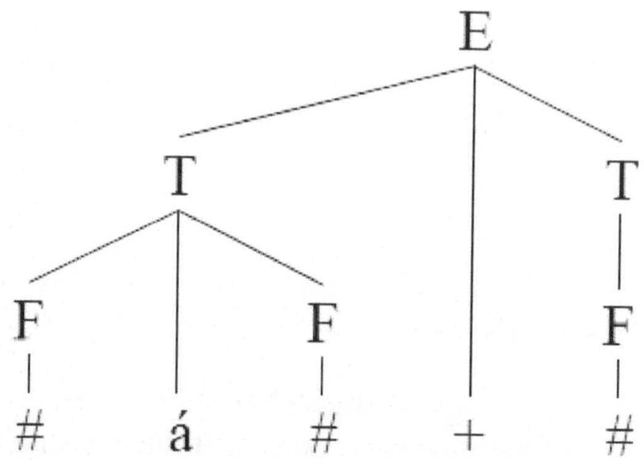

Exhibit 7.1: The unique parse tree for # · # + #

Top-down parsers begin with the start symbol as the goal of the analysis. in our example, "search for an E". The production for E tells us that we obtain an E if we find a sequence of T's separated by + or –. Hence we look for T's. The structure tree of an expression growsin this way as a sequence of goals from top (the root) to bottom (the leaves). While satisfying the goals (nonterminal symbols) the parser reads suitable symbols (terminal symbols) from left to right. in many practical cases a parser needs no backtrack. No backtracking is required if the current input symbol and the nonterminal to be expanded determine uniquely the production to be applied. A recursive- descent parser uses a set of recursive procedures to recognize itsinput with no backtracking.

Bottom-up methods build the structure tree from the leaves to the root. The text is reduced until the start symbol is obtained.

Syntax analysis of parenthesis-free expressions by counting

Syntax analysis can be very simple. Arithmetic expressionsin Polish notation are analyzed by counting. For sake of simplicity we assume that each operand in an arithmetic expression is denoted by the single character #. in order to decide whether a given string $c_1\ c_2 \dots c_n$ is a correct expression in postfix notation, we form an integer sequence $t_0, t_1, \dots, t_n$ according to the following rule:

$t_0 = 0.$

$t_{i+1} = t_i + 1$, if i > 0 and c_{i+1} is an operand.

$t_{i+1} = t_i - 1$, if i > 0 and c_{i+1} is an operator.

Example of a correct expression:

```
#   #   #   #   -   -   +   #   ·

c1  c2  c3  c4  c5  c6  c7  c8  c9
t0  t1  t2  t3  t4  t5  t6  t7  t8  t9

    0   1   2   3   4   3   2   1   2   1
```

Example of an incorrect expression (one operator is missing):

```
    #    #    #    +    ·    #    #    /
    c1   c2   c3   c4   c5   c6   c7   c8

t0  t1   t2   t3   t4   t5   t6   t7   t8
0   1    2    3    2    1    2    3    2
```

Theorem: The string $c_1\ c_2 \dots c_n$ over the alphabet $A = \{\#, +, -, \cdot, /\}$ is a syntactically correct expression in postfix notation if and only if the associated integer sequence $t_0, t_1, \dots$, tn satisfies the following conditions:

$$t_i > 0 \text{ for } 1 \leq i < n, \quad t_n = 1.$$

***Proof* ⇒:** Let c_1 c_2 ... c_n be a correct arithmetic expression in postfix notation. We prove by induction on the length n of the string that the corresponding integer sequence satisfies the conditions.

Base of induction: For n = 1 the only correct postfix expression is c_1 = #, and the sequence $t_0 = 0$, $t_1 = 1$ has the desired properties.

induction hypothesis: The theorem is correct for all expressions of length ≤ m.

induction step: Consider a correct postfix expression S of length m + 1 > 1 over the given alphabet A. Let s = $(s_i)_{0 \leq i \leq m}{}^{+1}$ be the integer sequence associated with S. Then s_is of the form S = T U Op, where 'Op' is an operator and T and U are correct postfix expressions of length j ≤ m and length k ≤ m, j + k = m. Let t = $(t_i)_{0 \leq i \leq j}$ and u = $(u_i)_{0 \leq i \leq k}$ be the integer sequences associated with T and U. We apply the induction hypothesis to T and U. The sequence s_is composed from t and u as follows:

$$\begin{array}{lllllllllll} s = & s_0, & s_1, & s_2, & \ldots, & s_j, & s_j+1, & s_j+2, & \ldots, & s_m, & s_{m+1} \\ & t_0, & t_1, & t_2, & \ldots, & t_j, & u_1+1, & u_2+1, & \ldots, & u_k+1, & 1 \\ & 0, & \ldots, & 1, & \ldots, & 2, & 1 & & & & \end{array}$$

Since t ends with 1, we add 1 to each element in u, and the subsequence therefore ends with $u_k + 1 = 2$. Finally, the operator 'O_p' decreases this element by 1, and s therefore ends with $s_m{}^{+1} = 1$. Since $t_i > 0$ for 1 ≤ i < j and ui > 0 for 1 ≤i < k, we obtain that $s_i > 0$ for 1 ≤ i < k + 1. Hence s has the desired properties, and we have proved one direction of the theorem.

***Proof* ⇐:** We prove by induction on the length n that a string c_1 c_2 ... c_n over A is a correct arithmetic expression in postfix notation if the associated integer sequence satisfies the conditions stated in the theorem.

Base of induction: For n = 1 the only sequence is $t_0 = 0$, $t_1 = 1$. it follows from the definition of the sequence that c_1 = #, which is a correct arithmetic expression in postfix notation.

induction hypothesis: The theorem is correct for all expressions of length ≤ m.

induction step: Let s = $(s_i)_{0 \leq i \leq m}{}^{+1}$ be the integer sequence associated with a string S = c_1 c_2 ... $c_m{}^{+1}$ of length m + 1 > 1 over the given alphabet A which satisfies the conditions stated in the theorem. Let j < m + 1 be the largest index with $s_j = 1$. Since s1 = 1 such an index j exists. Consider the substrings T = c_1 c_2 ... cj and U = cj cj^{+1} ... cm. The integer sequences $(s_i)_{0 \leq i \leq j}$ and $(s_i - 1)_{j \leq i \leq m}$ associated with T and U both satisfy the conditions stated in the theorem.

Hence we can apply the induction hypothesis and obtain that both T and U are correct postfix expressions. From the definition of the integer sequence we obtain that c_m^{+1} is an operand 'Op'. Since T and U are correct postfix expressions, $S = T\ U\ O_p$ is also a correct postfix expression, and the theorem is proved.

A similar proof shows that the syntactic structure of a postfix expression is unique. The integer sequence associated with a postfix expression is of practical importance: The sequence describes the depth of the stack during evaluation of the expression, and the largest number in the sequence is therefore the maximum number of storage cells needed.

Analysis by recursive descent

We return to the syntax of the simple arithmetic expressions of chapter 6 in the section "Example: syntax of simple expressions" (Exhibit 7.2). Using the expression # · (# – #) as an example, we show how these syntax diagrams are used to analyze any expressions by means of a technique called *recursive-descent parsing*. The progress of the analysis depends on the current state and the next symbol to be read: a lookahead of exactly one symbol suffices to avoid backtracking. in Exhibit 7.3 we move one step to the right after each symbol has been recognized, and we move vertically to step up or down in the recursion.

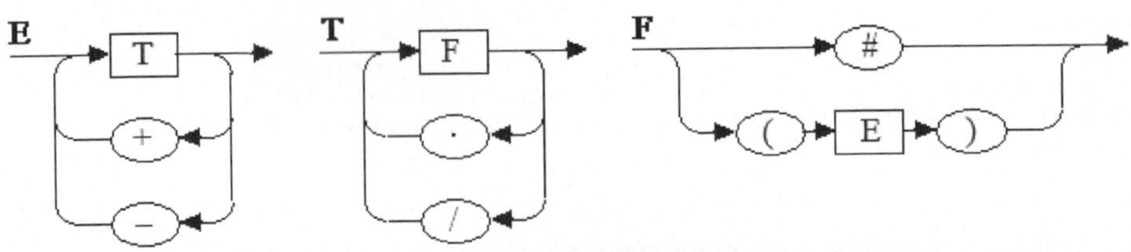

Exhibit 7.2: Standard syntax for simple arithmetic expressions (graphic does not match)

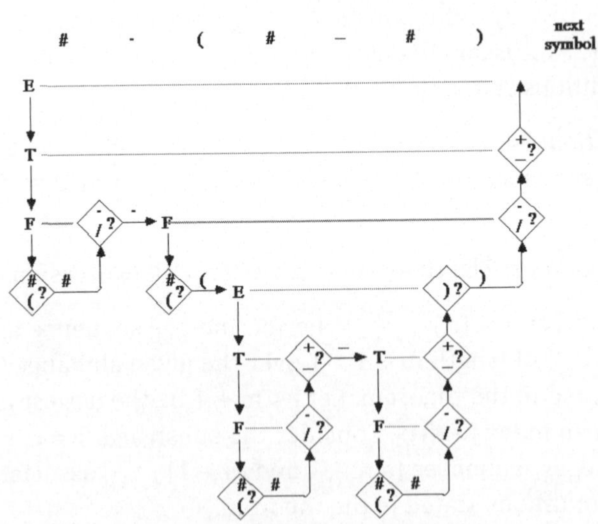

Exhibit 7.3: Trace of syntax analysis algorithm parsing the expression # · (# – #).

Turning syntax diagrams into a parser

in a programming language that allows recursion the three syntax diagrams for simple arithmetic expressions can be translated directly into procedures. A nonterminal symbol corresponds to a procedure call, a loop in the diagram generates a while loop, and a selection is translated into an if statement. When a procedure wants to delegate a goal it calls another, in cyclic order: E calls T calls F calls E, and so on. Procedures implementing such a recursive control structure are often called *recursive coroutines.*

The procedures that follow must be embedded into a program that provides the variable 'ch' and the procedures 'read' and 'error'. We assume that the procedure 'error' prints an error message and terminates the program. in a more sophisticated implementation, 'error' would return a message to the calling procedure (e.g. 'factor'). Then this error message is returned up the ladder of all recursive procedure calls active at the moment.

```
procedure expression;
begin
  term;
  while (ch = '+') or (ch = '–') do { read(ch);  term }
end;

procedure term;
begin
  factor;
  while (ch = '·') or (ch = '/') do { read(ch); factor }
end;

procedure factor;
begin
  if    ch = '(' then { read(ch); expression; if ch = ')' then read (ch) else error}
  elsif ch = '#' then read(ch)
  else error
end;
```

Before the first call of the procedure 'expression', a character has to be read into 'ch'. Furthermore, we assume that a correct expression is terminated by a period:

```
...
read(ch);  expression;  if  ch ≠ '.'  then  error;
...
```

Exercises

- ❖ Design recursive algorithms to translate the simple arithmetic expressions of chapter 6 in the section "Example: syntax of a simple expressions" into corresponding prefix and postfix expressions as defined in chapter 6 in the section "Parenthesis-free notation for arithmetic expressions". Same for the inverse translations.

- ❖ Using syntax diagrams and EBNF define a language of 'correctly nested parentheses expressions'. You have a bit of freedom (how much?) in defining exactly what is correctly nested and what is not, but obviously your definition must include expressions such as (), ((())), (()(())), and must exclude strings such as (,)(, ()) ()(.

- Design two parsing algorithms for your class of correctly nested parentheses expressions: one that Works by counting, the other through recursive descent.

Part-III

Objects, Algorithms, Programs

Computing with numbers and other objects

Since the introduction of computers four or five decades ago the meaning of the word *computation* has kept expanding. Whereas "computation" traditionally implied "numbers", today we routinely compute pictures, texts, and many other types of objects. When classified according to the types of objects being processed, three types of computer applications stand out prominently with respect to the influence they had on the development of computer science.

The first generation involved *numerical computing*, applied mainly to scientific and technical problems. Data to be processed consisted almost exclusively of numbers, or sets of numbers with a simple structure, such as vectors and matrices. Programs were characterized by long execution times but small sets of input and output data. Algorithms were more important than data structures, and many new numerical algorithms were invented. Lasting achievements of this first phase of computer applicationsinclude systematic study of numerical algorithms, error analysis, the concept of program libraries, and the first high-level programming languages, Fortran and Algol.

The second generation, hatched by the needs of commercial data processing, leads to the development of many new data structures. Business applications thrive on record keeping and updating, text and form processing, and report generation: there is not much computation in the numeric sense of the word, but a lot of reading, storing, moving, and printing of data. in other words, these applications are data intensive rather than computation intensive. By focusing attention on the problem of efficient management of large, dynamically varying data collections, this phase created one of the core disciplines of computer science: data structures, and corresponding algorithms for managing data, such as searching and sorting.

We are now in a third generation of computer applications, dominated by computing with geometric and pictorial objects. This change of emphasis was triggered by the advent of computers with bitmap graphics. in turn, this leads to the widespread use of sophisticated user interfaces that depend on graphics, and to a rapid increase in applications such as computer-aided design (CAD) and image processing and pattern recognition (in medicine, cartography, robot control). The young discipline of computational geometry has emerged in response to the growing importance of processing geometric and pictorial objects. it has created novel data structures and algorithms, some of which are presented in Parts V and VI.

Our selection of algorithmsin Part III reflects the breadth of applications whose history we have just sketched. We choose the simplest types of objects from each of these different domains of computation and some of the most concise and elegant algorithms designed to process them. The study of typical small programs is an essential part of programming. A large part of computer science consists of the knowledge of how typical problems can be solved; and the best way to gain such knowledge is to study the main ideas that make standard programs Work.

Algorithms and programs

Theoretical computer science treats *algorithm* as a formal concept, rigorously defined in a number of ways, such as Turing machines or lambda calculus. But in the context of programming, *algorithm* is typically used as an intuitive concept designed to help people express solutions to their problems. The formal counterpart of an algorithm is a procedure or program (fragment) that expresses the algorithm in a formally defined programming language. The process of formalizing an algorithm as a program typically requires many decisions: some superficial (e.g. what type of statement is chosen to set up a loop), some of great practical consequence (e.g. for a given range of values of n, is the algorithm's asymptotic complexity analysis relevant or misleading?).

We present algorithmsin whatever notation appears to convey the key ideas most clearly, and we have a clear preference for pictures. We present programsin an extended version of Pascal; readers should have little difficulty translating thisinto any programming language of their choice. Mastery of interesting small programs is the best way to get started in computer science. We encourage the reader to Work the examplesin detail.

The literature on algorithms. The development of new algorithms has been proceeding at a very rapid pace for several decades, and even a specialist can only stay abreast with the state of the art in some subfield, such as graph algorithms, numerical algorithms, or geometric algorithms. This rapid development is sure to continue unabated, particularly in the increasingly important field of parallel algorithms. The cutting edge of algorithm research is published in several journals that specialize in this research topic, including the *Journal of Algorithms* and

Algorithmica. This literature is generally accessible only after a student has studied a few textbooks on algorithms, such as [AHU 75], [Baa 88], [BB 88], [CLR 90], [GB 91], [HS 78], [Knu 73a], [Knu 81], [Knu 73b], [Man 89], [Meh 84a], [Meh 84b], [Meh 84c], [RND 77], [Sed 88], [Wil 86], and [Wir 86].

8

Truth Values, the Data type 'sET', and bit Acrobatics

Bits and boolean functions

The English mathematician George Boole (1815–1864) became one of the founders of symbolic logic when he endeavored to express logical argumentsin mathematical form. The goal of his 1854 book *The Laws Of Thought* was "to investigate the laws of those operations of the mind by which reasoning is performed; to give expression to them in the symbolic language of calculus. ..."

Truth values or *boolean values*, named in Boole's honor, possess the smallest possible useful domain: the binary domain, represented by yes/no, 1/0, true/false, T/F. in the late 1940s, as the use of binary arithmetic became standard and asinformation theory came to regard a two-valued quantity as the natural unit of information, the concise term *b*it was coined as an abbreviation of "binary digit". A bit, by any other name, is truly a primitive data element—at a sufficient level of detail, (almost) everything that happensin today's computers is bit manipulation. Just because bits are simple data quantities does not mean that processing them is necessarily simple, as we illustrate in this section by presenting some clever and efficient bit manipulation algorithms.

Boolean variables range over boolean values, and *boolean functions* take boolean arguments and produce boolean results. There are only four distinct boolean functions of a single boolean variable, among which 'not' is the most useful: it yields the complement of its argument (i.e. turns 0 into 1, and vice versa). The other three are the identity and the functions that yield the constants 0 and 1. There are 16 distinct boolean functions of two boolean variables, of which several are frequently used, in particular: 'and', 'or'; their negations 'nand', 'nor'; the exclusive-or 'xor'; and the implication '⊃'. These functions are defined as follows:

a	b	a and b	a or b	a nand b	a nor b	a xor b	$a \supset b$
0	0	0	0	1	1	0	1
0	1	0	1	1	0	1	1
1	0	0	1	1	0	1	0
1	1	1	1	0	0	0	1

Bits are the atomic data elements of today's computers, and most programming languages provide a data type 'boolean' and built-in operators for 'and', 'or', 'not'. To avoid the necessity for boolean expressions to be fully parenthesized, precedence relations are defined on these operators: 'not' takes precedence over 'and', which takes precedence over 'or'. Thus

x and not y or not x and y ⇔ ((x and (not y)) or ((not x) and y)).

What can you compute with boolean variables? Theoretically everything, since large finite domains can always be represented by a sufficient number of boolean variables: 16-bit integers, for example, use 16 boolean variables to represent the integer domain -2^{15}.. $2^{15}-1$. Boolean variables are often used for program optimization in practical problems where efficiency is important.

Swapping and crossovers: the versatile exclusive-or

Consider the swap statement x :=: y, which we use to abbreviate the cumbersome triple: t := x; x := y; y := t. On computers that provide bitwise boolean operations on registers, the swap operator :=: can be implemented efficiently without the use of a temporary variable.

The operator *exclusive-or*, often abbreviated as 'xor', is defined as

x xor y = x and not y or not x and y.

it yields true iff exactly one of its two argumentsis true.

The bitwise boolean operation z:= x op y on n-bit registers: x[1 .. n], y[1 .. n], z[1 .. n], is defined as

for i := 1 to n do z[i] := x[i] op y[i]

With a bitwise exclusive-or, the swap x :=: y can be programmed as

```
x := x xor y;   y := x xor y;   x := x xor y;
```

It still takes three statements, but no temporary variable. Given that registers are usually in short supply, and that a logical operation on registers is typically just as fast as an assignment, the latter code is preferable. Exhibit 8.1 traces the execution of this code on two 4-bit registers and shows exhaustively that the swap is performed correctly for all possible values of x and y.

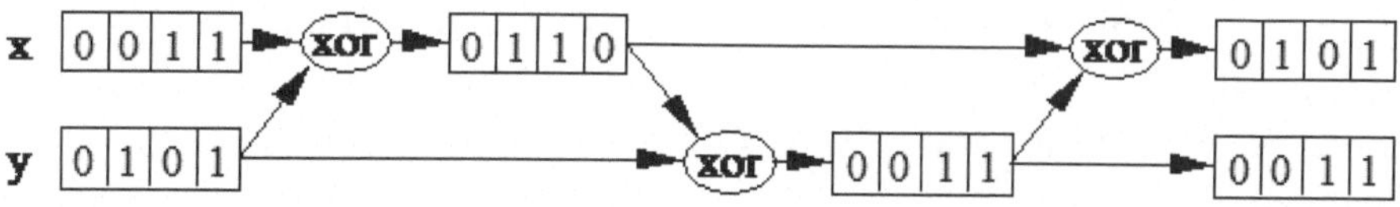

Exhibit 8.1: Trace of registers x and y under repeated exclusive-or operations.

Exercise: planar circuits without crossover of wires

The code above has yet another interpretation: How should we design a logical circuit that effects a logical crossover of two wires x and y while avoiding any physical crossover? If we had an 'xor' gate, the circuit diagram shown in Exhibit 8.2 would solve the problem. 'xor' gates must typically be realized as circuits built from simpler primitives, such as 'and', 'or', 'not'. Design a circuit consisting of 'and', 'or', 'not' gates only, which has the effect of crossing wires x and y while avoiding physical crossover.

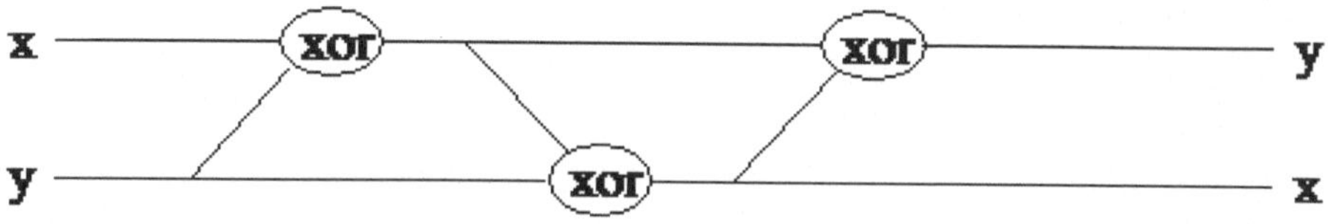

Exhibit 8.2: Three exclusive-or gates in series interchange values on two wires.

The bit sum or "population count"

A computer word is a fixed-length sequence of bits, call it a bit vector . Typical word lengths are 16, 32, or 64, and most instructions in most computers operate on all the bits in a word at the same time, in parallel. When efficiency is of great importance, it is worth exploiting to the utmost the bit parallelism built into the hardware of most computers. Today's programming languages often fail to refer explicitly to hardware features such as registers or wordsin memory, but it is usually possible to access individual bits if one knows the representation of integers or other data types. in this section we take the freedom to drop the constraint of *strong typing* built into Pascal and other modern languages. We interpret the content of a register or a word in memory as it suits the need of the moment: a bit string, an integer, or a set.

We are well aware of the dangers of such ambiguous interpretations: Programs become system and compiler dependent, and thus lose portability. If such ambiguity is localized in a single, small procedure, the danger may be kept under control, and the gain in efficiency may outweigh these drawbacks. in Pascal, for example, the type 'sE$_T$' is especially well suited to operate at the bit level. 'type s = set of (a, b, c)' consists of the 23 sets that can be formed from the three elements a, b, c. If the basic set M underlying the declaration of consists of n elements, then S has 2n elements. Usually, a value of type sisinternally represented by a vector of n contiguously allocated bits, one bit for each element of the set M. When computing with values of type S we operate on single bits using the boolean operators. The union of two sets of type s_is obtained by applying bitwise 'or', the intersection by applying bitwise 'and'. The complement of a set is obtained by applying bitwise 'not'.

type S = set of M

Example

M = {0, 1, ... , 7}

Set		Bit vector
		7 6 5 4 3 2 1 0 Elements
s_1	{0, 3, 4, 6}	0 1 0 1 1 0 0 1
s_2	{0, 1, 4, 5}	0 0 1 1 0 0 1 1
$s_1 \cup s_2$	{0, 1, 3, 4, 5, 6}	0 1 1 1 1 0 1 1
$s_1 \cap s_2$	{0, 4}	0 0 0 1 0 0 0 1
$\neg s_1$	{1, 2, 5, 7}	1 0 1 0 0 1 1 0

integers are represented on many small computers by 16 bits. We assume that a type '$w_1$6', for "word of length 16", can be defined. in Pascal, this might be

```
type  w16 = set of 0 .. 15;
```

A variable of type '$w_1$6' is a set of at most 16 elements represented as a vector of 16 bits.

Asking for the number of elements in a set s_is therefore the same as asking for the number of 1'sin the bit pattern that represents s. The operation that counts the number of elements in a set, or the number of 1'sin a word, is called the *population count* or bit *sum.* The bit sum is frequently used in inner loops of combinatorial calculations, and many a programmer has tried to make it as fast as possible. Let us look at four of these tries, beginning with the obvious.

Inspect every bit

```
function bitsum0(w: w16): integer;
var  i, c: integer;
begin
   c := 0;
   for  i := 0  to  15  do  { inspect every bit }
      if  i ∈ w {w[i] = 1}  then  c := c + 1;  { count the ones}
   return(c)
end;
```

Skip the zeros

Is there a faster way? The following algorithm looks mysterious and tricky. The expression w ∩ (w – 1) contains both an intersection operation '∩', which assumes that its operands are sets, and a subtraction, which assumes that w is an integer:

```
c := 0;
while  w ≠ 0  do  { c := c + 1;  w := w ∩ (w - 1) } ;
```

Such mixing makes sense only if we can rely on an implicit assumption on how sets and integers are represented as bit vectors. With the usual binary number representation, an example shows that when the body of the loop is executed once, the rightmost 1 of w is replaced by 0:

```
w                      1000100011001000

w - 1                  1000100011000111

w ∩ (w - 1)            1000100011000000
```

This clever code seems to look at the 1's only and skip over all the 0's: its loop is executed only as many times as there are 1'sin the word. This savings is worthwhile for long, sparsely populated words (few 1's and many 0's).

in the statement w := w ∩ (w – 1), w is used both as an integer (in w – 1) and as a set (as an operand in the intersection operation '∩'). Strongly typed languages, such as Pascal, do not allow such mixing of types. in the following function 'bitsum$_1$', the conversion routines 'w$_1$6toi' and 'itow$_1$6' are introduced to avoid this double interpretation of w. However, 'bitsum$_1$' is of interest only if such a type conversion requires no extra time (i.e. if one knows how sets and integers are represented internally).

```
function bitsum1(w: w16): integer;
var  c, i: integer;  w0, w1: w16;
begin
  w0 := w;   c := 0;
  while  w0 ≠ Ø  { empty set }  do  begin
    i := w16toi(w0);   { w16toi converts type w16 to integer }
    i := i - 1;
    w1 := itow16(i);   { itow16 converts type integer to w16 }
    w0 := w0 ∩ w1;   { intersection of two sets }
    c := c + 1
  end;
  return(c)
end;
```

Most languages provide some facility for permitting purely formal type conversions that result in no Work: 'EQUIVALENCE' statementsin Fortran, 'UNSPEC' in PL/1, variant recordsin Pascal. Such "conversions" are done merely by interpreting the contents of a given storage location in different ways.

Logarithmic bit sum

For a computer of word length n, the following algorithm computes the bit sum of a word w running through its loop only $\lceil \log_2 n \rceil$ times, as opposed to n times for '$bitsum_0$' or up to n times for '$bitsum_1$'. The following description holds for arbitrary n but is understood most easily if n = 2h.

The logarithmic bit sum Works on the familiar principle of divide-and-conquer. Let w denote a word consisting of n = 2h bits, and let S(w) be the bit sum of the bit string w. Split w into two halves and denote its left part by wL and its right part by wR. The bit sum obviously satisfies the recursive equation S(w) = S(wL) + S(wR). Repeating the same argument on the substrings wL and wR, and, in turn, on the substrings they create, we arrive at a process to compute S(w). This process terminates when we hit substrings of length 1 [i.e. substrings consisting of a single bit b; in this case we have S(b) = b]. Repeated halving leads to a recursive decomposition of w, and the bit sum is computed by a tree of n – 1 additions as shown below for n = 4 (Exhibit 8.3).

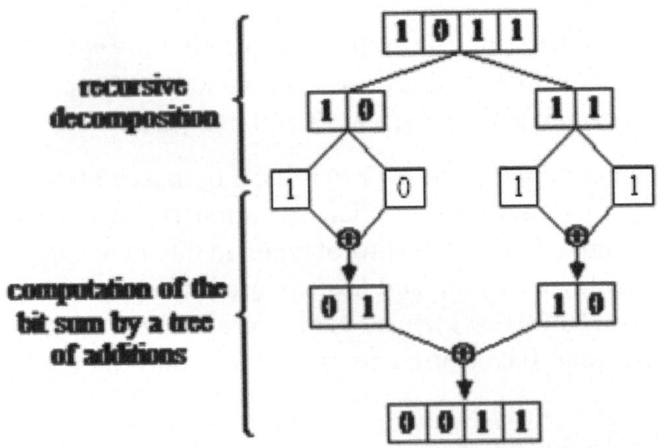

Exhibit 8.3: Logarithmic bit sum algorithm as a result of divide-and-conquer.

This approach of treating both parts of w symmetrically and repeated halving leads to a computation of depth h = $\lceil \log_2 n \rceil$. To obtain a logarithmic bit sum, we apply the additional trick of performing many additions in parallel. Notice that the total length of all operands on the same level is always n. Thus we can pack them into a single word and, if we arrange things cleverly, perform all the additions at the same level in one machine operation, an addition of two n-bit words.

Exhibit 8.4 shows how a number of the additions on short strings are carried out by a *single* addition on long strings. S(w) now denotes not only the bit sum but also its binary representation, padded with zeros to the left so as to have the appropriate length. Since the same algorithm is being applied to wL and wR, and since wL and wR are of equal length, exactly the same operations are performed at each stage on wL and its parts as on wR and its corresponding parts. Thus if the operations of addition and shifting operate on words of length n, a single one of these operations can be interpreted as performing many of the same operations on the shorter parts into which w has been split. This logarithmic speedup Works up to the word length of the computer. For n = 64, for example, recursive splitting generates six levels and translates into six iterations of the loop below.

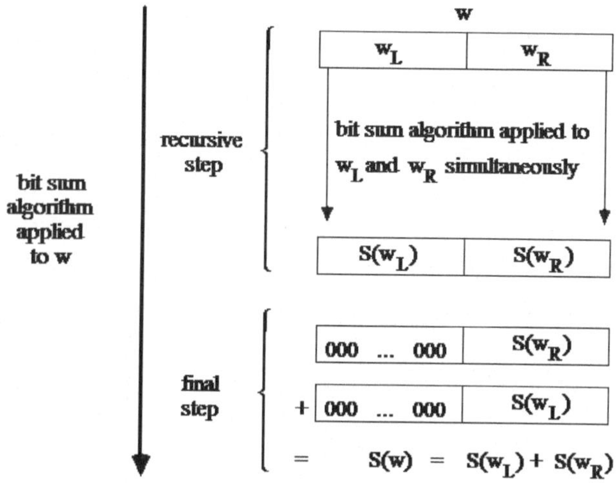

Exhibit 8.4: All processes generated by divide-and-conquer are performed in parallel on shared data registers.

The algorithm is best explained with an example; we use n = 8.

	w_7	w_6	w_5	w_4	w_3	w_2	w_1	w_0
w	1	1	0	1	0	0	0	1

First, extract the even-indexed bits w_6 w_4 w_2 w_0 and place a zero to the left of each bit to obtain weven. The newly inserted zeros are shown in small type.

		w_6		w_4		w_2		w_0
w_{even}	0	1	0	1	0	0	0	1

Next, extract the odd-indexed bits w_7 w_5 w_3 w_1, shift them right by one place into bit positions w_6 w_4 w_2 w_0, and place a zero to the left of each bit to obtain w_{odd}.

		w_7		w_5		w_3		w_1
w_{odd}	0	1	0	0	0	0	0	0

Then, numerically add w_{even} and w_{odd}, considered as integers written in base 2, to obtain w'.

	w'7	w'6	w'5	w'4	w'3	w'2	w'1	w'0
Weven	0	1	0	1	0	0	0	1
Wodd	0	1	0	0	0	0	0	0
w'	1	0	0	1	0	0	0	1

Next, we index not bits, but pairs of bits, from right to left: (w'_1 w'_0) is the zeroth pair, (w'_5 w'_4) is the second pair.

Extract the even-indexed pairs w'_5 w'_4 and w'_1 w'_0, and place a pair of zeros to the left of each pair to obtain w'_{even}.

			w'5	w'4			w'1	w'0
W'even	0	0	0	1	0	0	0	1

Next, extract the odd-indexed pairs w'_7 w'_6 and w'_3 w'_2, shift them right by two places into bit positions w'_5 w'_4 and w'_1 w'_0, respectively, and insert a pair of zeros to the left of each pair to obtain w'_{odd}.

			w'7	w'6			w'3	w'2
W'odd	0	0	1	0	0	0	0	0

Numerically, add w'_{even} and w'_{odd} to obtain w''.

	w''7	w''6	w''5	w''4	w''3	w''2	w''1	w''0
W''	0	0	1	1	0	0	0	1

Next, we index quadruples of bits, extract the quadruple w''_3 w''_2 w''_1 w''_0, and place four zeros to the left to obtain w''_{even}.

					w''3	w''2	w''1	w''0
W''even	0	0	0	0	0	0	0	1

Extract the quadruple w''_7 w''_6 w''_5 w''_4, shift it right four places into bit positions w''_3 w''_2 w''_1 w''_0, and place four zeros to the left to obtain w''_{odd}.

					w''7	w''6	w''5	w''4
W''odd	0	0	0	0	0	0	1	1

Finally, numerically add w"even and w"odd to obtain w''' = (00000100), which is the representation in base 2 of the bit sum of w (4 in this example). The following function implements this algorithm.

Logarithmic bit sum implemented for a 16-bit computer:

in 'bitsum$_2$' we apply addition and division operations directly to variables of type 'w$_1$6' without performing the type conversions that would be necessary in a strongly typed language such as Pascal.

```
function bitsum2(w: w16): integer;
const  mask[0] = '0101010101010101';
       mask[1] = '0011001100110011';
       mask[2] = '0000111100001111';
       mask[3] = '0000000011111111';
var  i, d: integer;  weven, wodd: w16;
begin
  d := 2;
  for  i := 0  to  3  do  begin
    weven := w ∩ mask[i];
    w := w / d;   { shift w right 2^i bits }
    d := d^2;
    wodd := w ∩ mask[i];
    w := weven + wodd
  end;
  return(w)
end;
```

Trade-off between time and space: the fastest algorithm

Are there still faster algorithms for computing the bit sum of a word? Is there an *optimal* algorithm? The question of optimality of algorithms is important, but it can be answered only in special cases. To show that an algorithm is optimal, one must specify precisely the class of algorithms allowed and the criterion of optimality. in the case of bit sum algorithms, such specifications would be complicated and largely arbitrary, involving specific details of how computers Work.

However, we can make a plausible argument that the following bit sum algorithm is the fastest possible, since it uses a table lookup to obtain the result in essentially one operation. The penalty for this speed is an extravagant use of memory space (2n locations), thereby making the algorithm impractical except for small values of n. The choice of an algorithm almost always involves trade-offs among various desirable properties, and the better an algorithm is from one aspect, the worse it may be from another.

The algorithm is based on the idea that we can pre-compute the solutions to all possible questions, store the results, and then simply look them up when needed. As an example, for n = 3, we would store the information

Word	Bit sum
0 0 0	0
0 0 1	1
0 1 0	1
0 1 1	2
1 0 0	1
1 0 1	2
1 1 0	2
1 1 1	3

What is the fastest way of looking up a word w in this table? Under assumptions similar to those used in the preceding algorithms, we can interpret w as an address of a memory cell that contains the bit sum of w, thus giving us an algorithm that requires only one memory reference.

Table lookup implemented for a 16-bit computer:

```
function bitsum3(w: w16): integer;
const  c: array[0 .. 65535] of integer = [0, 1, 1, 2, 1, 2, 2, 3,
… , 15, 16];
begin  return(c[w])  end;
```

In concluding this example, we notice the variety of algorithms that exist for computing the bit sum, each one based on entirely different principles, giving us a different trade-off between space and time. 'bitsum0' and 'bitsum_3' solve the problem by "brute force" and are simple to understand: 'bitsum0' looks at each bit and so requires much time; 'bitsum_3' stores the solution for each separate case and thus requires much space. The logarithmic bit sum algorithm is an elegant compromise: efficient with respect to both space and time, it merely challenges the programmer's wits.

Exercises

- Show that there are exactly 16 distinct boolean functions of two variables.
- Show that each of the boolean functions 'nand' and 'nor' is universal in the following sense: Any boolean function f(x, y) can be written as a nested expression involving only 'nands', and it can also be written using only 'nors'. Show that no other boolean function of two variables is universal.
- Consider the logarithmic bit sum algorithm, and show that *any* strategy for splitting w (not just the halving split) requires n – 1 additions.

9

Ordered Sets

Sets of elements processed on a computer are always ordered according to some criterion. in the preceding example of the “population count” operation, a set is ordered arbitrarily and implicitly simply because it is mapped onto linear storage; a programmer using that set can ignore any order imposed by the implementation and access the set through functions that hide irrelevant details. in most cases, however, the order imposed on a set is not accidental, but is prescribed by the problem to be solved and/or the algorithm to be used. in such cases the programmer explicitly deals with issues of how to order a set and how to use any existing order to advantage.

Searching in ordered sets is one of the most frequent tasks performed by computers: whenever we operate on a data item, that item must be selected from a set of items. Searching is also an ideal ground for illustrating basic concepts and techniques of programming.

At times, ordered sets need to be rearranged (permuted). The chapter “Sorting and its complexity” is dedicated to the most frequent type of rearrangement: permuting a set of elements into ascending order. Here we discuss another type of rearrangement: reordering a set according to a given permutation.

Sequential search

Consider the simple case where a fixed set of n data elements is given in an array A:

```
const  n = ... ;   { n > 0 }
type  index = 0 .. n;   elt = ... ;
var  A: array[1 .. n] of elt;     or     var A: array[0 .. n] of elt;
```

Sequential or linear search is the simplest technique for determining whether A contains a given element x. it is a trivial example of an in*cremental algo*rithm, which processes a set of data one element at a time. If the search for x is successful, we return an index i, 1 ≤ i ≤ n, to point to x. The convention that i = 0 signals unsuccessful search is convenient and efficient, asit encodes all possible outcomes in a single parameter.

```
function find(x: elt): index;
var  i: index;
begin
   i := n;
   while  (i > 0)  { can access A }  cand  (A[i] ≠ x)  { not yet
found }  do
(1)    { (1 ≤ i ≤ n) ∧ (∀ k, i ≤ k: A[k] ≠ x) }
       i := i - 1;

(2)    { (∀k, i < k: A[k] ≠ x) ∧ ((i= 0) ∧ ((1 ≤ i ≤ n) ∧ (A[i] = x))) }
       return(i)
end;
```

The 'cand' operator used in the termination condition is the *conditional* 'and'. Evaluation proceeds from left to right and stops as soon as the value of the boolean expression is determined: If i > 0 yields 'false', we immediately terminate evaluation of the boolean expression without accessing A[i], thus avoiding an out-of-bounds error.

We have included two assertions, (1) and (2), that express the main points necessary for a formal proof of correctness: mainly, that each iteration of the loop extends by one element the subarray known *not* to contain the search argument x. Assertion (1) is trivially true after the initialization i := n, and remains true whenever the body of the while loop is about to be executed. Assertion (2) states that the loop terminatesin one of two ways:

- i = 0 signals that the entire array has been scanned unsuccessfully.
- x has been found at index i.

A formal correctness proof would have to include an argument that the loop doesindeed terminate—a simple argument here, since i isinitialized to n, decreases by 1 in each iteration, and thus will become 0 after a finite number of steps.

The loop is terminated by a Boolean expression composed of two terms: reaching the end of the array, i = 0, and testing the current array element, A[i] = x. The second term is unavoidable, but the first one can be spared by making sure that x is always found before the index i drops off the end of the array. This is achieved by extending the array by one cell A[0] and placing the search argument x in it as a *sentinel*. If no true element x stops the scan of the array, the sentinel will. Upon exit from the loop, the value of i reveals the outcome of the search, with the convention that 0 signals an unsuccessful search:

```
function find(x: elt): index;
var  i: index;
begin
   A[0] := x;   i := n;
   while  A[i] ≠ x   do   i := i - 1;
   return(i)
end;
```

How efficient is sequential search? An unsuccessful search always scans the entire array. If all n array elements have equal probability of being searched for, the average number of iterations of the while loop in a successful search is

$$\frac{1}{n}(1+2+K+n)=\frac{n+1}{2}.$$

This algorithm needs time proportional to n in the *average* and the *worst case.*

Binary search

If the data elements stored in the array A are ordered according to the order relation ≤ defined on their domain, that is the search for an element x can be made much faster because a comparison of x with any array element A[m] provides more information than it doesin the unordered case. The result x ≠ A[m] excludes not only A[m], but also all elements on one or the other side of A[m], depending on whether x is greater or smaller than A[m] (Exhibit 9.1).

$\forall$ k, 1 ≤ k < n: A[k] ≤ A[k + 1]

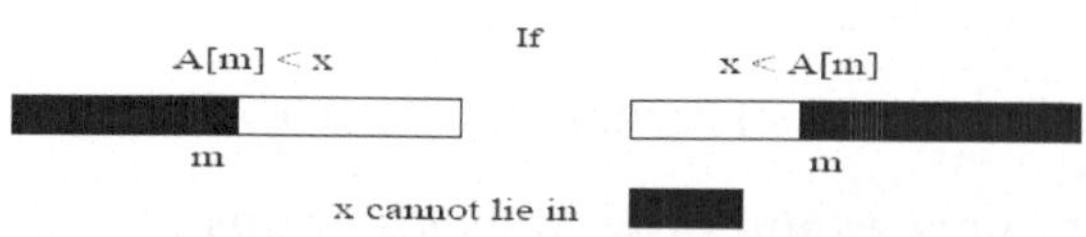

Exhibit 9.1: Binary search identifies regions where the search argument is guaranteed to be absent.

The following function exploits this additional information:

```
  const  n = ... ;   { n > 0 }
  type  index = 1 .. n;  elt = ... ;
  var  A: array[1 .. n] of elt;

  function find(x: elt;  var m: index): boolean;
  var  u, v: index;
  begin
     u := 1;  v := n;
     while  u ≤ v  do  begin
(1)     { (u ≤ v) ∧ (∀ k, 1 ≤ k < u: A[k] < x) ∧(∀ k, v < k ≤ n: A[k] > 
x) }
        m := any value such that u ≤ m ≤ v ;
        if     x < A[m]      thenv := m - 1
        elsif    x > A[m]   then u := m + 1
(2)        else   {x = A[m] }     return(true)
     end;
(3)     { (u = v + 1) ∧(∀ k, 1 ≤ k < u: A[k]< x) ∧  (∀ k, v < k ≤ n:
A[k] > x) }
     return(false)
  end;
```

u and v bound the interval of uncertainty that might contain x. Assertion (1) states that A[1], ... , A[u – 1] are known to be smaller than x; A[v + 1], ... , A[n] are known to be greater than x. Assertion (2), before exit from the function, states that x has been found at index m. in assertion (3), u = v + 1 signals that the interval of uncertainty has shrunk to become empty. If there exists more than one match, this algorithm will find one of them.

This algorithm is correct independently of the choice of m but is most efficient when m is the midpoint of the current search interval:

```
m := (u + v) div 2;
```

With this choice of m each comparison either finds x or eliminates half of the remaining elements. Thus at most $\lceil \log_2 n \rceil$ iterations of the loop are performed in the worst case.

Exercise: binary search

The array

```
var A: array [1 .. n] of integer;
```

contains n integersin ascending order: A[1] ≤ A[2] ≤ ... ≤ A[n].

- ❖ Write a recursive binary search that returns 0 if x is not in A, and an index i such that A[i] = x if x isin A.

```
function rbs (x, u, v: integer): integer;
```

- ❖ What is the maximal depth of recursive calls of 'rbs' in terms of n?
- ❖ Describe the advantages and disadvantages of this recursive binary search as compared to the iterative binary search.

Exercise: searching in a partially ordered two-dimensional array

Consider the n by m array:

```
var  A: array[1 .. n, 1 .. m] of integer;
```

and assume that the integersin each row and in each column are in ascending order; that is,

```
A[i, j] ≤ A[i, j + 1] for i = 1, ... , n and j = 1, ... , m - 1;
A[i, j] ≤ A[i + 1, j] for i = 1, ... , n - 1 and j = 1, ... , m.
```

- ❖ Design an algorithm that determines whether a given integer x is stored in the array A. Describe your algorithm in words and figures. *Hint:* Start by comparing x with A[1, m] (Exhibit 9.2).

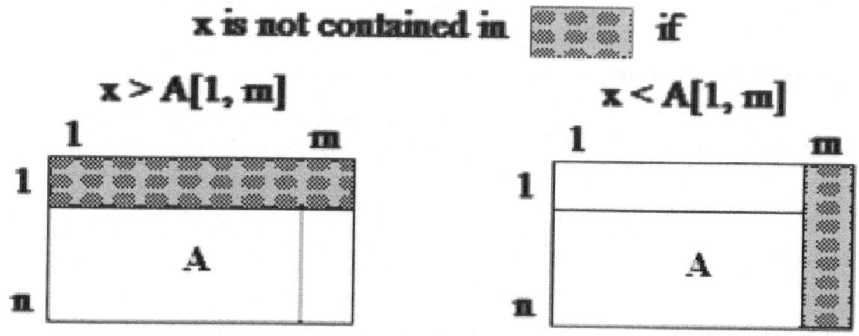

Exhibit 9.2: Another example of the idea of excluded regions.

- ❖ Implement your algorithm by a

```
function IsInArray (x: integer): boolean;
```

- ❖ Show that your algorithm is correct and terminates, and determine its worst case time complexity.

Solution

(a) The algorithm compares x first with A[1, m]. If x is smaller than A[1, m], then x cannot be contained in the last column, and the search process is continued by comparing x with A[1, m – 1]. If x is greater than A[1, m], then x cannot be contained in the first row, and the search process is continued by comparing x with A[2, m]. Exhibit 9.3 shows part of a typical search process.

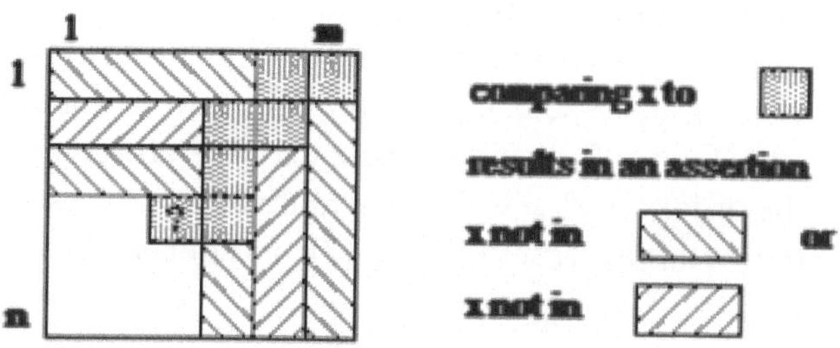

Exhibit 9.3: Excluded regions combine to leave only a staircase-shaped strip to examine.

```
(b)   function IsInArray(x: integer): boolean;
      var  r, c: integer;
      begin
         r := 1;  c := m;
         while  (r ≤ n) and (c ≥ 1)  do
         {1}  if x < A[r, c] then c := c - 1
              elsif  x > A[r, c] then r := r + 1
      else { x = A[r, c] }  {2}  return(true);
    {3}  return(false)
  end;
```

❖ At positions {1}, {2}, and {3}, the invariant

$\forall$ i, 1 ≤ i ≤ n,$\forall$ j, 1 ≤ j ≤ m:

(j > c ⇒ x ≠ A[i, j]) ∧ (i < r ⇒ x ≠ A[i, j] (*)

states that the hatched rows and columns of A do not contain x. At {2},

(1 ≤ r ≤ n) ∧ (1 ≤ c ≤ m) ∧ (x = A[r, c])

states that r and c are within index range and x has been found at (r, c). At {3},

(r = n + 1)∨(c = 0)

states that r or c are outside the index range. This coupled with (*) implies that x is not in A:

(r = n + 1) ∨ (c = o) ⇒ $\forall$ i, 1 ≤ i ,≤ n, $\forall$ j. 1 ≤ j ≤ m: x ≠ A[i, j].

Each iteration through the loop either decreases c by one or increases r by one. If x is not contained in the array, either c becomes zero or r becomes greater than n after a finite number of steps, and the algorithm terminates. in each step, the algorithm eliminates either a row from the top or a column from the right. in the worst case it Worksits way from the upper right corner to the lower left corner in n + m – 1 steps, leading to a complexity of Θ(n + m).

In-place permutation

Representations of a permutation. Consider an array D[1 .. n] that holds n data elements of type 'elt'. These are ordered by their position in the array and must be rearranged according to a specific permutation given in another array. Exhibit 9.4 shows an example for n = 5. Assume that a, b, c, d, e, stored in this order, are to be rearranged in the order c, e, d, a, b. This permutation is represented naturally by either of the two permutation arrays t (to) or f (from) declared as

```
var  t, f: array[1 .. n] of 1 .. n;
```

The exhibit also shows a third representation of the same permutation: the decomposition of this permutation into cycles. The element in D[1] movesinto D[4], the one in D[4] into D[3], the one in D[3] into D[1], closing a cycle that we abbreviate as (1 4 3), or (4 3 1), or (3 1 4). There is another cycle (2 5), and the entire permutation is represented by (1 4 3) (2 5).

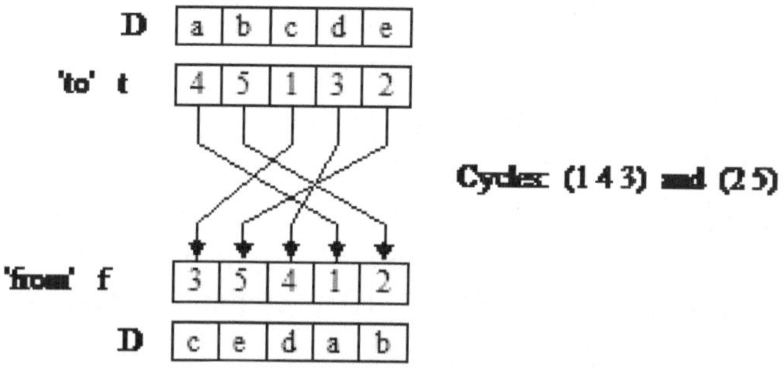

Exhibit 9.4: A permutation and its representations in terms of 'to', 'from', and cycles.

The cycle representation is intuitively most informative, as it directly reflects the decomposition of the problem into independent sub-problems, and both the 'to' and 'from' information is easily extracted from it. But 'to' and 'from' dispense with parentheses and lead to more concise programs.

Consider the problem of executing this permutation in *place*: Both the given data and the result are stored in the same array D, and only a (small) constant amount of auxiliary storage may be used, independently of n. Let us use the example of in-place permutation to introduce a notation that is frequently convenient, and to illustrate how the choice of primitive operations affects the solution.

A *multiple assignment* statement will do the job, using either 'to' or 'from':

// (1 ≤ i ≤ n) **{** D[t[i]] := D[i] **}**

or

// (1 ≤ i ≤ n) **{** D[i]} := D[f[i]] **}**

The characteristic properties of a multiple assignment statement are:

- The left-hand side is a sequence of variables, the right-hand side is a sequence of expressions, and the two sequences are matched according to length and type. The value of the i-th expression on the right is assigned to the i-th variable on the left.
- All the expressions on the right-hand side are evaluated using the original values of all variables that occur in them, and the resulting values are assigned "simultaneously" to the variables on the left-hand side. We use the sign // to designate concurrent or parallel execution.

Few of today's programming languages offer multiple assignments, in particular those of variable length used above. Breaking a multiple assignment into single assignments usually forces the programmer to introduce temporary variables. As an example, notice that the direct sequentialization:

for i := 1 to n do D[t[i]] := D[i]

or

for i := 1 to n do D[i] := D[f[i]]

is faulty, as some of the elements in D will be overwritten before they can be moved. Overwriting can be avoided at the cost of nearly doubling memory requirements by allocating an array A[1 .. n] of data elements for temporary storage:

for i := 1 to n do A[t[i]] := D[i];

for i := 1 to n do D[i] := A[i];

This, however, is not an in-place computation, as the amount of auxiliary storage grows with n. it is unnecessarily inefficient: There are elegant in-place permutation algorithms based on the conventional primitive of the single assignment statement. They all assume that the permutation array may be destroyed as the permutation is being executed. If the representation of the permutation must be preserved, additional storage is required for bookkeeping, typically of a size proportional to n. Although this additional space may be as little as n bits, perhapsin order to distinguish the elements processed from those yet to be moved, such an algorithm is not technically in place.

Nondeterministic algorithms. Problems of rearrangement always appear to admit many different solutions—a phenomenon that is most apparent when one considers the multitude of sorting algorithms in the literature. The reason is clear: When n elements must be moved, it may not matter much which elements are moved first and which ones later. Thusit is useful to look for *nondeterministic algorithms* that refrain from specifying the precise sequence of all actions taken, and instead merely iterate *condition* ⇒ *action* statements, with the meaning "wherever *condition* applies perform the corresponding *action*". These algorithms are nondeterministic because each of several distinct conditions may apply at lots of different places, and we may "fire" any action that is currently enabled. Adding sequential control to a nondeterministic algorithm turns it into a deterministic algorithm. Thus a nondeterministic algorithm corresponds to a class of deterministic ones that share common invariants, but differ in the order in which steps are executed. The correctness of a nondeterministic algorithm implies the correctness of all its sequential instances. Thus it is good algorithm design practice to develop a correct nondeterministic algorithm first, then turn it into a deterministic one by ordering execution of its steps with the goal of efficiency in mind.

Deterministic sequential algorithms come in a variety of forms depending on the choice of primitive (assignment or swap), data representation ('to' or 'from'), and technique. We focus on the latter and consider two techniques: cycle rotation and cycle clipping. *Cycle rotation* follows naturally from the idea of decomposing a permutation into cycles and processing one cycle at a time, using temporary storage for a single element. it fits the 'from' representation somewhat more efficiently than the 'to' representation, as the latter requires a swap of two elements where the former uses an assignment. *Cycle clipping* uses the primitive 'swap two elements' so effectively as a step toward executing a permutation that it needs no temporary storage for elements. Because no temporary storage is tied up, it is not necessary to finish processing one cycle before starting on the next one –elements can be clipped from their cycles in any order. Clipping Works efficiently with either representation, but is easier to understand with 'to'. We present cycle rotation with 'from' and cycle clipping with 'to' and leave the other two algorithms as exercises.

Cycle rotation

A search for an in-place algorithm naturally leads to the idea of processing a permutation one cycle at a time: every element we place at its destination bumps another one, but we avoid holding an unbounded number of bumped elementsin temporary storage by rotating each cycle, one element at a time. This Works best using the 'from' representation. The following loop rotates the cycle that passes through an arbitrary index i:

Rotate the cycle starting at index i, updating f:

```
j := i;{ initialize a two-pronged fork to travel along the cycle }
p := f[j];  { p is j's predecessor in the cycle }
A := D[j];  { save a single element in an auxiliary variable A }
while  p ≠ i  do  { D[j] := D[p];  f[j] := j;  j := p;  p := f[j]} ;
D[j] := A;  { reinsert the saved element into the former cycle ... }
f[j] := j;  { ... but now it is a fixed point }
```

This code Works trivially for a cycle of length 1, where p = f[i] = i guards the body of the loop from ever being executed. The statement f[j] := j in the loop is unnecessary for rotating the cycle. its purpose is to identify an element that has been placed at its final destination, so this code can be iterated for $1 \le i \le n$ to yield an in-place permutation algorithm. For the sake of efficiency we add two details: (1) We avoid unnecessary movements A := D[j]; D[j] := A of a possibly voluminous element by guarding cycles of length 1 with the test 'i ≠ f[i]', and (2) we terminate the iteration at n – 1 on the grounds that when n – 1 elements of a permutation are in their correct place, the n-th one is also. Using the code above, this leads to

for i := 1 to n – 1 do if i ≠ f[i] then rotate the cycle starting at index i, updating f

Exercise

Implement cycle rotation using the 'to' representation. *Hint:* Use the swap primitive rather than element assignment.

Cycle clipping

Cycle clipping is the key to elegant in-place permutation using the 'to' representation. At each step, we clip an arbitrary element d out of an arbitrary cycle of length > 1, thus reducing the latter's length by 1. As shown in Exhibit 9.5, we place d at its destination, where it forms a cycle of length 1 that needs no further processing. The element it displaces, c, can find a (temporary) home in the cell vacated by d. it is probably out of place there, but no more so than it was at its previous home; its time will come to be relocated to its final destination. Since we have permuted elements, we must update the permutation array to reflect accurately the permutation yet to be performed. This is a local operation in the vicinity of the two elements that were swapped, somewhat like tightening a belt by one notch—all but two of the elementsin the clipped cycle remain unaffected. The Exhibit below shows an example. in order to execute the permutation (1 4 3) (2 5), we clip d from its cycle (1 4 3) by placing d at its destination D[3], thus bumping c into the vacant cell D[4]. This amounts to representing the cycle (1 4 3) as a product of two shorter cycles: the swap (3 4), which can be done right away, and the cycle (1 4) to be executed later. The cycle (2 5) remains unaffected. The ovalsin Exhibit 9.5 indicate that corresponding entries of D and t are moved together. Exhibit 9.6 shows what happens to a cycle clipped by a swap

```
// { t[i], D[i]  :=:  t[t[i]], D[t[i]] }
```

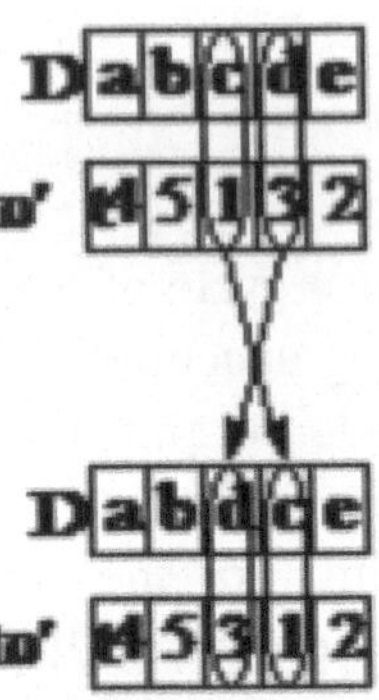

Exhibit 9.5: Clipping one element out of a cycle of a permutation.

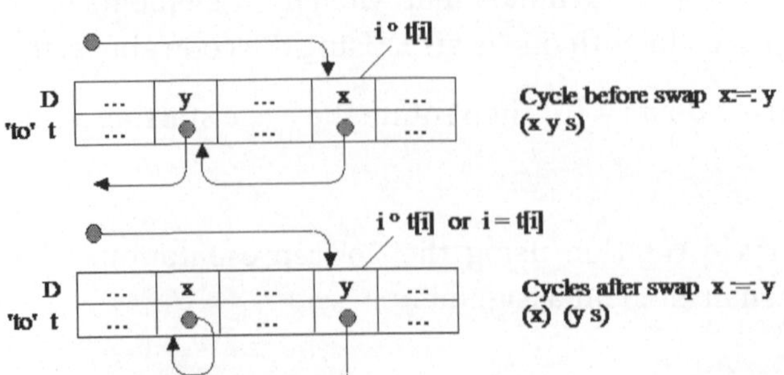

Exhibit 9.6: Effect of a swap caused by the condition i ≠ t[i].

Cycles of length 1 are left alone, and the absence of cycles of length > 1 signals termination. Thus the following *condition* ⇒ *action* statement, iterated as long as the condition i ≠ t[i] can be met, executes a permutation represented in the array t:

```
∃ i:i ≠ t[i]   ⇒   // { t[i], D[i]   :=:   t[t[i]], D[t[i]] }
```

We use the multiple swap operator // { :=: } with the meaning: evaluate all four expressions using the original values of all the variables involved, then perform all four assignments simultaneously. it can be implemented using six single assignments and two auxiliary variables, one of type 1 .. n, the other of type 'elt'. Each swap places (at least) one element into its final position, say j, where it is guarded from any further swaps by virtue of j = t[j]. Thus the nondeterministic algorithm above executes at most n – 1 swaps: When n – 1 elements are in final position, the n-th one is also.

The conditions on i can be checked in any order, as long as they are checked exhaustively, for example:

```
{ (0)   (1 ≤ j < 0) ⇒ j = t[j] }
for  i := 1   to   n - 1   do
   { (1)   (1 ≤ j < i) ⇒ j = t[j] }
   while  i ≠ t[i]   do   // { t[i], D[i]   :=:   t[t[i]], D[t[i]] }
   { (2)   (1 ≤ j ≤ i) ⇒ j = t[j] }
{ (3)   (1 ≤ j ≤ n - 1) ⇒ j = t[j] }
```

For each value of i, i is the leftmost position of the cycle that passes through i. As the while loop reduces this cycle to cycles of length 1, all swaps involve i and t[i] > i, as asserted by the invariant (1) (1 ≤ j < i) ⇒ j = t[j], which precedes the while loop. At completion of the while loop, the assertion is strengthened to include i, as stated in invariant (2) (1 ≤ j ≤ i) ⇒ j = t[j]. This reestablishes (1) for the next higher value of i. The vacuously true assertion serves as the basis of this proof by induction. The final assertion (3) is just a restatement of assertion (2) for the last value of i. Since t[1] ... t[n] is a permutation of 1 ...n, (3) implies that n = t[n].

Exercise: cycle clipping using the 'from' representation

The nondeterministic algorithm expressed as a multiple assignment

// (1 ≤ i ≤ n) { D[i]} := D[f[i]] }

is equally as valid for the 'from' representation asits analog

// (1 ≤ i ≤ n) { D[t[i]] := D[i] }

was for the 'to' representation. But in contrast to the latter, the former cannot be translated into a simple iteration of the *condition* ⇒ *action* statement:

∃ i: i ≠ f[i] ⇒ // { f[i], D[i] :=: f[f[i]], D[f[i]] }

Why not? Can you salvage the idea of cycle clipping using the 'from' representation

Exercises

1. Write two functions that implement sequential search, one with sentinel as shown in the first section, "Sequential search" the other without sentinel. Measure and compare their running time on random arrays of various sizes
2. Measure and compare the running times of sequential search and binary search on random arrays of size n, for n = 1 to n = 100. Sequential search is obviously faster for small values of n, and binary search for large n, but where is the crossover? Explain your observations.

10

Strings

Most programming languages support simple operations on strings (e.g. comparison, concatenation, extraction, searching). Searching for a specified pattern in a string (text) is the computational kernel of most string processing operations. Several efficient algorithms have been developed for this potentially time-consuming operation. The approach presented here is very general; it allows searching for a pattern that consists not only of a single string, but a set of strings. The cardinality of this set influences the storage space needed, but not the time. it leads us to the concept of a *finite-state machine* (fsm).

Recognizing a pattern consisting of a single string

Problem: Given a (long) string $z = z_1 z_2 \ldots z_n$ of n characters and a (usually much shorter) string $p = p_1 p_2 \ldots pm$ of m characters (the pattern), find all (nonoverlapping) occurrences of p in z. By sliding a window of length m from left to right along z and examining most characters z_i m times we solve the problem using $m \cdot n$ comparisons. By constructing a finite-state machine from the pattern p it suffices to examine each character z_i exactly once, as shown in Exhibit 10.1. Each state corresponds to a prefix of the pattern, starting with the empty prefix and ending with the complete pattern. The input symbols are the input characters $z_1, z_2, \ldots, z_n$ of z. in the j-th step the input character z_j leads from a state corresponding to the prefix $p_1 p_2 \ldots p_i$ to

- the state with prefix $p_1 p_2 \ldots pi\ p_i^{+1}$ if $z_j = p_i^{+1}$
- a different state (often the empty prefix, λ) if $z_j \neq p_i^{+1}$

Example

p = barbara (Exhibit 10.1).

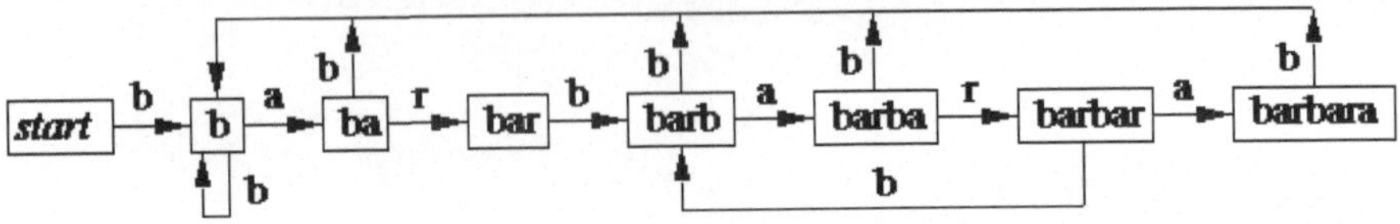

Exhibit 10.1: State diagram showing some of the transitions. All other state transitions lead back to the initial state.

Notice that the pattern 'barbara', although it sounds repetitive, cannot overlap with any part of itself. Constructing a finite-state machine for such a pattern is straightforward. But consider a self-overlapping pattern such as 'barbar', or 'abracadabra', or 'xx', where the first k > 0 characters are identical with the last: The text 'barbarbar' contains two overlapping occurrences of the pattern 'barbar', and 'xxxx' contains three occurrences of 'xx'. A finite-state machine constructed in an analogous fashion as the one used for 'barbara' always finds the first of several overlapping occurrences but might miss some of the later ones. As an exercise, construct finite-state machines that detect all occurrences of self-overlapping patterns.

Recognizing a set of strings: a finite-state-machine interpreter

Finite-state machines (fsm, also called "finite automata") are typically used to recognize patterns that consist of a *set* of strings. An adequate treatment of this more general problem requires introducing some concepts and terminology widely used in computer science.

Given a finite set A of input symbols, the *alphabet*, A* denotes the (infinite) set of all (finite) strings over A, including the nullstring λ. Any subset $L \subseteq A^*$, finite or infinite, is called a set of strings, or a *language*, over A. **Recognizing** *a language* L refers to the ability to examine any string $z \in A^*$, one symbol at a time from left to right, and deciding whether or not $z \in L$.

A *deterministic* finite-state machine M is essentially given by a finite set S of *states*, a finite alphabet A of in*put symbols*, and a *transition function* f: $S \times A \rightarrow S$. The state diagram depicts the states and the inputs, which lead from one state to another; thus a finite-state machine maps strings over A into sequences of states.

When treating any specific problem, it is typically useful to expand this minimal definition by specifying one or more of the following additional concepts. An init*ial state* s_0 S, a subset $F \subseteq S$ of *final* or *accepting states*, a finite alphabet B of *output symbols* and an *output function* g: $S \rightarrow B$, which can be used to assign certain actions to the states in S. We use the concepts of initial state s_0 and of accepting states F to define the notion "**Recognizing** a set of strings":

A set $L \subseteq A^*$ of strings is *recognized* or *accepted* by the finite-state machine M = (S, A, f, s_0, F) iff all the stringsin L, and no others, lead M from s_0 to some state $s \in F$.

Example

Exhibit 10.3 shows the state diagram of a finite-state machine that recognizes parameter lists as defined by the syntax diagramsin Exhibit 10.2. L (letter) stands for a character a .. z, D (digit) for a digit 0 .. 9.

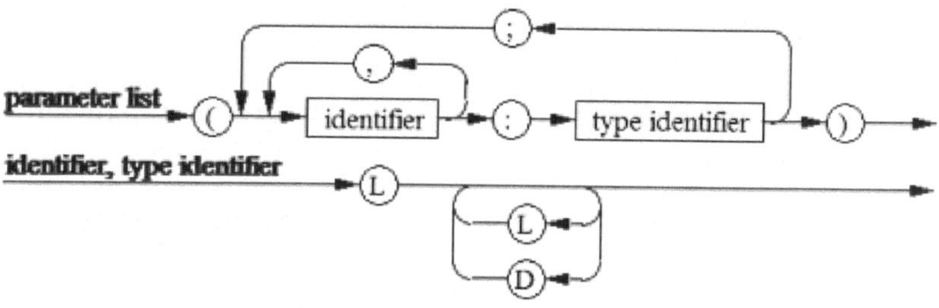

Exhibit 10.2: Syntax diagram of simple parameter lists.

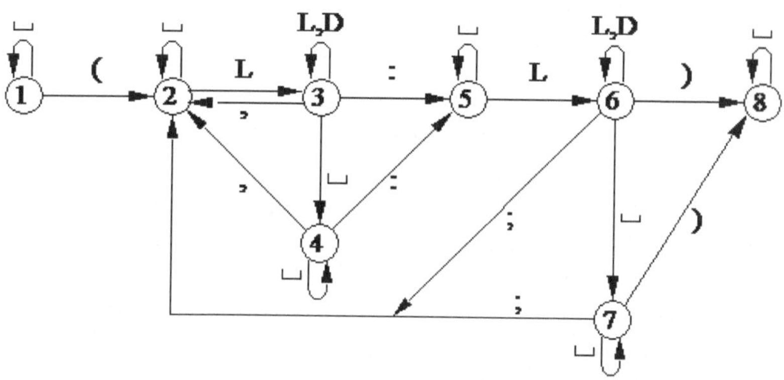

Exhibit 10.3: State diagram of finite-state machine to accept parameter lists. The starting state is '1', the single accepting state is '8'.

A straightforward implementation of a finite-state machine interpreter uses a transition matrix T to represent the state diagram. From the current state s the input symbol c leads to the next state T[s, c]. it is convenient to introduce an error state that captures all illegal transitions. The transition matrix T corresponding to Exhibit 10.3 looks as follows:

```
L represents a character a .. z.
D represents a digit 0 .. 9.
! represents all characters that are not explicitly mentioned.
```

	␣	(	)	:	,	;	L	D	!	
0	0	0	0	0	0	0	0	0	0	error state
1	1	2	0	0	0	0	0	0	0	skip blank
2	2	0	0	0	0	0	3	0	0	left parenthesis read
3	4	0	0	5	2	0	3	3	0	reading variable identifier
4	4	0	0	5	2	0	0	0	0	skip blank
5	5	0	0	0	0	0	6	0	0	colon read
6	7	0	8	0	0	2	6	6	0	reading type identifier
7	7	0	8	0	0	2	0	0	0	skip blank
8	8	0	0	0	0	0	0	0	0	right parenthesis read

The following is a suitable environment for programming a finite-state-machine interpreter:

```
const  nstate = 8; { number of states, without error state }
type state = 0 .. nstate; { 0 = error state, 1 = initial state }
   inchar = ' ' .. '¨'; { 64 consecutive ASCII characters }
   tmatrix = array[state, inchar] of state;
var  T: tmatrix;
```

After initializing the transition matrix T, the procedure 'silentfsm' interprets the finite-state machine defined by

T. it processes the sequence of input characters and jumps around in the state space, but it produces no output.

```
procedure silentfsm(var T: tmatrix);
var  s: state;  c: inchar;
begin
   s := 1;  { initial state }
   while  s ≠ 0  do  { read(c);  s := T[s, c] }

end;
```

The simple structure of 'silentfsm' can be employed for a useful finite-state-machine interpreter in which initialization, error condition, input processing and transitionsin the state space are handled by procedures or functions 'initfsm', 'alive', 'processinput', and 'transition' which have to be implemented according to the desired behavior. The terminating procedure 'terminate' should print a message on the screen that confirms the correct termination of the input or shows an error condition.

```
procedure fsmsim(var T: tmatrix);
var  ... ;
begin
   initfsm;
   while  alive  do  { processinput;  transition };
   terminate
end;
```

Exercise: finite-state recognizer for multiples of 3

Consider the set of strings over the alphabet {0, 1} that represent multiples of 3 when interpreted as binary numbers, such as: 0, 00, 11, 00011, 110. Design two finite-state machines for **Recognizing** this set:

- *Left to right:* Mlr reads the strings from most significant bit to least significant.
- *Right to left:* Mrl reads the strings from least significant bit to most significant.

Solution

Left to right: Let r_k be the number represented by the k leftmost bits, and let b be the (k + 1)-st bit, interpreted as an integer. Then $r_k^{+1} = 2 \cdot r_k + b$. The states correspond to r_k mod 3 (Exhibit 10.4). Starting state and accepting state: 0'.

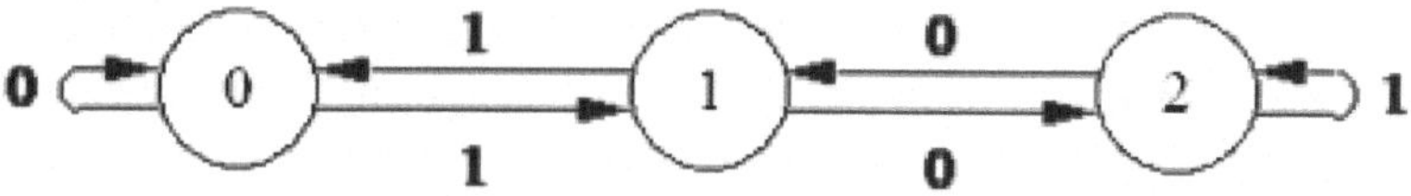

Exhibit 10.4: Finite-state machine computes remainder modulo 3 left to right.

Right to left: $r_k^{+1} = b \cdot 2^k + r_k$. Show by induction that the powers of 2 are alternatingly congruent to 1 and 2 modulo 3 (i.e. 2_k mod 3 = 1 for k even, 2_k mod 3 = 2 for k odd). Thus we need a modulo 2 counter, which appearsin Exhibit 10.5 as two rows of three states each. Starting state: 0. Accepting states: 0 and 0'.

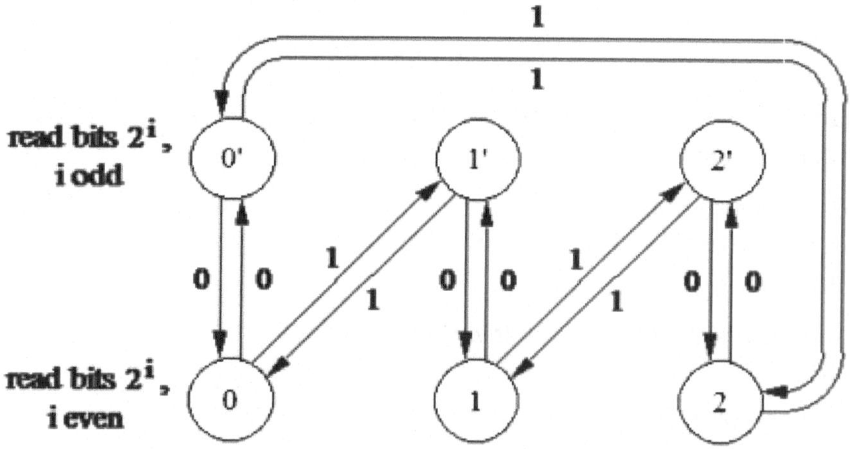

Exhibit 10.5: Finite-state machine computes remainder modulo 3 right to left.

Exercises and programming projects

Draw the state diagram of several finite-state machines, each of which searches a string z for all occurrences of an interesting pattern with repetitive parts, such as 'abaca' or 'Caracas'.

Draw the state diagram of finite-state machines that detect all occurrences of a self-overlapping pattern such as 'abracadabra', 'barbar', or 'xx'.

Finite-state recognizer for various days:

Design a finite-state machine for automatic recognition of the set of nine words:

```
'monday','tuesday','wednesday','thursday',
'friday', 'saturday', 'sunday', 'day', 'daytime'
```

in a text. The underlying alphabet consists of the lowercase letters 'a' .. 'z' and the blank. Draw the state diagram of the finite-state machine; identify the initial state and indicate accepting states by a double circle. it suffices to recognize membership in the set without **Recognizing** each word individually.

Implementation of a pattern recognizer:

Some useful procedures and functions require no parameters, hence most programming languagesincorporate the concept of an empty parameter list. There are two reasonable syntax conventions about how to write the headers of parameterless procedures and functions:

```
(1)  procedure p;   function f: T;
(2)  procedure p();function f(): T;
```

Examples: Pascal uses convention (1); Modula-2 allows both (1) and (2) for procedures, but only (2) for function procedures.

For each convention (1) and (2), modify the syntax diagram in Exhibit 10.2 to allow empty parameter lists, and draw the state diagrams of the corresponding finite-state machines.

Standard Pascal defines parameter lists by means of the syntax diagram shown in Exhibit 10.6.

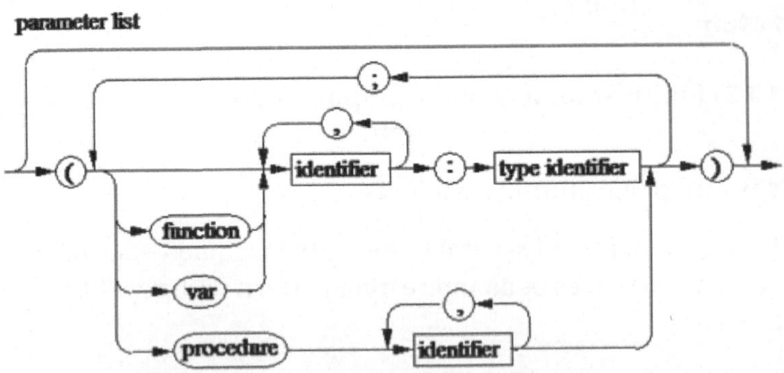

Exhibit 10.6: Syntax diagram for standard Pascal parameter lists.

Draw a state diagram for the corresponding finite-state machine. For brevity's sake, consider the reserved words 'function', 'var' and 'procedure' to be atomic symbols rather than strings of characters.

11

Matrices and Graphs: Transitive Closure

In any systematic presentation of data objects, it is useful to distinguish *primitive* or *atomic objects* from *composite* or *structured objects.* in each of the preceding chapters we have seen both types: A bit, a character, or an identifier is usually considered primitive; a word of bits, a string of characters, an array of identifiers is naturally treated as composite. Before proceeding to the most common primitive objects of computation, numbers, let us discuss one of the most important types of structured objects, matrices. Even when matrices are filled with the simplest of primitive objects, bits, they generate interesting problems and useful algorithms.

Paths in a graph

Syntax diagrams and state diagrams are examples of a type of object that aboundsin computer science: A *graph* consists of *nodes* or *vertices*, and of *edges* or *arcs* that connect a pair of nodes. Nodes and edges often have additional information attached to them, such as labels or numbers. If we wish to treat graphs mathematically, we need a definition of these objects.

Directed graph. Let N be the set of n elements {1, 2, ... , n} and E a binary relation: E N ⊆ ξ N, also denoted by an arrow, →. Consider N to be the set of nodes of a directed graph G, and E the set of arcs (directed edges). A directed graph G may be represented by its *adjacency matrix* A (Exhibit 11.1), an n ξ n boolean matrix whose elements A[i, j] determine the existence of an arc from i to j:

A[i, j] = true iff i → j.

An arc is a path of length 1. From A we can derive all paths of any length. This leads to a relation denoted by a double arrow, ⇒, called the *transitive closure* of E:

i ⇒ j, iff there exists a path from i to j

(i.e. a sequence of arcs$_i$ → i_1, i_1 → i_2, i_2 → i_3, ... , i_k → j). We accept paths of length 0 (i.e. i ⇒ i for all $_i$). This relation ⇒ is represented by a matrix C= A∗(Exhibit 11.1):

C[i, j] = true iff i ⇒ j.

C stands for *connectivity* or *reachability matrix;* C = A* is also called *transitive hull* or *transitive closure*, since it is the smallest transitive relation that "encloses" E.

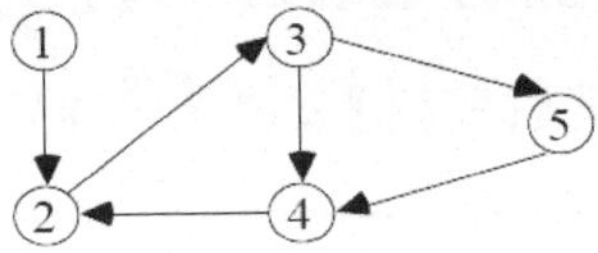

A

	1	2	3	4	5
1		T			
2			T		
3				T	T
4		T			
5				T	

C

	1	2	3	4	5
1	T	T	T	T	T
2		T	T	T	T
3		T	T	T	T
4		T	T	T	T
5		T	T	T	T

Exhibit 11.1: Example of a directed graph with its adjacency and connectivity matrix.

(Undirected) graph. If the relation E ⊆ N ξ N is *symmetric* [i.e. for every ordered pair (i, j) of nodesit also contains the opposite pair (j, $_i$)] we can identify the two arcs (i, j) and (j, $_i$) with a single *edge*, the unordered pair (i, j). Books on graph theory typically start with the definition of *undirected* graphs (graphs, for short), but we treat them as a special case of directed graphs because the latter occur much more often in computer science. Whereas graphs are based on the concept of an edge between two nodes, *directed* graphs embody the concept of one-way *arcs* leading *from* a node *to* another one.

Boolean matrix multiplication

Let A, B, C be n ξ n boolean matrices defined by

```
type nnboolean: array[1 .. n, 1 .. n] of boolean;
var  A, B, C: nnboolean;
```

The boolean matrix multiplication C = A · B is defined as and implemented by

$$C[i, j] = \underset{1 \le k \le n}{\text{OR}} (A[i, k] \text{ and } B[k, j])$$

and implemented by

```
procedure mmb(var a, b, c: nnboolean);
var  i, j, k: integer;
begin
   for  i := 1  to  n  do
      for  j := 1  to  n  do  begin
         c[i, j] := false;
         for  k := 1  to  n  do  c[i, j] := c[i, j] or (a[i, k] and
b[k, j])  (*)
      end
end;
```

Remar$_k$: Remember (in the section, "Pascal and its dialects: Lingua franca of computer science") that we usually assume the boolean operations 'or' and 'and' to be conditional (i.e. their arguments are evaluated only as far as necessary to determine the value of the expression). An extension of this simple idea leads to an alternative way of coding boolean matrix multiplication that speeds up the innermost loop above for large values of n. Explain why the following code is equivalent to (∗∗):

```
k:=1;
while  not c[i, j] and (k ≤ n)  do  { c[i, j] := a[i, k] and b[k,
j]; k := k + 1 }
```

Multiplication also defines powers, and this gives us a first solution to the problem of computing the transitive closure. If Al^{+1} denotes the L-th power of A, the formula

$$A^{L+1}[i,j] = \underset{1 \le k \le n}{OR} (A^{L}[i,k] \text{ and } A[k,j])$$

has a clear interpretation: There exists a path of length L + 1 from i to j iff, for some node k, there exists a path of length L from i to k and a path of length 1 (a single arc) from k to j. Thus A2 represents all paths of length 2; in general, A^L represents all paths of length L, for L ≥ 1:

$A^L[i, j]$ = true iff there exists a path of length L from i to j.

Rather than dealing directly with the adjacency matrix A, it is more convenient to construct the matrix A' = A or I. The identity matrix I has the values 'true' along the diagonal, 'false' everywhere else. Thusin A' all diagonal elements A'[i, i] = true. Then A'L describes all paths of length ≤ L (instead of exactly equal to L), for L ≥ 0. Therefore, the transitive closure is $A* = A'^{(n-1)}$

The efficiency of an algorithm is often measured by the number of "elementary" operations that are executed on a given data set. The execution time of an elementary operation [e.g. the binary boolean operators (and, or) used above] does not depend on the operands. To estimate the number of elementary operations performed in boolean matrix multiplication as a function of the matrix size n, we concentrate on the leading terms and neglect the lesser terms. Let us use asymptotic notation in an intuitive way; it is defined formally in Part IV.

The number of operations (and, or), executed by procedure 'mmb' when multiplying two boolean n ξ n matrices is $\Theta(n^3)$ since each of the nested loops$_i$siterated n times. Hence the cost for computing $A'^{(n-1)}$ by repeatedly multiplying with A' is $\Theta(n^4)$. This algorithm can be improved to $\Theta(n^3 \cdot \log n)$ by repeatedly squaring: A'^2, A'^4, A'^8, ... , A'^k where k is the smallest power of 2 with $k \geq n - 1$. it is not necessary to compute exactly $A'^{(n-1)}$. instead of A'^{13}, for example, it suffices to compute A'^{16}, the next higher power of 2, which contains all paths of length at most 16. in a graph with 14 nodes, this set is equal to the set of all paths of length at most 1.

Warshall's algorithm

In search of a faster algorithm we consider other ways of iterating over the set of all paths. instead of iterating over paths of growing length, we iterate over an increasing number of nodes that may be used along a path from node i to node j. This idea leads to an elegant algorithm due to Warshall [War 62]:

Compute a sequence of matrices $B_0, B_1, B_2, \ldots, B_n$:

$B_0[i, j] = A'[i, j]$ = true iff $i = j$ or $i \rightarrow j$.

$B_1[i, j]$ = true iff $i \Rightarrow j$ using at most node 1 along the way.

$B_2[i, j]$ = true iff $i \Rightarrow j$ using at most nodes 1 and 2 along the way

...

$B_k[i, j]$ = true iff $i \Rightarrow j$ using at most nodes 1, 2, ... , k along the way.

The matrices B_0, B_1, ... express the existence of paths that may touch an increasing number of nodes along the way from node i to node j; thus Bn talks about unrestricted paths and is the connectivity matrix $C = B_n$.

An iteration step $B_{k-1} \rightarrow B_k$ is computed by the formula

$$B_k[i, j] = B_{k-1}[i, j] \text{ or } (B_{k-1}[i, k] \text{ and } B_{k-1}[k, j]).$$

The cost for performing one step is $\Theta(n^2)$, the cost for computing the connectivity matrix is therefore $\Theta(n^3)$. A comparison of the formula for Warshall's algorithm with the formula for matrix multiplication shows that the n-ary 'OR' has been replaced by a binary 'or'.

At first sight, the following procedure appears to execute the algorithm specified above, but a closer look reveals that it executes something else: the assignment in the innermost loop computes new values that are used immediately, instead of the old ones.

```
procedure warshall(var a: nnboolean);
var i, j, k: integer;
begin
  for k := 1 to n do
    for i := 1 to n do
      for j := 1 to n do
        a[i, j] := a[i, j] or (a[i, k] and a[k, j])
        { this assignment mixes values of the old and new matrix }
end;
```

A more thorough examination, however, shows that this "naively" programmed procedure computes the correct result in-place more efficiently than would direct application of the formulas for the matrices Bk. We encourage you to verify that the replacement of old values by new ones leavesintact all values needed for later steps; that is, show that the following equalities hold:

$$B_k[i, k] = B_{k-1}[i, k] \text{ and } B_k[k, j] = B_{k-1}[k, j].$$

Exercise: distancesin a directed graph, Floyd's algorithm

Modify Warshall's algorithm so that it computes the shortest distance between any pair of nodesin a directed graph where each arc is assigned a length ≥ 0. We assume that the data is given in an n ξ n array of reals, where d[i, j] is the length of the arc between node i and node j. If no arc exists, then d[i, j] is set to ∞, a constant that is the largest real number that can be represented on the given computer. Write a procedure 'dist' that Works on an array d of type

```
type  nnreal = array[1 .. n, 1 .. n] of real;
```

Think of the meaning of the boolean operations 'and' and 'or' in Warshall's algorithm, and find arithmetic operations that play an analogous role for the problem of computing distances. Explain your reasoning in words and pictures.

Solution

The following procedure 'dist' implements Floyd's algorithm [Flo 62]. We assume that the length of a nonexistent arc is ∞, that x + ∞ = ∞, and that min(x, ∞) = x for all x.

```
procedure dist(var d: nnreal);
var  i, j, k: integer;
begin
  for  k := 1  to  n  do
    for  i := 1  to  n  do
      for  j := 1  to  n  do
        d[i, j] := min(d[i, j], d[i, k] + d[k, j])
end;
```

Exercise: shortest paths

In addition to the distance d[i, j] of the preceding exercise, we wish to compute a shortest path from i to j (i.e. one that realizes this distance). Extend the solution above and write a procedure 'shortestpath' that returnsits result in an array 'next' of type:

```
  type  nnn = array[1 .. n, 1 .. n] of 0 .. n;
next[i,j] contains the next node after i on a shortest path from i to
j, or 0 if no such path exists.
```

Solution

```
procedure shortestpath(var d: nnreal; var next: nnn);
var  i, j, k: integer;
begin
   for  i := 1  to  n  do
    for  j := 1  to  n  do
      if  d[i, j] ≠ ∞  then  next[i, j] := j  else  next[i, j] :=
0;
   for  k := 1  to  n  do
     for  i := 1  to  n  do
       for  j := 1  to  n  do
         if  d[i, k] + d[k, j] < d[i, j]  then
           { d[i, j] := d[i, k] + d[k, j];  next[i, j] := next[i, k]
}
end;
```

It is easy to prove that next[i, j] = 0 at the end of the algorithm iff d[i, j] = ∞ (i.e. there is no path from i to j).

Minimum spanning tree in a graph

Consider a *weighted* graph G = (V, E, w), where V = {v_1, ..., v_n} is the set of vertices, E = {E_1, ... , e_m} is the set of edges, each edge e_i is an unordered pair (v_j, v_k) of vertices, and w: E → R assigns a real number to each edge, which we call its weight. We consider only *connected* graphs G, in the sense that any pair (v_j, v_k) of vertices is connected by a sequence of edges. in the following example, the edges are labeled with their weight (Exhibit 11.2).

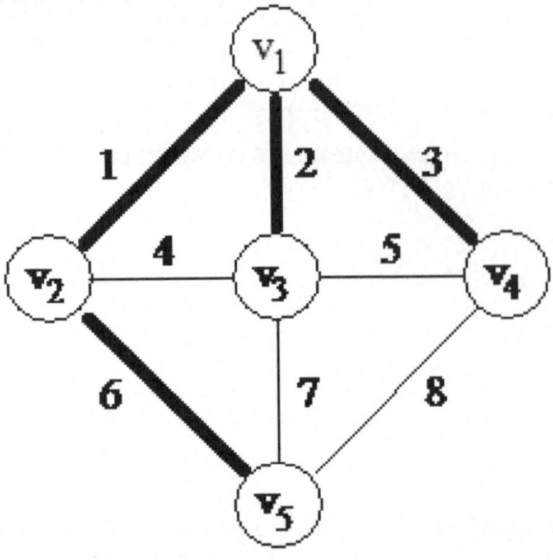

Exhibit 11.2: Example of a minimum spanning tree.

A *tree* T is a connected graph that contains no circuits: any pair (vj, vk) of verticesin T is connected by a unique sequence of edges. A *spanning tree* of a graph G is a subgraph T of G, given by its set of edges ET ⊆ E, that is a tree and satisfies the additional condition of being maximal, in the sense that no edge in E \ ET can be added to T without destroying the tree property. *Observation:* a connected graph G has at least one spanning tree. The *weight* of a spanning tree is the sum of the weights of all its edges. A *minimum spanning tree* is a spanning tree of minimal weight. in Exhibit 11.2, the bold edges form the minimal spanning tree.

Consider the following two algorithms:

Grow:

$E_T := \emptyset$; *{ initialize to empty set }* while T is not a spanning tree do $E_T :=$ ET ∪ {a min cost edge that does not form a circuit when added to ET}

Shrink:

$E_T := E$; *{ initialize to set of all edges }* while T is not a spanning tree do $E_T :=$ ET \ {a max cost edge that leaves T connected after its removal}

Claim: The "growing algorithm" and "shrinking algorithm" determine a minimum spanning tree.

If T is a spanning tree of G and e = (vj, vk) ∉ ET, we define Ckt(e, T), "the circuit formed by adding e to T" as the set of edgesin ET that form a path from vj to vk. in the example of Exhibit 11.2 with the spanning tree shown in bold edges we obtain Ckt((v4, v5), T) = {(v4, v1), (v1, v2), (v2, v5)}.

Exercise

Show that for each edge e ∉ ET there exists exactly one such circuit. Show that for any e ∉ ET and any t ∉ Ckt(e, T) the graph formed by (ET \ {t}) ∪ {e} is still a spanning tree.

A local minimum spanning tree of G is a spanning tree T with the property that there exist no two edges e ∉ ET , t ∉ Ckt(e, T) with w(e) < w(t).

Consider the following 'exchange algorithm', which computes a local minimum spanning tree:

Exchange:

T := any spanning tree;

while there exists e $\notin E_T$, t ∈ Ckt(e, T) with w(e) < w(t) do

$E_T := (E_T \setminus \{t\}) \cup \{e\}$; *{ exchange }*

Theorem: A local minimum spanning tree for a graph G is a minimum spanning tree.

For the proof of this theorem we need:

Lemma: If T' and T" are arbitrary spanning trees for G, T' ≠ T", then there exist e" $\notin E_T'$, e' $\notin E_T''$, such that e"∈ Ckt(e', T") and e' ∈ Ckt(e", T').

Proof: Since T' and T" are spanning trees for G and T' ≠ T", there exists e" $\in E_T'' \setminus E_T'$. Assume that Ckt(e", T') ⊆T". Then e" and the edges in Ckt(e", T') form a circuit in T'' that contradicts the assumption that T'' is a tree. Hence there must be at least one e' ∈ Ckt(e'', T') \ E_T''.

Assume that for all e' ∈ Ckt(e'', T') \ E_T'' we have e'' ∈ Ckt(e', T''). Then forms a circuit in T'' that contradicts the proposition that T'' is a tree. Hence there must be at least one e' ∈ Ckt(e'', T') \ E_T'' with e'' ∈ Ckt(e', T'').

$$\{e''\} \cup (Ckt(e'', T') \cap E_{T''}) \cup \bigcup_{e' \in Ckt(e'', T') \setminus E_{T''}} Ckt(e', T'')$$

Proof of the Theorem: Assume that T' is a local minimum spanning tree. Let T'' be a minimum spanning tree.

If T' ≠ T'' the lemma implies the existence of e' ∈ Ckt(e'', T') \ E_T'' and e'' ∈ Ckt(e', T'') \ E_T'.

If w(e') < w(e''), the graph defined by the edges (E_T'' \ {e''}) ∪ {e'} is a spanning tree with lower weight than T''.

Since T'' is a minimum spanning tree, this is impossible and it follows that w(e') ≥w (e''). (**)

If w(e') > w(e''), the graph defined by the edges (E_T' \ {e'}) ∪ {e''} is a spanning tree with lower weight than T'.

Since T'' is a local minimum spanning tree, this is impossible and it follows that w(e') ≤ w(e''). (****)

From (*) and (**) it follows that w(e') = w(e'') must hold. The graph defined by the edges (E_T'' \ {e''}) ∪ {e'} is still a spanning tree that has the same weight as T''. We replace T'' by this new minimum spanning tree and continue the replacement process. Since T' and T'' have only finitely many edges the process will terminate and T'' will become equal to T'. This proves that T'' is a minimum spanning tree.

The theorem implies that the tree computed by 'Exchange' is a minimum spanning tree.

Exercises

Consider how to extend the transitive closure algorithm based on boolean matrix multiplication so that it computes (a) distances and (b) a shortest path.

Prove that the algorithms 'Grow' and 'Shrink' compute local minimum spanning trees. Thus they are minimum spanning trees by the theorem of the section entitled "Minimum spanning tree in a graph".

12

Integers

Operations on integers

Five basic operations account for the lion's share of integer arithmetic:

+ – · div mod

The product 'x · y', the quotient 'x div y', and the remainder 'x mod y' are related through the following *div-mod identity*:

- (x div y) · y + (x mod y) = x for y ≠ 0.

Many programming languages provide these five operations, but unfortunately, 'mod' tends to behave differently not only between different languages but also between different implementations of the same language. How come have we not learned in school what the remainder of a division is?

The div-mod identity, a cornerstone of number theory, defines 'mod' assuming that all the other operations are defined. it is mostly used in the context of nonnegative integers $x \geq 0$, $y > 0$, where everything is clear, in particular the convention $0 \leq x \bmod y < y$. One half of the domain of integers consists of negative numbers, and there are good reasons for extending all five basic operations to the domain of all integers (with the possible exception of y = 0), such as:

- Any operation with an undefined result hinders the portability and testing of programs: if the "forbidden" operation does get executed by mistake, the computation may get into non-repeatable states. Example: from a practical point of view it is better not to leave 'x div 0' undefined, as is customary in mathematics, but to define the result as '= overflow', a feature typically supported in hardware.

- Some algorithms that we usually consider in the context of nonnegative integers have natural extensions into the domain of all integers (see the following sections on 'gcd' and modular number representations).

Unfortunately, the attempt to extend 'mod' to the domain of integers runs into the problem mentioned above:

How should we define 'div' and 'mod'? Let's follow the standard mathematical approach of listing desirable properties these operations might possess. in addition to the "sacred" div-mod identity (1) we consider:

❖ Symmetry of div: (−x) div y = x div (−y) = −(x div y).

The most plausible way to extend 'div' to negative numbers.

❖ A constraint on the possible values assumed by 'x mod y', which, for y > 0, reduces to the convention of nonnegative remainders:

$0 \leq x \bmod y < y$.

This is important because a standard use of 'mod' is to partition the set of integers into y residue classes. We consider a weak and a strict requirement:

(3') Number of residue classes = |y|: for given y and varying x, 'x mod y' assumes exactly |y| distinct values. (3") in addition, we ask for nonnegative remainders: $0 \leq x \bmod y < |y|$.

Pondering the consequences of these desiderata, we soon realize that 'div' cannot be extended to negative arguments by means of symmetry. Even the relatively innocuous case of positive denominator y > 0 makes it impossible to preserve both (2) and (3"), as the following failed attempt shows:

$((-3) \text{ div } 2) \cdot 2 \;+\; ((-3) \bmod 2) \quad ?\!=\!? \quad -3$ Preserving (1)

$(-(3 \text{ div } 2)) \cdot 2 \;+\; 1 \quad ?\!=\!? \quad -3$ and using (2) and (3")

$(-1) \cdot 2 \;+\; 1 \;\neq\; -3$ ... fails!

Even the weak condition (3'), which we consider essential, is incompatible with (2). For y = −2, it follows from and (2) that there are three residue classes modulo (−2): x mod (−2) yields the values 1, 0, −1; for example, 1 mod (−2) = 1, 0 mod (−2) = 0, (−1) mod (−2) = −1.

❖ This does not go with the fact that 'x mod 2' assumes only the two values 0, 1. Since a reasonable partition into residue classes is more important than the superficially appealing symmetry of 'div', we have to admit that (2) was just wishful thinking.

Without giving any reasons, [Knu 73a] (see the chapter "Reducing a task to given primitives; programming motion) defines 'mod' by means of the div-mod identity (1) as follows:

$x \bmod y = x - y \cdot \lfloor x / y \rfloor$, if $y \neq 0$; $x \bmod 0 = x$;

Thus he implicitly defines x div y = $\lfloor x / y \rfloor$, where $\lfloor z \rfloor$, the "floor" of z, denotes the largest integer ≤ z; the "ceiling" $\lceil z \rceil$ denotes the smallest integer ≥ z. Knuth extends the domain of 'mod' even further by defining "x mod 0 = x". With the exception of this special case y = 0, Knuth's definition satisfies (3'): Number of residue classes = |y|. The definition does not satisfy (3"), but a slightly more complicated condition. For given y ≠ 0, we have 0 ≤ x mod y < y, if y > 0; and 0 ≥ x mod y > y, if y < 0. Knuth's definition of 'div' and 'mod' has the added advantage that it holds for real numbers as well, where 'mod' is a useful operation for expressing the periodic behavior of functions [e.g. tan x = tan (x mod π)].

Exercise: another definition of 'div' and 'mod'

❖ Show that the definition

$$\mathbf{x\ div\ y} := \begin{cases} \lfloor x/y \rfloor & \text{if } y > 0 \\ \lceil x/y \rceil & \text{if } y < 0 \end{cases}$$

In conjunction with the div-mod identity (1) meets the strict requirement (3").

Solution

x \ y	-5	-4	-3	-2	-1	1	2	3	4	5
5	-1	-1	-1	-2	-5	5	2	1	1	1
4	0	-1	-1	-2	-4	4	2	1	1	0
3	0	0	-1	-1	-3	3	1	1	0	0
2	0	0	0	-1	-2	2	1	0	0	0
1	0	0	0	0	-1	1	0	0	0	0
0	0	0	0	0	0	0	0	0	0	0
-1	1	1	1	1	1	-1	-1	-1	-1	-1
-2	1	1	1	1	2	-2	-1	-1	-1	-1
-3	1	1	1	2	3	-3	-2	-1	-1	-1
-4	1	1	2	2	4	-4	-2	-2	-1	-1
-5	1	2	2	3	5	-5	-3	-2	-2	-1

x div y

x \ y	-5	-4	-3	-2	-1	1	2	3	4	5
5	0	1	2	1	0	0	1	2	1	0
4	4	0	1	0	0	0	0	1	0	4
3	3	3	0	1	0	0	1	0	3	3
2	2	2	2	0	0	0	0	2	2	2
1	1	1	1	1	0	0	1	1	1	1
0	0	0	0	0	0	0	0	0	0	0
-1	4	3	2	1	0	0	1	2	3	4
-2	3	2	1	0	0	0	0	1	2	3
-3	2	1	0	1	0	0	1	0	1	2
-4	1	0	2	0	0	0	0	2	0	1
-5	0	3	1	1	0	0	1	1	3	0

x mod y

Exercise

Fill out comparable tables of values for Knuth's definition of 'div' and 'mod'.

Solution

x \ y	-5	-4	-3	-2	-1	1	2	3	4	5
5	-1	-2	-2	-3	-5	5	2	1	1	1
4	-1	-1	-2	-2	-4	4	2	1	1	0
3	-1	-1	-1	-2	-3	3	1	1	0	0
2	-1	-1	-1	-1	-2	2	1	0	0	0
1	-1	-1	-1	-1	-1	1	0	0	0	0
0	0	0	0	0	0	0	0	0	0	0
-1	0	0	0	0	1	-1	-1	-1	-1	-1
-2	0	0	0	1	2	-2	-1	-1	-1	-1
-3	0	0	1	1	3	-3	-2	-1	-1	-1
-4	0	1	1	2	4	-4	-2	-2	-1	-1
-5	1	1	1	2	5	-5	-3	-2	-2	-1

x div y

x \ y	-5	-4	-3	-2	-1	1	2	3	4	5
5	0	-3	-1	-1	0	0	1	2	1	0
4	-1	0	-2	0	0	0	0	1	0	4
3	-2	-1	0	-1	0	0	1	0	3	3
2	-3	-2	-1	0	0	0	0	2	2	2
1	-4	-3	-2	-1	0	0	1	1	1	1
0	0	0	0	0	0	0	0	0	0	0
-1	-1	-1	-1	-1	0	0	1	2	3	4
-2	-2	-2	-2	0	0	0	0	1	2	3
-3	-3	-3	0	-1	0	0	1	0	1	2
-4	-4	0	-1	0	0	0	0	2	0	1
-5	0	-1	-2	-1	0	0	1	1	3	0

x mod y

The Euclidean algorithm

A famous algorithm for computing the greatest common divisor (gcd) of two natural numbers appears in Book 7 of Euclid's Elements (ca. 300 BC). it is based on the identity gcd(u, v) = gcd(u – v, v), which can be used for u > v to reduce the size of the arguments, until the smaller one becomes 0.

We use these properties of the greatest common divisor of two integers u and v > 0: gcd(u, 0) = u By convention this also holds for u = 0.

gcd(u, o) = u	By convention this also holds for u = o.
gcd(u, v) = gcd(v, u)	Permutation of arguments, important for the termination of the following procedure.
gcd(u, v) = gcd(v, u – q · v)	For any integer q.

The formulas above translate directly into a recursive procedure:

```
function gcd(u, v: integer): integer;
begin
  if  v = 0  then  return(u)  else  return(gcd(v, u mod v))
end;
```

A test for the relative size of u and v is unnecessary. If initially u < v, the first recursive call permutes the two arguments, and thereafter the first argument is always larger than the second.

This simple and concise solution has a relatively high implementation cost. A stack, introduced to manage the recursive procedure calls, consumes space and time. in addition to the operations visible in the code (test for equality, assignment, and 'mod'), hidden stack maintenance operations are executed. There is an equally concise iterative version that requires a bit more thinking and writing, but is significantly more efficient:

```
function gcd(u, v: integer): integer;
var  r: integer;
begin
  while  v ≠ 0  do  { r := u mod v;  u := v;  v := r };
  return(u)
end;
```

The prime number sieve of Eratosthenes

The oldest and best-known algorithm of type *sieve* is named after Eratosthenes (ca. 200 BC). A set of elements is to be separated into two classes, the "good" ones and the "bad" ones. As is often the case in life, bad elements are easier to find than good ones. A sieve process successively eliminates elements that have been recognized as bad; each element eliminated helps in identifying further bad elements. Those elements that survive the epidemic must be good.

Sieve algorithms are often applicable when there is a striking asymmetry in the complexity or length of the proofs of the two assertions "p is a good element" and "p is a bad element". This theme occurs prominently in the complexity theory of problems that appear to admit only algorithms whose time requirement grows faster than polynomially in the size of the input (NP completeness). Let us illustrate this asymmetry in the case of prime numbers, for which Eratosthenes' sieve is designed. in this analogy, "prime" is "good" and "nonprime" is "bad".

A prime is a positive integer greater than 1 that is divisible only by 1 and itself. Thus primes are defined in terms of their *lack* of an easily verified property: a prime has no factors other than the two trivial ones. To prove that 1 675 307 419 is not prime, it suffices to exhibit a pair of factors:

$$1\,675\,307\,419 = 1\,234\,567 \cdot 1\,357.$$

This verification can be done by hand. The proof that $2^{17} - 1$ is prime, on the other hand, is much more elaborate. in general (without knowledge of any special property this particular number might have) one has to verify, for each and every number that qualifies as a candidate factor, that it is not a factor. This is obviously more time consuming than a mere multiplication.

Exhibiting factors through multiplication is an example of what is sometimes called a "one-way" or "trapdoor" function: the function is easy to evaluate (just one multiplication), but its inverse is hard. in this context, the inverse of multiplication is not division, but rather factorization. Much of modern cryptography relies on the difficulty of factorization.

The prime number sieve of Eratosthenes Works as follows. We mark the smallest prime, 2, and erase all of its multiples within the desired range 1 .. n. The smallest remaining number must be prime; we mark it and erase its multiples. We repeat this process for all numbers up to $\sqrt{n}$: If an integer $c < n$ can be factored, $c = a \cdot b$, then at least one of the factors is $<\sqrt{n}$.

```
{ sieve of Eratosthenes marks all the primes in 1 .. n }
const  n = ... ;
var  sieve: packed array [2 .. n] of boolean;
         p, sqrtn, i: integer;
...
begin
   for  i := 2  to  n  do  sieve[i] := true;   { initialize the
sieve }
   sqrtn := trunc(sqrt(n));
   { it suffices to consider as divisors the numbers up to √ n }
   p := 2;
   while  p ≤ sqrtn  do  begin
      i := p · p;
      while  i ≤ n  do  { sieve[i] := false;  i := i + p };
      repeat  p := p + 1  until  sieve[p];
   end;
end;
```

Large integers

The range of numbers that can be represented directly in hardware is typically limited by the word length of the computer. For example, many small computers have a word length of 16 bits and thus limit integers to the range $-2^{15} \leq a < +2^{15} = 32768$. When the built-in number system is insufficient, a variety of software techniques are used to extend its range. They differ greatly with respect to their properties and intended applications, but all of them come at an additional cost in memory and, above all, in the time required for performing arithmetic operations. Let us mention the most common techniques.

Double-length or double-precision integers. Two words are used to hold an integer that squares the available range as compared to integers stored in one word. For a 16-bit computer we get 32-bit integers, for a 32- bit computer we get 64-bit integers. Operations on double-precision integers are typically slower by a factor of 2 to 4.

Variable precision integers. The idea above is extended to allocate as many words as necessary to hold a given integer. This technique is used when the size of intermediate results that arise during the course of a computation is unpredictable. it calls for list processing techniques to manage memory. The time of an operation depends on the size of its arguments: linearly for addition, mostly quadratically for multiplication.

Packed BCD integers. This is a compromise between double precision and variable precision that comes from commercial data processing. The programmer defines the maximal size of every integer variable used, typically by giving the maximal number of decimal digits that may be needed to express it. The compiler allocates an array of bytes to this variable that contains the following information: maximal length, current length, sign, and the digits. The latter are stored in BCD (binary-coded decimal) representation: a decimal digit is coded in 4 bits, two of them are packed into a byte. Packed BCD integers are expensive in space because most of the time there is unused allocated space; and even more so in time, due to digit-by-digit arithmetic. They are unsuitable for lengthy scientific/technical computations, but OK for I/O-intensive data processing applications.

Modular number systems: the poor man's large integers

Modular arithmetic is a special-purpose technique with a narrow range of applications, but is extremely efficient where it applies—typically in combinatorial and number-theoretic problems. it handles addition, and particularly multiplication, with unequaled efficiency, but lacks equally efficient algorithms for division and comparison. Certain combinatorial problems that require high precision can be solved without divisions and with few comparisons; for these, modular numbers are unbeatable.

Chinese Remainder Theorem: Let $m_1, m_2, \dots, m_k$ be pairwise relatively prime positive integers, called *moduli*. Let $m = m_1 \cdot m_2 \cdot \dots \cdot m_k$ be their product. Given k positive integers $r_1, r_2, \dots, r_k$, called *residues*, with $0 \le r_i <$ mi for $1 \le i \le r_k$, there exists exactly one integer r, $0 \le r < m$, such that r mod $m_i = r_i$ for $1 \le i \le k$.

The Chinese remainder theorem is used to represent integers in the range $0 \le r < m$ uniquely as k-tuples of their residues modulo mi. We denote this number representation by

$r \sim [r_1, r_2, \dots, r_k]$.

The practicality of modular number systems is based on the following fact: The arithmetic operations (+ , – , ·) on integers r in the range $0 \le r < m$ are represented by the same operations, applied componentwise to k-tuples [r 1, $r_2, \dots, r_k$]. A modular number system replaces a single +, –, or · in a large range by k operations of the same type in small ranges.

```
If r ~ [r1, r2, ... , rk], s ~ [s1, s2, ... , sk], t ~ [t1, t2, ... , tk],
then:
   (r + s)mod m = t ⇔ (ri + si) mod mi = ti for 1 ≤ i ≤ k,
   (r - s)mod m = t ⇔ (ri - si) mod mi = ti for 1 ≤ i ≤ k,
   (r · s)mod m = t ⇔ (ri · si) mod mi = ti for 1 ≤ i ≤ k.
```

Example

$m_1 = 2$ and $m_2 = 5$, hence $m = m_1 \cdot m_2 = 2 \cdot 5 = 10$. in the following table the numbers r in the range 0.. 9 are represented as pairs modulo 2 and modulo 5.

r	0	1	2	3	4	5	6	7	8	9
r mod 2	0	1	0	1	0	1	0	1	0	1
r mod 5	0	1	2	3	4	0	1	2	3	4

Let r = 2 and s = 3, hence r · s = 6. in modular representation: r ~ [0, 2], s ~ [1, 3], hence r · s ~ [0, 1].

A useful modular number system is formed by the moduli

$m_1 = 99, m_2 = 100, m_3 = 101$, hence $m = m_1 \cdot m_2 \cdot m_3 = 999900$.

Nearly a million integers in the range $0 \le r < 999900$ can be represented. The conversion of a decimal number to its modular form is easily computed by hand by adding and subtracting pairs of digits as follows:

r mod 99: Add pairs of digits, and take the resulting sum mod 99.

r mod 100: Take the least significant pair of digits.

r mod 101: Alternatingly add and subtract pairs of digits, and take the result mod 101.

The largest integer produced by operations on components is $100^2 \sim 2^{13}$; it is smaller than $2^{15} = 32768 \sim 32\text{k}$ and thus causes no overflow on a computer with 16-bit arithmetic.

Example

r = 123456

r mod 99 = (56 + 34 + 12) mod 99 = 3

r mod 100 = 56

r mod 101 = (56 – 34 + 12) mod 101 = 34

r ~ [3, 56, 34]

s = 654321

s mod 99 = (21 + 43 + 65) mod 99 = 30

s mod 100 = 21

s mod 101 = (21 – 43 + 65) mod 101 = 43

s ~ [30, 21, 43]

r + s ~ [3, 56, 34] + [30, 21, 43] = [33, 77, 77]

Modular arithmetic has some shortcomings: division, comparison, overflow detection, and conversion to decimal notation trigger intricate computations.

Exercise: Fibonacci numbers and modular arithmetic

The sequence of Fibonacci numbers

0, 1, 1, 2, 3, 5, 8, 13, 21, 34, 55, 89, 144, 233, ...

is defined by

$x_0 = 0, x_1 = 1, x_n = x_{n-1} + x_{n-2}$ for $n \geq 2$.

Write (a) a recursive function (b) an iterative function that computes the n-th element of this sequence. Using modular arithmetic, compute Fibonacci numbers up to 108 on a computer with 16-bit integer arithmetic, where the largest integer is $2^{15} - 1 = 32767$.

Using moduli $m_1 = 999$, $m_2 = 1000$, $m_3 = 1001$, what is the range of the integers that can be represented uniquely by their residues $[r_1, r_2, r_3]$ with respect to these moduli?

Describe in words and formulas how to compute the triple $[r_1, r_2, r_3]$ that uniquely represents a number r in this range.

Modify the function in (b) to compute Fibonacci numbers in modular arithmetic with the moduli 999, 1000, and 1001. Use the declaration

```
type  triple = array [1 .. 3] of integer;
```

and write the procedure

```
procedure modfib(n: integer; var r: triple);
```

Solution

```
(a)       function fib(n: integer): integer;
      begin
        if  n ≤ 1  then  return(n)  else  return(fib(n - 1) + fib(n - 2))
      end;

(b)       function fib(n: integer): integer;
      var  p, q, r, i: integer;
      begin
        if  n ≤ 1  then  return(n)

   else  begin
      p := 0;   q := 1;
      for  i := 2  to  n  do  { r := p + q;  p := q;  q := r };
      return(r)
   end
end;
```

Random numbers

The colloquial meaning of the term *at random* often implies "unpredictable". But *random numbers* are used in scientific/technical computing in situations where unpredictability is neither required nor desirable. What is needed in simulation, in sampling, and in the generation of test data is *not* unpredictability but certain statistical properties. A *random number generator* is a program that generates a sequence of numbers that passes a number of specified statistical tests. Additional requirements include: it runs fast and uses little memory; it is portable to computers that use a different arithmetic; the sequence of random numbers generated can be reproduced (so that a test run can be repeated under the same conditions).

in practice, random numbers are generated by simple formulas. The most widely used class, linear congruential generators, given by the formula

$r_{i+1} = (a \cdot r_i + c) \bmod m$

are characterized by three integer constants: the multiplier a, the increment c, and the modulus m. The sequence is initialized with a seed r0.

All these constants must be chosen carefully. Consider, as a bad example, a formula designed to generate random days in the month of February:

$r_0 = 0, r_{i+1} = (2 \cdot r_i + 1) \bmod 28.$

it generates the sequence 0, 1, 3, 7, 15, 3, 7, 15, 3, ... Since $0 \le r_i < m$, each generator of the form above generates a sequence with a prefix of length < m which is followed by a period of length ≤ m. in the example, the prefix 0, 1 of length 2 is followed by a period 3, 7, 15 of length 3. Usually we want a long period. Results from number theory assert that a period of length m is obtained if the following conditions are met:

- m is chosen as a prime number.
- (a – 1) is a multiple of m.
- m does not divide c.

Example

```
  r0 = 0,  ri+1 = (8 · ri + 1) mod 7
generates a sequence: 0, 1, 2, 3, 4, 5, 6, 0, ... with a period of length
7.
```

Shall we accept this as a sequence of random integers, and if not, why not? Should we prefer the sequence 4, 1, 6, 2, 3, 0, 5, 4, ... ?

For each application of random numbers, the programmer/analyst has to identify the important statistical properties required. Under normal circumstances these include:

No periodicity over the length of the sequence actually used. *Example:* to generate a sequence of 100 random weekdays ∈ {Su, Mo, ... , Sat}, do not pick a generator with modulus 7, which can generate a period of length at most 7; pick one with a period much longer than 100.

A desired distribution, most often the uniform distribution. If the range 0 .. m – 1 is partitioned into k equally sized intervals $I_1, I_2, \ldots, I_k$, the numbers generated should be uniformly distributed among these intervals; this must be the case not only at the end of the period (this is trivially so for a generator with maximal period m), but for any initial part of the sequence.

Many well-known statistical tests are used to check the quality of random number generators. The *run test* (the lengths of monotonically increasing and monotonically decreasing subsequences must occur with the right frequencies); the *gap test* (given a test interval called the "gap", how many consecutively generated numbers fall outside?); the *permutation test* (partition the sequence into subsequences of t elements; there are t! possible relative orderings of elements within a subsequence; each of these orderings should occur about equally often).

Exercise: visualization of random numbers

Write a program that lets its user enter the constants a, c, m, and the seed r0 for a linear congruential generator, then displays the numbers generated as dots on the screen: A pair of consecutive random numbers is interpreted as the (x, y)-coordinates of the dot. You will observe that most generators you enter have obvious flaws: our visual system is an excellent detector of regular patterns, and most regularities correspond to undesirable statistical properties.

The point made above is substantiated in [PM 88].

The following simple random number generator and some of its properties are easily memorized: $r_0 = 1$, $r_i^{+1} = 125 \cdot r_i \bmod 8192$.

- $8192 = 2^{13}$, hence the remainder mod 8192 is represented by the 13 least significant bits.
- $125 = 127 - 2 = (1111101)$ in binary representation.
- Arithmetic can be done with 16-bit integers without overflow and without regard to the representation of negative numbers.
- The numbers r_k generated are exactly those in the range $0 \le r_k < 8192$ with $r_k \bmod 4 = 1$ (i.e. the period has length 211 = 2048).
- Its statistical properties are described in [Kru 69], [Knu 81] contains the most comprehensive treatment of the theory of random number generators.
- As a conclusion of this brief introduction, remember an important rule of thumb:

  Never choose a random number generator at random!

Exercises

- Work out the details of implementing double-precision, variable-precision, and BCD integer arithmetic, and estimate the time required for each operation as compared to the time of the same operation in single precision. For variable precision and BCD, introduce the length L of the representation as a parameter.
- The least common multiple (lcm) of two integers u and v is the smallest integer that is a multiple of u and v. Design an algorithm to compute lcm(u, v).

- The prime decomposition of a natural number n > 0 is the (unique) multiset PD(n) = [p1, p2, ... , pk] of primes pi whose product is n. A multiset differs from a set in that elements may occur repeatedly (e.g. PD(12) = [2, 2, 3]). Design an algorithm to compute PD(n) for a given n > 0.
- Work out the details of modular arithmetic with moduli 9, 10, 11.
- Among the 95 linear congruential random number generators given by the formula $r_{i+1} = a \cdot r_i \bmod m$, with prime modulus m = 97 and 1 < a < 97, find out how many get disqualified "at first sight" by a simple visual test. Consider that the period of these RNGs is at most 97.

13

Reals

Floating-point numbers

Real numbers, those declared to be of type REAL in a programming language, are represented as floating-point numbers on most computers. A floating-point number z is represented by a (signed) mantissa m and a (signed) exponent e with respect to a base b: z=± m·b±e (e.g. $z=+0.11 \cdot 2^{-1}$). This section presents a very brief introduction to floating-point arithmetic. We recommend [Gol91] as a comprehensive survey.

Floating-point numbers can only approximate real numbers, and in important ways, they behave differently. The major difference is due to the fact that any floating-point number system is a *finite number system*, as the mantissa m and the exponent e lie in a bounded range. Consider, as a simple example, the following number system:

$z = \pm 0.B_1 b_2 \cdot 2\pm e$, where B_1, b_2, and e may take the values 0 and 1.

The number representation is *not unique*: The same real number may have many different representations, arranged in the following table by numerical value (lines) and constant exponent (columns).

1.5	$+0.11 \cdot 2^{+1}$		
1.0	$+0.10 \cdot 2^{+1}$		
0.75		$+0.11 \cdot 2^{\pm 0}$	
0.5	$+0.01 \cdot 2^{+1}$	$+0.10 \cdot 2^{\pm 0}$	
0.375			$+0.11 \cdot 2^{-1}$
0.25		$+0.01 \cdot 2^{\pm 0}$	$+0.10 \cdot 2^{-1}$
0.125			$+0.01 \cdot 2^{-1}$
0.	$+0.00 \cdot 2^{+1}$	$+0.00 \cdot 2^{\pm 0}$	$+0.00 \cdot 2^{-1}$

The table is symmetric for negative numbers. Notice the cluster of representable numbers around zero. There are only 15 different numbers, but 2^5= 32 different representations.

Exercise: a floating-point number system

Consider floating-point numbers represented in a 6-bit "word" as follows: The four bits b b_2 B_1 B_0 represent a signed mantissa, the two bits e e_0 a signed exponent to the base 2. Every number has the form x=b b_2 B_1 B_0·2^{ee0}.

Both the exponent and the mantissa are integers represented in *2's complement form*. This means that the integer values –2..1 are assigned to the four different representations e e0 as shown:

v	e	e_o
0	0	0
1	0	1
−2	1	0
−1	1	1

- Complete the following table of the values of the mantissa and their representation, and write down a formula to compute v from b b_2 B_1 B_0.

v	b	b_2	b_1	b_o
0	0	0	0	0
1	0	0	0	1
…				
7	0	1	1	1
−8	1	0	0	0
…				
−1	1	1	1	1

- How many different number representations are there in this floating-point system?
- How many different numbers are there in this system? Draw all of them on an axis, each number with all its representations.

On a byte-oriented machine, floating-point numbers are often represented by 4 bytes =32 bits: 24 bits for the signed mantissa, 8 bits for the signed exponent. The mantissa m is often interpreted as a fraction $0 \le m < 1$, whose precision is bounded by 23 bits; the 8-bit exponent permits scaling within the range $2^{-128} \le 2^e \le 2^{127}$. Because 32- and 64-bit floating-point number systems are so common, often coexisting on the same hardware, these number systems are often identified with "single precision" and "double precision", respectively. in recent years an IEEE standard format for-single precision floating-point numbers has emerged, along with standards for higher precisions: double, single extended, and double extended.

IEEE standard single-precision floating-point format:

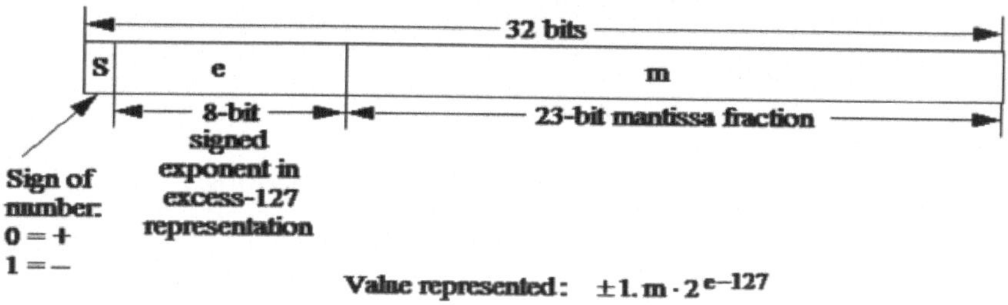

The following example shows the representation of the number

$^{+1}.011110 \ldots 0 \cdot 2^{-54}$

in the IEEE format:

0	0 1 0 0 1 0 0 1	0 1 1 1 1 0 ... 0

Some dangers

Floating-point computation is fraught with problems that are hard to analyze and control. Unexpected results abound, as the following examples show. The first two use a binary floating-point number system with a signed 2- bit mantissa and a signed 1-bit exponent. Representable numbers lie in the range

$-0.11 \cdot 2^{+1} \leq z \leq +0.11 \cdot 2^{+1}$.

Example: $y + x = y$ and $x \neq 0$

it suffices to choose $|x|$ small as compared to $|y|$; for example,

$x = 0.01 \cdot 2^{-1}$, $y = 0.10 \cdot 2^{+1}$.

The addition forces the mantissa of x to be shifted to the right until the exponents are equal (i.e. x is represented as $0.0001 \cdot 2^{+1}$). Even if the sum is computed correctly as $0.1001 \cdot 2^{+1}$ in an accumulator of double length, storing the result in memory will force rounding: $x + y = 0.10 \cdot 2^{+1} = y$.

Example: Addition is not associative: $(x + y) + z \neq x + (y + z)$

The following values for x, y, and z assign different values to the left and right sides. Left side: $(0.10 \cdot 2^{+1} + 0.10 \cdot 2^{-1}) + 0.10 \cdot 2^{-1} = 0.10 \cdot 2^{+1}$

Right side: $0.10 \cdot 2^{+1} + (0.10 \cdot 2^{-1} + 0.10 \cdot 2^{-1}) = 0.11 \cdot 2^{+1}$

A useful rule of thumb helps prevent the loss of significant digits: Add the small numbers before adding the large ones.

Example: ((x + y)2 – x2 – 2xy) / y2 = 1?

Let's evaluate this expression for large |x| and small |y| in a floating-point number system with five decimal digits.

$x = 100.00, y = .01000$

$x + y = 100.01$

$(x + y)^2 = 10002.0001$, rounded to five digits yields 10002.

$x^2 = 10000.$

$(x + y)^2 - x^2 = 2.????$ (four digits have been lost!)

$2xy = 2.0000$

$(x + y)^2 - x^2 - 2xy = 2.???? - 2.0000 = 0.?????$

Now five digits have been lost, and the result is meaningless.

Example: numerical instability

Recurrence relations for sequences of numbers are prone to the phenomenon of *numerical* instability. Consider the sequence

$x_0 = 1.0, x_1 = 0.5, x_{n+1} = 2.5 \cdot x_n - x_{n-1}.$

We first solve this linear recurrence relation in closed form by trying $x_i=r^i$ for $r \neq 0$. This leads to $r^{n+1} = 2.5 \cdot r^n - r^{n-1}$, and to the quadratic equation $0 = r^2 - 2.5 \cdot r + 1$, with the two solutions $r = 2$ and $r = 0.5$.

The general solution of the recurrence relation is a linear combination:

$x_i = a \cdot 2^i + b \cdot 2^{-i}.$

The starting values $x_0 = 1.0$ and $x_1 = 0.5$ determine the coefficients a=0 and b=1, and thus the sequence is given exactly as $x_i = 2^{-i}$. If the sequence $x_i = 2^{-i}$ is computed by the recurrence relation above in a floating-point number system with one decimal digit, the following may happen:

$x_2 = 2.5 \cdot 0.5 - 1 = 0.2$ (rounding the exact value 0.25),

$x_3 = 2.5 \cdot 0.2 - 0.5 = 0$ (represented exactly with one decimal digit),

$x_4 = 2.5 \cdot 0 - 0.2 = -0.2$ (represented exactly with one decimal digit),

$x_5 = 2.5 \cdot (-0.2) - 0 = -0.5$ represented exactly with one decimal digit),

$x_6 = 2.5 \cdot (-0.5) - (-0.2) = -1.05$ (exact) $= -1.0$ (rounded),

$x_7 = 2.5 \cdot (-1) - (-0.5) = -2.0$ (represented exactly with one decimal digit),

$x_8 = 2.5 \cdot (-2)-(-1) = -4.0$(represented exactly with one decimal digit).

As soon as the first rounding error has occurred, the computed sequence changes to the alternative solution $x_i = a \cdot 2_i$, as can be seen from the doubling of consecutive computed values.

Exercise: floating-point number systems and calculations

- Consider a floating-point number system with two ternary digits t1, t2 in the mantissa, and a ternary digit e in the exponent to the base 3. Every number in this system has the form x = .t 1 t2 · 3e, where t_1, t_2, and 3^e assume a value chosen among {0,1,2}. Draw a diagram that shows all the different numbers in this system, and for each number, all of its representations. How many representations are there? How many different numbers?

- Recall the series

$$\frac{1}{1-x} = 1 + x + x^2 + x^3 + \ldots$$

which holds for $|x| < 1$, for example,

$$\frac{1}{1-\frac{1}{2}} = 2 = 1 + \frac{1}{2} + \frac{1}{4} + \frac{1}{8} + \ldots.$$

Use this formula to express 1/0.7 as a series of powers.

Horner's method

A polynomial of n-th degree (e.g. n = 3) is usually represented in the form

$a_3 \cdot x_3 + a_2 \cdot x^2 + a_1 \cdot x + a_0$

but is better evaluated in nested form,

$((a_3 \cdot x + a_2) \cdot x + a_1) \cdot x + a_0$.

The first formula needs n multiplications of the form ai · xi and, in addition, n–1 multiplications to compute the powers of x. The second formula needs only n multiplications in total: The powers of x are obtained for free as a side effect of the coefficient multiplications.

The following procedure assumes that the (n^{+1}) coefficients ai are stored in a sufficiently large array a of type 'coeff':

```
type  coeff = array[0 .. m] of real;

function horner(var a: coeff;  n: integer;  x: real): real;
var  i: integer;  h: real;
begin
  h := a[n];
  for  i := n - 1  downto  0  do  h := h · x + a[i];
  return(h)
end;
```

Bisection

Bisection is an iterative method for solving equations of the form f(x) = 0. Assuming that the function f : R ⁇ R is continuous in the interval [a, b] and that f(a) · f(b) < 0, a root of the equation f(x) = 0 (a zero of f) must lie in the interval [a, b] (Exhibit 13.1). Let m be the midpoint of this interval. If f(m) = 0, m is a root. If f(m) · f(a) < 0, a root must be contained in [a, m], and we proceed with this subinterval; if f(m) · f(b) < 0, we proceed with [m, b]. Thus at each iteration the in*terval of uncertai*nty that must contain a root is half the size of the interval produced in the previous iteration. We iterate until the interval is smaller than the tolerance within which the root must be determined.

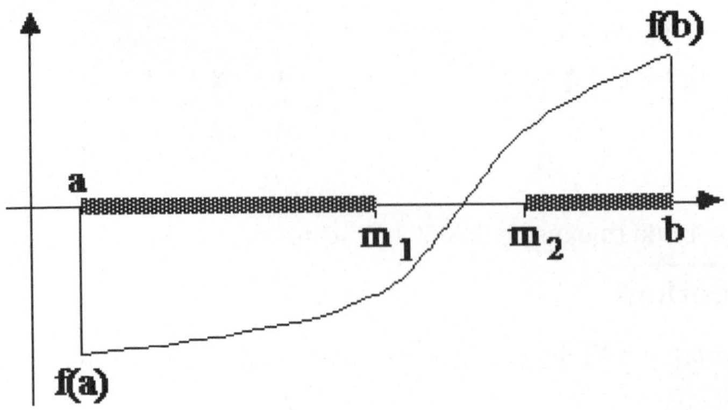

Exhibit 13.1: As in binary search, bisection excludes half of the interval under consideration at every step.

```
function bisect(function f: real;  a, b: real): real;
const  epsilon = 10^-6;
var  m: real;  faneg: boolean;
begin
  faneg := f(a) < 0.0;
  repeat
    m := (a + b) / 2.0;
    if  (f(m) < 0.0) = faneg  then  a := m  else  b := m
  until  |a - b| < epsilon;
  return(m)
```

A *sequence* x1, x^2, x_3,... converging to x *converges linearly* if there exist a constant c and an index i_0 such that for all I > i0: $|x_{i+1} - x| \leq c \cdot |xi - x|$. An *algor*ithm is said to converge linearly if the sequence of approximations constructed by this algorithm converges linearly. in a linearly convergent algorithm each iteration adds a constant number of significant bits. For example, each iteration of bisection halves the interval of uncertainty in each iteration (i.e. adds one bit of precision to the result). Thus bisection converges linearly with c = 0.5. A sequence x_1, x^2, x_3,... *converges quadratically* if there exist a constant c and an index i0 such that for all i > i0: $|xi^{+1} - x| \leq c \cdot |xi - x|2$.

Newton's method for computing the square root

Newton's method for solving equations of the form f(x) = 0 is an example of an algorithm with quadratic convergence. Let f: R → R be a continuous and differentiable function. An approximation x_{i+1} is obtained from xi by approximating f(x) in the neighborhood of xi by its tangent at the point (x_i, f(x_i)), and computing the intersection of this tangent with the x-axis (Exhibit 13.2). Hence

$$x_{i+1} = x_i - \frac{f(x_i)}{f'(x_i)}.$$

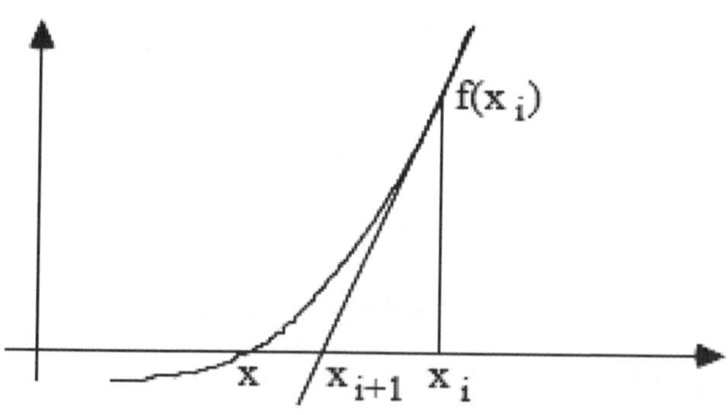

Exhibit 13.2: Newton's iteration approximates a curve locally by a tangent.

Newton's method is not guaranteed to converge (*Exercise:* construct counterexamples), but when it converges, it does so quadratically and therefore very fast, since each iteration doubles the number of significant bits.

To compute the square root x = $\sqrt{a}$ of a real number a > 0 we consider the function f(x) = x^2 – a and solve the equation x^2– a = 0. With f'(x)= 2 · x we obtain the iteration formula:

$$x_{i+1} = x_i - \frac{x_i^2 - a}{2x_i} = \frac{1}{2}\left(x_i + \frac{a}{x_i}\right).$$

The formula that relates xi and xi^{+1} can be transformed into an analogous formula that determines the propagation of the relative error:

$$R_i = \frac{x_i - x}{x}.$$

Since

$$x_{i+1} = \frac{1}{2}\left(x_i + \frac{a}{x_i}\right) = \frac{x_i^2 + x^2}{2x_i} = x + \frac{(x_i - x)^2}{2x_i},$$

we obtain for the relative error:

$$\frac{x_{i+1} - x}{x} = \frac{(x_i - x)^2}{2x \cdot x_i} = \frac{1}{2 \cdot \frac{x_i}{x}} \cdot \left(\frac{x_i - x}{x}\right)^2.$$

Using

$$R_i = \frac{x_i - x}{x} \quad \text{and} \quad 1 + R_i = \frac{x_i}{x}$$

we get a recurrence relation for the relative error:

$$R_{i+1} = \frac{R_i^2}{2 \cdot (1 + R_i)}.$$

If we start with $x_0 > 0$, it follows that $1+R_0 > 0$. Hence we obtain $r_1 > r_2 > r_3 > \ldots > 0$.

As soon as R_i becomes small (i.e. $R_i \ll 1$), we have $1 + R_i \approx 1$, and we obtain

$R_{i+1} \approx 0.5 \cdot R_i^2$

Newton's method converges quadratically as soon as x_i is close enough to the true solution. With a bad initial guess $R_i \gg 1$ we have, on the other hand, $1 + R_i \approx R_i$, and we obtain $R_{i+1} \approx 0.5 \cdot R_i$ (i.e. the computation appears to converge linearly until $R_i \ll 1$ and proper quadratic convergence starts).

Thus it is highly desirable to start with a good initial approximation x_0 and get quadratic convergence right from the beginning. We assume normalized binary floating-point numbers (i.e. $a = m \cdot 2^e$ with $0.5 \le m < 1$). A good approximation of is obtained by choosing any mantissa c with $0.5 \le c < 1$ and halving the exponent:

In order to construct this initial approximation x0, the programmer needs read and write access not only to a "real number" but also to its components, the mantissa and exponent, for example, by procedures such as

```
procedure mantissa(z: real): integer;
procedure exponent(z: real): integer;
procedure buildreal(mant, exp: integer): real;
```

Today's programming languages often lack such facilities, and the programmer is forced to use backdoor tricks to construct a good initial approximation. If x0 can be constructed by halving the exponent, we obtain the following upper bounds for the relative error:

$r_1 < 2^{-2}, r_2 < 2^{-5}, r_3 < 2^{-11}, R_4 < 2^{-23}, R_5 < 2^{-47}, R_6 < 2^{-95}$.

It is remar$_k$able that four iterations suffice to compute an exact square root for 32-bit floating-point numbers, where 23 bits are used for the mantissa, one bit for the sign and eight bits for the exponent, and that six iterations will do for a "number cruncher" with a word length of 64 bits. The starting value x 0 can be further optimized by choosing c carefully. it can be shown that the optimal value of c for computing the square root of a real number is $c = 1/\sqrt{2} \approx 0.707$.

Exercise: square root

Consider a floating-point number system with two decimal digits in the mantissa: Every number has the form $x = \pm .d_1 d_2 \cdot 10^{\pm e}$.

How many different *number representations* are there in this system?

How many different *numbers* are there in this system? Show your reasoning.

Compute $\sqrt{50} \cdot 102$ in this number system using Newton's method with a starting value x0 = 10. Show every step of the calculation. Round the result of any operation to two digits immediately.

Solution

- A number representation contains two sign bits and three decimal digits, hence there are 22 · 103 = 4000 distinct number representations in this system.
- There are three sources of redundancy:
- Multiple representations of zero
- Exponent +0 equals exponent –0
- Shifted mantissa: ±.d0 · 10 ±e=±.0d · 10 ±e + 1

A detailed count reveals that there are 3439 different numbers.

Zero has $2^2 \cdot 10 = 40$ representations, all of the form $\pm.00 \cdot 10^{\pm e}$, with two sign bits and one decimal digit e to be freely chosen. Therefore, $r_1 = 39$ must be subtracted from 4000.

If e = 0, then $\pm.d_1d_2 \cdot 10^{+0} = \pm.d_1d_2 \cdot 10^{-0}$. We assume furthermore that $d_1d_2 \neq 00$. The case $d_1d_2 = 00$ has been covered above. Then there are $2 \cdot 99$ such pairs. Therefore, $r_2 = 198$ must be subtracted from 4000.

If $d_2 = 0$, then $\pm.d_10 \cdot 10^{\pm e} = \pm.0d_1 \cdot 10^{\pm e+1}$. The case $d_1 = 0$ has been treated above. Therefore, we assume that $d_1 \neq 0$. Since ±e can assume the 18 different values −9, −8, ... , −1, +0, +1, ... +8, there are $2 \cdot 9 \cdot 18$ such pairs. Therefore, $r_3 = 324$ must be subtracted from 4000.

There are $4000 - r_1 - r_2 - r_3 = 3439$ different numbers in this system.

❖ Computing Newton's square root algorithm:

$x_0 = 10$

$x_1 = .50 \cdot (10 + 50/10) = .50 \cdot (10 + 5) = .50 \cdot 15 = 7.5$

$x_2 = .50 \cdot (7.5 + 50/7.5) = .50 \cdot (7.5 + 6.6) = .50 \cdot 14 = 7$

$x_3 = .50 \cdot = .50 \cdot (7 + 50/7) = (7 + 7.1) = .50 \cdot 14 = 7$

Exercises

❖ Write up all the distinct numbers in the floating-point system with number representations of the form $z=0.B_1b_2 \cdot 2E_1E_2$, where B_1, b_2 and E_1, E_2 may take the values 0 and 1, and mantissa and exponent are represented in 2's complement notation.

❖ Provide simple numerical examples to illustrate floating-point arithmetic violations of mathematical identities.

14

Straight Lines and Circles

Points are the simplest geometric objects; straight lines and line segments come next. Together, they make up the lion's share of all primitive objects used in two-dimensional geometric computation (e.g. in computer graphics). Using these two primitives only, we can approximate any curve and draw any picture that can be mapped onto a discrete raster. If we do so, most queries about complex figures get reduced to basic queries about points and line segments, such as: is a given point to the left, to the right, or *on* a given line? Do two given line segments intersect? As simple as these questions appear to be, they must be handled efficiently and carefully. Efficiently because these basic primitives of geometric computations are likely to be executed millions of times in a single program run. Carefully because the ubiquitous phenomenon of *degenerate configurations* easily traps the unwary programmer into overflow or meaningless results.

Intersection

The problem of deciding whether two line segments intersect is unexpectedly tricky, as it requires a consideration of three distinct nondegenerate cases, as well as half a dozen degenerate ones. Starting with degenerate objects, we have cases where one or both of the line segments degenerate into points. The code below assumes that line segments of length zero have been eliminated. We must also consider nondegenerate objects in degenerate configurations, as illustrated in Exhibit 14.1. Line segments A and B intersect (strictly). C and D, and E and F, do not intersect; the intersection point of the infinitely extended lines lies on C in the first case, but lies neither on E nor on F in the second case. The next three cases are degenerate: G and H intersect barely (i.e. in an endpoint); I and J overlap (i.e. they intersect in infinitely many points); K and L do not intersect. Careless evaluation of these last two cases is likely to generate overflow.

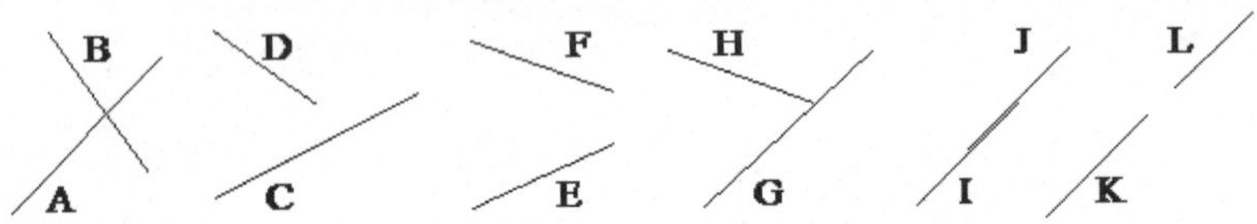

Exhibit 14.1: Cases to be distinguished for the segment intersection problem.

Computing the intersection point of the infinitely extended lines is a naive approach to this decision problem that leads to a three-step process:

Check whether the two line segments are parallel (a necessary precaution before attempting to compute the intersection point). If so, we have a degenerate configuration that leads to one of three special cases: not collinear, collinear nonoverlapping, collinear overlapping

Compute the intersection point of the extended lines (this step is still subject to numerical problems for lines that are almost parallel).

Check whether this intersection point lies on both line segments.

If all we want is a yes/no answer to the intersection question, we can save the effort of computing the intersection point and obtain a simpler and more robust procedure based on the following idea: two line segments intersect strictly iff the two endpoints of each line segment lie on opposite sides of the infinitely extended line of the other segment.

Let L be a line given by the equation $h(x, y) = a \cdot x + b \cdot y + c = 0$, where the coefficients have been normalized such that $a_2 + b_2 = 1$. For a line L given in this Hessean normal form, and for any point $p = (x, y)$, the function h evaluated at p yields the signed distance between p and L: $h(p) > 0$ if p lies on one side of L, $h(p) < 0$ if p lies on the other side, and $h(p) = 0$ if p lies on L. A line segment is usually given by its endpoints (x_1, y_1) and (x_2, y_2), and the Hessean normal form of the infinitely extended line L that passes through (x_1, y_1) and (x_2, y_2) is

$$h(x,y) = \frac{(y_2 - y_1)\cdot(x - x_1) - (x_2 - x_1)\cdot(y - y_1)}{L_{12}} = 0,$$

where

$$L_{12} = \sqrt{(x_2 - x_1)^2 + (y_2 - y_1)^2} > 0$$

is the length of the line segment, and $h(x, y)$ is the distance of $p = (x, y)$ from L. Two points p and q lie on opposite sides of L iff $h(p) \cdot h(q) < 0$ (Exhibit 14.2). $h(p) = 0$ or $h(q) = 0$ signals a degenerate configuration. Among these, $h(p) = 0$ and $h(q) = 0$ iff the segment (p, q) is collinear with L.

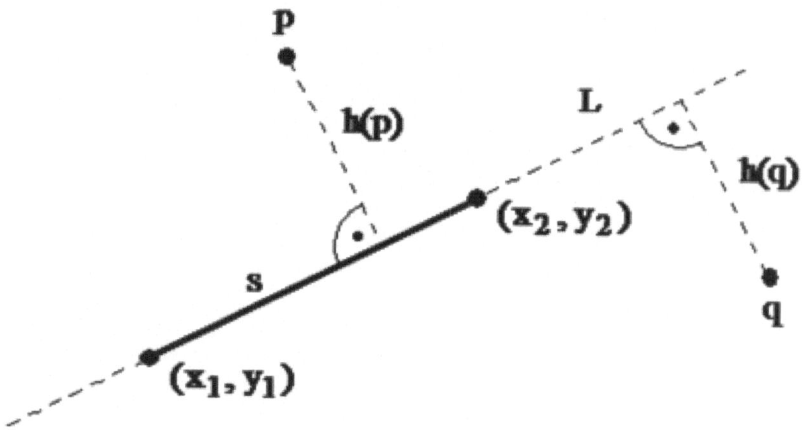

Exhibit 14.2: Segment s, its extended line L, and distance to points p, q as computed by function h.

```
type point   =  record  x, y: real  end;
     segment  =  record  p1, p2: point  end;

function d(s: segment; p: point): real;
{ computes h(p) for the line L determined by s }
var dx, dy, L12: real;
begin
  dx := s.p2.x - s.p1.x;  dy := s.p2.y - s.p1.y;
  L12 := sqrt(dx · dx + dy · dy);
  return((dy · (p.x - s.p1.x) - dx · (p.y - s.p1.y)) / L12)
end;
```

To optimize the intersection function, we recall the assumption $L_{12} > 0$ and notice that we do not need the actual distance, only its sign. Thus the function d used below avoids computing L_{12}. The function 'intersect' begins by checking whether the two line segments are collinear, and if so, tests them for overlap by intersecting the intervals obtained by projecting the line segments onto the x-axis (or onto the y-axis, if the segments are vertical). Two intervals [a, b] and [c, d] intersect iff min(a, b) ≤ max(c, d) and min(c, d) ≤ max(a, b). This condition could be simplified under the assumption that the representation of segments and intervals is ordered "from left to right" (i.e. for interval [a, b] we have a ≤ b). We do not assume this, as line segments often have a natural direction and cannot be "turned around".

```
function d(s: segment;  p: point): real;
begin
  return((s.p2.y - s.p1.y)  · (p.x - s.p1.x) - (s.p2.x - s.p1.x)  ·
(p.y - s.p1.y))
end;

function overlap(a, b, c, d: real): boolean;
begin  return((min(a, b) ≤ max(c, d)) and (min(c, d) ≤ max(a, b)))
end;

function intersect(s1, s2: segment): boolean;
var  d11, d12, d21, d22: real;
begin
  d11  := d(s1, s2.p1);  d12  := d(s1, s2.p2);
  if  (d11 = 0) and (d12 = 0)  then  { s1 and s2 are collinear }
    if  s1.p1.x = s1.p2.x  then  { vertical }
      return(overlap(s1.p1.y, s1.p2.y, s2.p1.y, s2.p2.y))
    else  { not vertical }
      return(overlap(s1.p1.x, s1.p2.x, s2.p1.x, s2.p2.x))
  else  begin  { s1 and s2 are not collinear }
    d21  := d(s2, s1.p1);  d22  := d(s2, s1.p2);
    return((d11  · d12 ≤ 0) and (d21  · d22 ≤ 0))
  end
end;
```

In addition to the degeneracy issues we have addressed, there are numerical issues of near-degeneracy that we only mention. The length L_{12} is a *condition number* (i.e. an indicator of the computation's accuracy). As Exhibit 14.3 suggests, it may be numerically impossible to tell on which side of a short line segment L a distant point p lies.

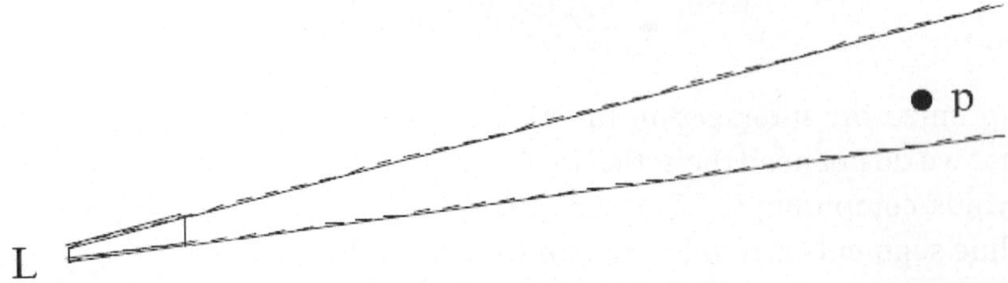

Exhibit 14.3: A point's distance from a segment amplifies the error of the "which side" computation.

Conclusion: A geometric algorithm must check for degenerate configurations explicitly—the code that handles configurations "in general position" will not handle degeneracies.

Clipping

The widespread use of windows on graphic screens makes clipping one of the most frequently executed operations: Given a rectangular window and a configuration in the plane, draw that part of the configuration which lies within the window. Most configurations consist of line segments, so we show how to clip a line segment given by its endpoints (x1, y_1) and (x^2, y_2) into a window given by its four corners with coordinates {left, right} × {top, bottom}.

The position of a point in relation to the window is described by four boolean variables: ll (to the left of the left border), rr (to the right of the right border), bb (below the lower border), tt (above the upper border):

```
type  wcode = set of (ll, rr, bb, tt);
```

A point inside the window has the code ll = rr = bb = tt = false, abbreviated 0000 (Exhibit 14.4).

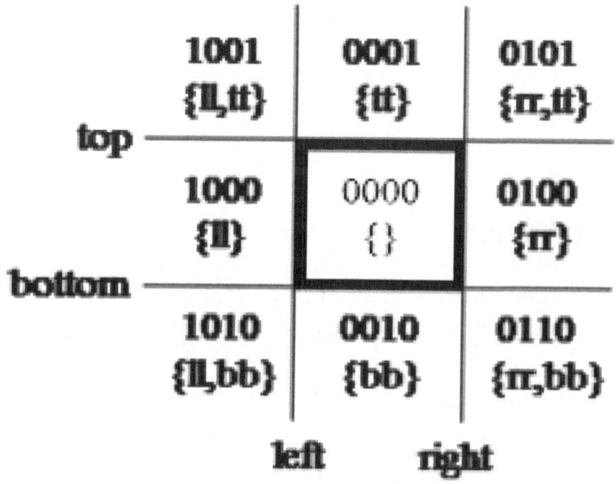

Exhibit 14.4: The clipping window partitions the plane into nine regions.

The procedure 'classify' determines the position of a point in relation to the window:

```
  procedure classify(x, y: real; var c: wcode);
  begin
    c := Ø;   { empty set }
    if  x < left  then  c := {ll}  elsif  x > right  then  c := {rr};
    if  y < bottom  then  c := c ∪ {bb}  elsif  y > top  then c := c ∪
{tt}
  end;
```

The procedure 'clip' computes the endpoints of the clipped line segment and calls the procedure 'showline' to draw it:

```
procedure clip(x1, y1, x2, y2: real);
```

```
var  c, c1, c2: wcode;   x, y: real;   outside: boolean;
begin  { clip }
  classify(x1, y1, c1);  classify(x2, y2, c2);  outside := false;
  while  (c1 ≠ Ø) or (c2 ≠ Ø)  do
    if  c1 ∩ c2 ≠ Ø  then
      { line segment lies completely outside the window }
      { c1 := Ø;  c2 := Ø;  outside := true }
    else  begin
      c := c1;
      if  c = Ø  then  c := c2;
      if  ll ∈ c  then  { segment intersects left }
        { y := y1 + (y2 - y1) · (left - x1) / (x2 - x1);  x := left }
      elsif  rr ∈ c  then  { segment intersects right }
        { y := y1 + (y2 - y1) · (right - x1) / (x2 - x1); x := right }
      elsif  bb ∈ c  then  { segment intersects bottom }
        { x := x1 + (x2 - x1) · (bottom - y1) / (y2 - y1); y := bottom }
      elsif  tt ∈ c  then  { segment intersects top }
        { x := x1 + (x2 - x1) · (top - y1) / (y2 - y1);  y := top };
      if  c = c1 then { x1 := x;  y1 := y;  classify(x, y, c1) }
          else { x2 := x;  y2 := y;  classify(x, y, c2) }
    end;
  if  not outside  then  showline(x1, y1, x2, y2)
end;  { clip }
```

Drawing digitized lines

A raster graphics screen is an integer grid of pixels, each of which can be turned on or off. Euclidean geometry does not apply directly to such a discretized plane. Any designer using a CAD system will prefer Euclidean geometry to a discrete geometry as a model of the world. The problem of how to approximate the Euclidean plane by an integer grid turns out to be a hard question: How do we map Euclidean geometry onto a digitized space in such a way as to preserve the rich structure of geometry as much as possible? Let's begin with simple instances: How do you map a straight line onto an integer grid, and how do you draw it efficiently? Exhibit 14.5 shows reasonable examples.

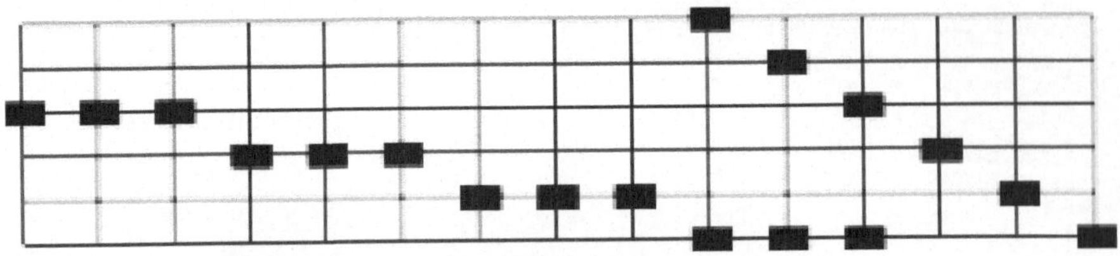

Exhibit 14.5: Digitized lines look like staircases.

Consider the slope $m = (y_2 - y_1) / (x_2 - x_1)$ of a segment with endpoints $p_1 = (x_1, y_1)$ and $p_2 = (x_2, y_2)$. If $|m| \le 1$ we want one pixel blackened on each x coordinate; if $|m| \ge 1$, one pixel on each y coordinate; these two requirements are consistent for diagonals with $|m| = 1$. Consider the case $|m| \le 1$. A unit step in x takes us from point (x, y) on the line to (x + 1, y + m). So for each x between x_1 and x_2 we paint the pixel (x, y) closest to the mathematical line according to the formula $y = \text{round}(y_1 + m \cdot (x - x_1))$. For the case $|m| > 1$, we reverse the roles of x and y, taking a unit step in y and incrementing x by 1/m. The following procedure draws line segments with $|m| \le 1$ using unit steps in x.

```
procedure line(x1, y1, x2, y2: integer);
var  x, sx: integer;  m: real;
begin
  PaintPixel(x1, y1);
  if  x1 ≠ x2  then  begin
    x := x1;  sx := sgn(x2 - x1);  m := (y2 - y1) / (x2 - x1);
    while  x ≠ x2  do
      { x := x + sx;  PaintPixel(x, round(y1 + m · (x - x1))) }
  end
end;
```

This straightforward implementation has a number of disadvantages. First, it uses floating-point arithmetic to compute integer coordinates of pixels, a costly process. in addition, rounding errors may prevent the line from being reversible: *reversibility* means that we paint the same pixels, in reverse order, if we call the procedure with the two endpoints interchanged. Reversibility is desirable to avoid the following blemishes: that a line painted twice, from both ends, looks thicker than other lines; worse yet, that painting a line from one end and erasing it from the other leaves spots on the screen. A weaker constraint, which is only concerned with the result and not the process of painting, is easy to achieve but is less useful.

Weak reversibility is most easily achieved by ordering the points p_1 and p_2 lexicographically by x and y coordinates, drawing every line from left to right, and vertical lines from bottom to top. This solution is inadequate for animation, where the direction of drawing is important, and the sequence in which the pixels are painted is determined by the application—drawing the trajectory of a falling apple from the bottom up will not do. Thus interactive graphics needs the stronger constraint.

Efficient algorithms, such as Bresenham's [Bre 65], avoid floating-point arithmetic and expensive multiplications through *incremental* computation: Starting with the current point p_1, a next point is computed as a function of the current point and of the line segment parameters. it turns out that only a few additions, shifts, and comparisons are required. In the following we assume that the slope m of the line satisfies $|m| \le 1$. Let

$\Delta x = x_2 - x_1$, $sx = \text{sign}(\Delta x)$, $\Delta y = y_2 - y_1$, $sy = \text{sign}(\Delta y)$.

Assume that the pixel (x, y) is the last that has been determined to be the closest to the actual line, and we now want to decide whether the next pixel to be set is (x + sx, y) or (x + sx, y + sy). Exhibit 14.6 depicts the case sx = 1 and sy = 1.

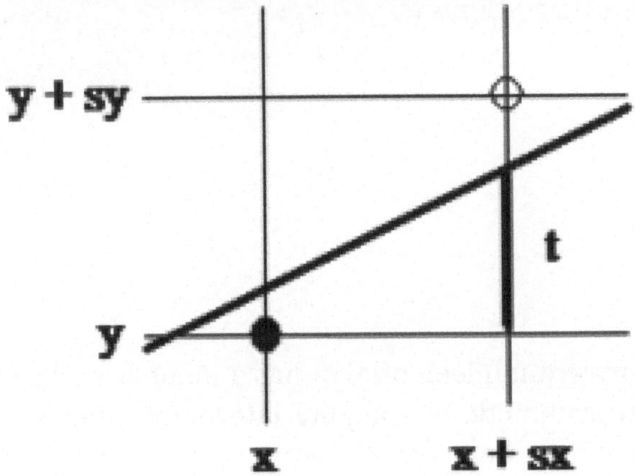

Exhibit 14.6: At the next coordinate x + sx, we identify and paint the pixel closest to the line.

Let t denote the absolute value of the difference between y and the point with abscissa x + sx on the actual line.

Then t is given by

$$t = sy \cdot \left(y_1 + \frac{\Delta y}{\Delta x} \cdot (x + sx - x_1) - y \right).$$

The value of t determines the pixel to be drawn:

$$t < \frac{1}{2} \Leftrightarrow 2 \cdot t - 1 < 0: \qquad \text{draw}\,(x + sx, y).$$

$$t > \frac{1}{2} \Leftrightarrow 2 \cdot t - 1 > 0: \qquad \text{draw}\,(x + sx, y + sy).$$

As the following example shows, reversibility is not an automatic consequence of the geometric fact that two points determine a unique line, regardless of correct rounding or the order in which the two endpoints are presented. A problem arises when two grid points are equally close to the straight line (Exhibit 14.7).

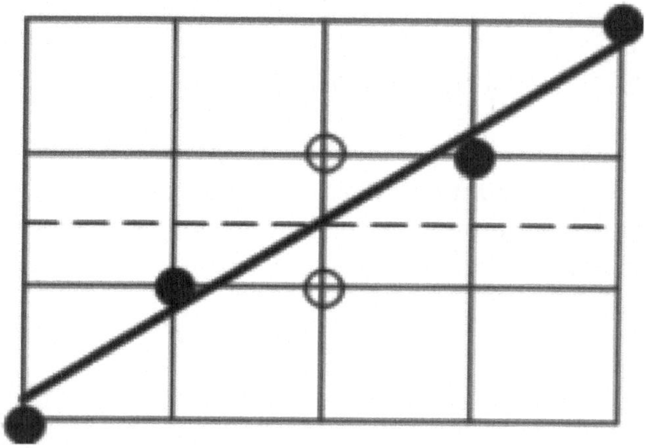

Exhibit 14.7: Breaking the tie among equidistant grid points.

If the tie is not broken in a consistent manner (e.g. by always taking the upper grid point), the resulting algorithm fails to be reversible:

$$t = \frac{1}{2} \quad \Leftrightarrow \quad 2 \cdot t - 1 = 0: \qquad \text{draw} \begin{cases} (x + sx, y) & \text{if } sy = -1, \\ (x + sx, y + sy) & \text{if } sy = 1. \end{cases}$$

All the variables introduced in this problem range over the integers, but the ratio $\frac{(\Delta y)}{(\Delta x)}$ appears to introduce rational expressions. This is easily remedied by multiplying everything with Δx. We define the decision variable d as $d = |\Delta x| \cdot (2 \cdot t - 1) = sx \cdot \Delta x \cdot (2 \cdot t - 1)$. (∗)

Let d_i denote the decision variable which determines the pixel $(x^{(i)}, y^{(i)})$ to be drawn in the i-th step. Substituting t and inserting $x = x^{(i-1)}$ and $y = y^{(i-1)}$ in (⍰) we obtain

$d_i = sx \cdot sy \cdot (2 \cdot \Delta x \cdot y_1 + 2 \cdot (x^{(i-1)} + sx - x_1) \cdot \Delta y - 2 \cdot \Delta x \cdot y^{(i-1)} - \Delta x \cdot sy)$ and

$d_{i+1} = sx \cdot sy \cdot (2 \cdot \mathbf{\Delta x} \cdot y_1 + 2 \cdot (x^{(i)} + sx - x_1) \cdot \Delta y - 2 \cdot \Delta x \cdot y^{(i)} - \Delta x \cdot sy)$.

Subtracting d_i from d_{i+1}, we get

$d_{i+1} - d_i = sx \cdot sy \cdot (2 \cdot (x^{(i)} - x^{(i-1)}) \cdot \Delta y - 2 \cdot \Delta x \cdot (y^{(i)} - y^{(i-1)}))$.

Since $x^{(i)} - x^{(i-1)} = sx$, we obtain

$d_{i+1} = d_i + 2 \cdot sy \cdot \Delta y - 2 \cdot sx \cdot \Delta x \cdot sy \cdot (y^{(i)} - y^{(i-1)})$.

If $d_i < 0$, or di = 0 and sy = −1, then $y^{(i)} = y^{(i-1)}$, and therefore

$d_{i+1} = d_i + 2 \cdot |\Delta y|$.

If $d_i > 0$, or $d_i = 0$ and sy = 1, then $y^{(i)} = y^{(i-1)} + sy$, and therefore

$d_{i+1} = d_i + 2 \cdot |\Delta y| - 2 \cdot |\Delta x|$.

This iterative computation of d_{i+1} from the previous d_i lets us select the pixel to be drawn. The initial starting value for d_1 is found by evaluating the formula for di, knowing that $(x^{(0)}, y^{(0)}) = (x_1, y_1)$. Then we obtain

$$d_1 = 2 \cdot |\Delta y| - |\Delta x|.$$

The arithmetic needed to evaluate these formulas is minimal: addition, subtraction and left shift (multiplication by 2). The following procedure implements this algorithm; it assumes that the slope of the line is between –1 and 1.

```
procedure BresenhamLine(x1, y1, x2, y2: integer);
var  dx, dy, sx, sy, d, x, y: integer;
begin
  dx := |x2 - x1|;  sx := sgn(x2 - x1);
  dy := |y2 - y1|;  sy := sgn(y2 - y1);
  d := 2 · dy - dx;  x := x1;  y := y1;
  PaintPixel(x, y);
  while  x ≠ x2  do  begin
    if  (d > 0) or ((d = 0) and (sy = 1))  then  { y := y + sy;-
2·dx};
    x := x + sx;  d := d + 2 · dy;
    PaintPixel(x, y)
  end
end;
```

The riddle of the braiding straight lines

Two straight lines in a plane intersect in at most one point, right? Important geometric algorithms rest on this well-known theorem of Euclidean geometry and would have to be reexamined if it were untrue. Is this theorem true for *computer lines*, that is, for data objects that represent and approximate straight lines to be processed by a program? Perhaps yes, but mostly no.

Yes. it is possible, of course, to program geometric problems in such a way that every pair of straight lines has at most, or exactly, one intersection point. This is most readily achieved through symbolic computation. For example, if the intersection of L_1 and L_2 is denoted by an expression 'intersect(L_1, L_2)' that is never evaluated but simply combined with other expressions to represent a geometric construction, we are free to postulate that 'intersect(L_1, L_2)' is a point.

No. For reasons of efficiency, most computer applications of geometry require the immediate numerical evaluation of every geometric operation. This calculation is done in a discrete, finite number system in which case the theorem need not be true. This fact is most easily seen if we Work with a discrete plane of pixels, and we represent a straight line by the set of all pixels touched by an ideal mathematical line. Exhibit 14.8 shows three digitized straight lines in such a square grid model of plane geometry. Two of the lines intersect in a common interval of three pixels, whereas two others have no pixel in common, even though they obviously intersect.

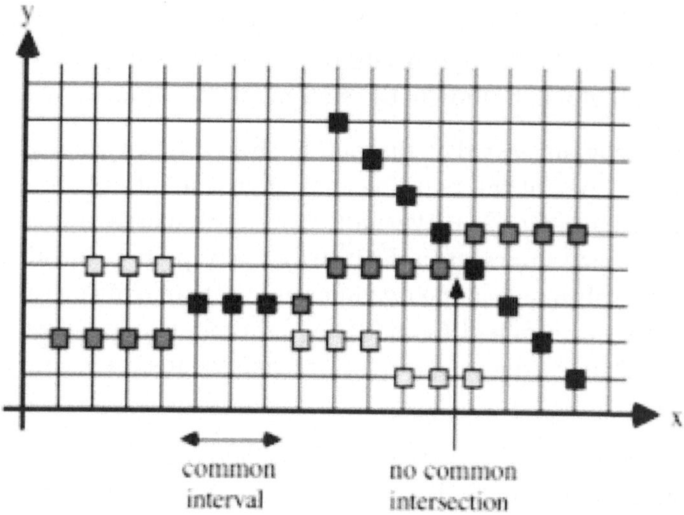

Exhibit 14.8: Two intersecting lines may share none, one, or more pixels.

With floating-point arithmetic the situation is more complicated; but the fact remains that the Euclidean plane is replaced by a discrete set of points embedded in the plane—all those points whose coordinates are representable in the particular number system being used. Experience with numerical computation, and the hazards of rounding errors, suggests that the question "in how many points can two straight lines intersect?" admits the following answers:

- There is no intersection—the mathematically correct intersection cannot be represented in the number system.
- A set of points that lie close to each other: for example, an interval.
- Overflow aborts the calculation before a result is computed, even if the correct result is representable in the number system being used.

Exercise: two lines intersect in how many points?

Construct examples to illustrate these phenomena when using floating-point arithmetic. Choose a suitable system G of floating-point numbers and two distinct straight lines

$a_i \cdot x + b_i \cdot y + c_i = 0$ with $a_i, b_i, c_i \in G$, i=1, 2,

such that, when all operations are performed in G:

There is no point whose coordinates x, y ∈ G satisfy both linear equations.

There are many points whose coordinates x, y ∈ G satisfy both linear equations.

There is exactly one point whose coordinates x, y ∈ G satisfy both linear equations, but the straightforward computation of x and y leads to overflow.

As a consequence of (a) it follows that the definition "two lines intersect they share a common point" is inappropriate for numerical computation. Formulate a numerically meaningful definition of the statement "two line segments intersect".

Exercise (b) may suggest that the points shared by two lines are neighbors. Pictorially, if the slopes of the two lines are almost identical, we expect to see a blurred, elongated intersection. We will show that worse things may happen: two straight lines may intersect in arbitrarily many points, and these points are separated by intervals in which the two lines alternate in lying on top of each other. Computer lines may be braided! To understand this phenomenon, we need to clarify some concepts: What exactly is a straight line represented on a computer? What is an intersection?

There is no one answer, there are many! Consider the analogy of the mathematical concept of real numbers, defined by axioms. When we approximate real numbers on a computer, we have a choice of many different number systems (e.g. various floating-point number systems, rational arithmetic with variable precision, interval arithmetic). These systems are typically not defined by means of axioms, but rather in terms of concrete representations of the numbers and algorithms for executing the operations on these numbers. Similarly, a *computer line* will be defined in terms of a concrete representation (e.g. two points, a point and a slope, or a linear expression). All we obtain depends on the formulas we use and on the basic arithmetic to operate on these representations. The notion of a straight line can be formalized in many different ways, and although these are likely to be mathematically equivalent, they will lead to data objects with different behavior when evaluated numerically. Performing an operation consists of evaluating a formula. Substituting a formula by a mathematically equivalent one may lead to results that are topologically different, because equivalent formulas may exhibit different sensitivities toward rounding errors.

Consider a computer that has only integer arithmetic, i.e. we use only the operations +, –, ·, div. Let Z be the set of integers. Two straight lines gi (i = 1, 2) are given by the following equations:

$$a_i \cdot x + b_i \cdot y + c_i = 0 \text{ with } a_i, b_i, c_i \in Z;\ b_i \neq 0.$$

We consider the problem of whether two given straight lines intersect in a given point x0. We use the following method: Solve the equations for y [i. e. $y = E_1(x)$ and $y = E_2(x)$] and test whether $E_1(x0)$ is equal to $E_2(x0)$.

Is this method suitable? First, we need the following definitions:

$$\mathbf{sign(x)} := \begin{cases} 1 & \text{if } x > 0, \\ 0 & \text{if } x = 0, \\ -1 & \text{if } x < 0. \end{cases}$$

$x \in Z$ is a *turn* for the pair (E_1, E_2) iff

$\text{sign}(E_1(x) - E_2(x)) \neq \text{sign}(E_1(x + 1) - E_2(x + 1))$.

An algorithm for the intersection problem is correct if there are at most two turns.

The intuitive idea behind this definition is the recognition that rounding errors may force us to deal with an intersection interval rather than a single intersection point; but we wish to avoid separate intervals. The definition above partitions the x-axis into at most three disjoint intervals such that in the left interval the first line lies above or below the second line, in the middle interval the lines "intersect", and in the right interval we have the complementary relation of the left one (Exhibit 14.9).

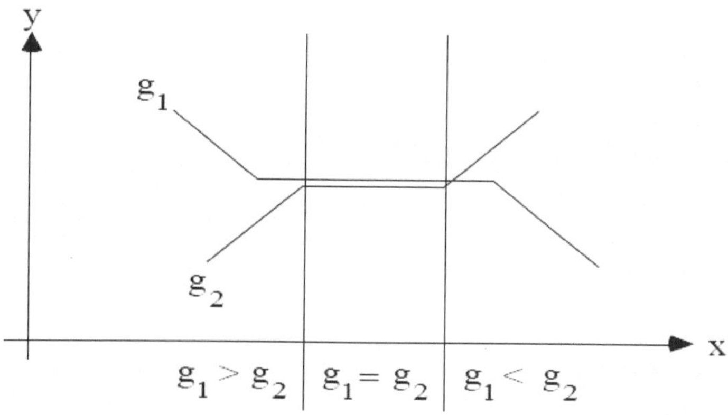

Exhibit 14.9: Desirable consistency condition for intersection of nearly parallel lines.

Consider the straight lines:

$3 \cdot x - 5 \cdot y + 40 = 0$ and $2 \cdot x - 3 \cdot y + 20 = 0$

which lead to the evaluation formulas

$$y = \frac{3 \cdot x + 40}{5} \quad \text{and} \quad y = \frac{2 \cdot x + 20}{3}.$$

Our naive approach compares the expressions

$(3 \cdot x + 40)$ div 5 and $(2 \cdot x + 20)$ div 3.

Using the definitions it is easy to calculate that the turns are

7, 8, 10, 11, 12, 14, 15, 22, 23, 25, 26, 27, 29, 30.

The straight lines have become step functions that intersect many times. They are braided (Exhibit 14.10).

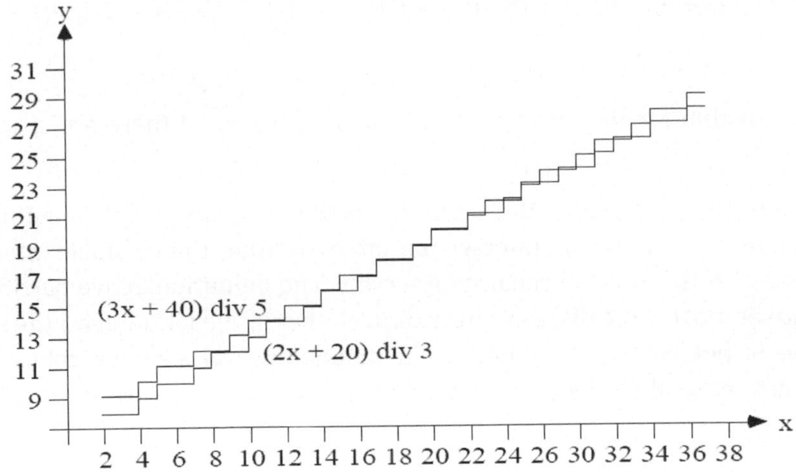

Exhibit 14.10: Braiding straight lines violate the consistency condition of Exhibit 14.9.

Exercise: show that the straight lines

$x - 2 \cdot y = 0$

$k \cdot x - (2 \cdot k + 1) \cdot y = 0$ for any integer $k > 0$ have $2 \cdot k - 1$ turns in the first quadrant.

Is braiding due merely to integer arithmetic? Certainly not: rounding errors also occur in floating-point arithmetic, and we can construct even more pathological behavior. As an example, consider a floating-point arithmetic with a two-decimal-digit mantissa. We perform the evaluation operation:

$$y = -\frac{a \cdot x + c}{b}$$

and truncate intermediate results immediately to two decimal places. Consider the straight lines (Exhibit 14.11)

$4.3 \cdot x - 8.3 \cdot y = 0,$

$1.4 \cdot x - 2.7 \cdot y = 0.$

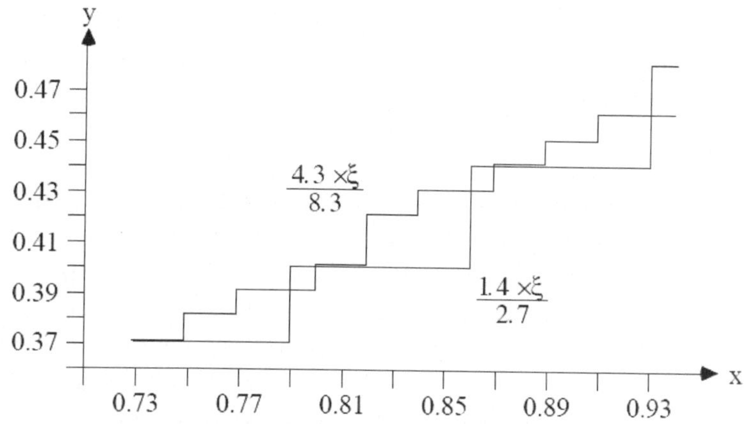

Exhibit 14.11: Example to be verified by manual computation.

These examples were constructed by intersecting straight lines with almost the same slope—a numerically ill- conditioned problem. While Working with integer arithmetic, we made the mistake of using the error-prone 'div' operator. The comparison of rational expressions does not require division.

$$\frac{-c_1 - a_1 \cdot x_0}{b_1} = \frac{-c_2 - a_2 \cdot x_0}{b_2}$$

Let $a_1 \cdot x + B_1 \cdot y + c_1 = 0$ and $a_2 \cdot x + b_2 \cdot y + c_2 = 0$ be two straight lines. To find out whether they intersect at x_0, we have to check whether the equality holds. This is equivalent to $b_2 \cdot c_1 - B_1 \cdot c_2 = x_0 \cdot (a_2 \cdot B_1 - a1 \cdot b_2)$.

The last formula can be evaluated without error if sufficiently large integer arguments are allowed. Another way to evaluate this formula without error is to limit the size of the operands. For example, if a_i, b_i, c_i, and x_0 are n-digit binary numbers, it suffices to be able to represent 3n-digit binary numbers and to compute with n-digit and 2n- digit binary numbers.

These examples demonstrate that programming even a simple geometric problem can cause unexpected difficulties. Numerical computation forces us to rethink and redefine elementary geometric concepts.

Digitized circles

The concepts, problems and techniques we have discussed in this chapter are not at all restricted to dealing with straight lines—they have their counterparts for any kind of digitized spatial object. Straight lines, defined by linear formulas, are the simplest nontrivial spatial objects and thus best suited to illustrate problems and solutions. In this section we show that the incremental drawing technique generalizes in a straightforward manner to more complex objects such as circles.

The basic parameters that define a circle are the center coordinates (x_c, y_c) and the radius r. To simplify the presentation we first consider a circle with radius r centered around the origin. Such a circle is given by the equation

$x^2 + y^2 = r^2$.

Efficient algorithms for drawing circles, such as Bresenham's [Bre 77], avoid floating-point arithmetic and expensive multiplications through *incremental* computation: A new point is computed depending on the current point and on the circle parameters. Bresenham's circle algorithm was conceived for use with pen plotters and therefore generates all points on a circle centered at the origin by incrementing all the way around the circle. We present a modified version of his algorithm which takes advantage of the *eight-way symmetry* of a circle. If (x, y) is a point on the circle, we can easily determine seven other points lying on the circle (Exhibit 14.12). We consider only the 45° segment of the circle shown in the figure by incrementing from x = 0 to x = y = r /, and use eight-way symmetry to display points on the entire circle.

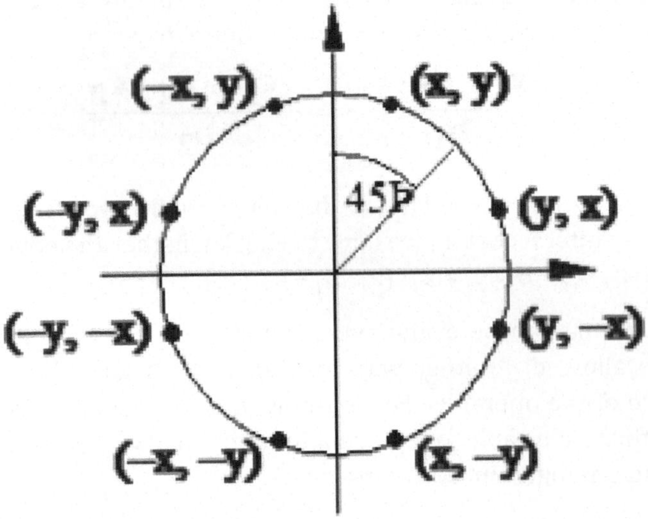

Exhibit 14.12: Eightfold symmetry of the circle.

Assume that the pixel p = (x, y) is the last that has been determined to be closest to the actual circle, and we now want to decide whether the next pixel to be set is p1 = (x + 1, y) or p_2 = (x + 1, y – 1). Since we restrict ourselves to the 45° circle segment shown above these pixels are the only candidates. Now define

$$d' = (x + 1)^2 + y^2 - r^2$$

$$d'' = (x + 1)^2 + (y - 1)^2 - r^2$$

which are the differences between the squared distances from the center of the circle to p_1 (or p_2) and to the actual circle. If $|d'| \leq |d''|$, then p_1 is closer (or equidistant) to the actual circle; if $|d'| > |d''|$, then p_2 is closer. We define the decision variable d as correctly selects the pixel that is closest to the actual circle. Exhibit 14.13 shows a small part of the pixel grid and illustrates the various possible ways [(1) to (5)] how the actual circle may intersect the vertical line at x + 1 in relation to the pixels p1 and p2.

$d = d' + d''$. (*)

We will show that the rule

If $d \leq 0$ then select p_1 else select p_2.

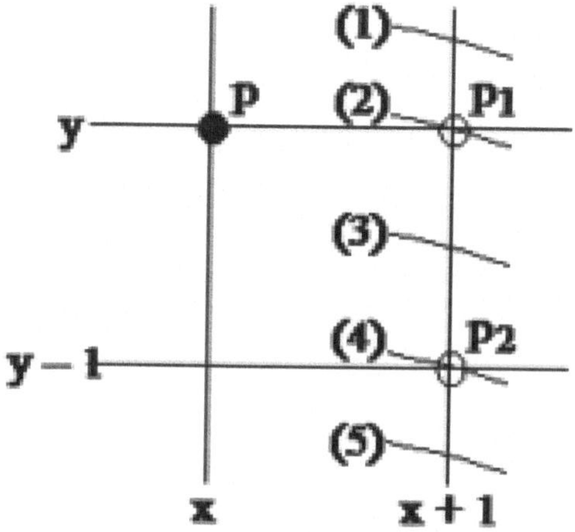

Exhibit 14.13: For a given octant of the circle, if pixel p is lit, only two other pixels p1 and p2 need be examined.

In cases (1) and (2) p2 lies inside, p_1 inside or on the circle, and we therefore obtain $d' \leq 0$ and $d'' < 0$. Now $d < 0$, and applying the rule above will lead to the selection of p_1. Since $|d'| \leq |d''|$ this selection is correct. in case (3) p_1 lies outside and p_2 inside the circle and we therefore obtain $d' > 0$ and $d'' < 0$. Applying the rule above will lead to the selection of p_1 if $d \leq 0$, and p_2 if $d > 0$. This selection is correct since in this case $d \leq 0$ is equivalent to $|d'| \leq |d''|$. in cases (4) and (5) p1 lies outside, p2 outside or on the circle and we therefore obtain $d' > 0$ and $d'' \geq 0$. Now $d > 0$, and applying the rule above will lead to the selection of p_2. Since $|d'| > |d''|$ this selection is correct.

Let d_i denote the decision variable that determines the pixel $(x^{(i)}, y^{(i)})$ to be drawn in the i-th step. Starting with $(x^{(0)}, y^{(0)}) = (0, r)$ we obtain

$d_1 = 3 - 2 \cdot r$.

If $d_i \leq 0$, then $(x^{(i)}, y^{(i)}) = (x^{(i)} + 1, y^{(i-1)})$, and therefore

$d_{i+1} = d_i + 4 \cdot x_{i-1} + 6.$

If $d_i > 0$, then $(x^{(i)}, y^{(i)}) = (x^{(i)} + 1, y^{(i-1)} - 1)$, and therefore

$d_{i+1} = d_i + 4 \cdot (x_{i-1} - y_{i-1}) + 10.$

This iterative computation of di^{+1} from the previous di lets us select the correct pixel to be drawn in the (i + 1)-th step. The arithmetic needed to evaluate these formulas is minimal: addition, subtraction, and left shift (multiplication by 4). The following procedure 'BresenhamCircle' which implements this algorithm draws a circle with center (x_c, y_c) and radius r. it uses the procedure 'CirclePoints' to display points on the entire circle. in the cases x = y or r = 1 'CirclePoints' draws each of four pixels twice. This causes no problem on a raster display.

```
procedure BresenhamCircle(xc, yc, r: integer);

  procedure CirclePoints(x, y: integer);
  begin
    PaintPixel(xc + x, yc + y); PaintPixel(xc - x, yc + y);
    PaintPixel(xc + x, yc - y); PaintPixel(xc - x, yc - y);
    PaintPixel(xc + y, yc + x); PaintPixel(xc - y, yc + x);
    PaintPixel(xc + y, yc - x); PaintPixel(xc - y, yc - x)
  end;
var  x, y, d: integer;
begin
  x := 0;  y := r;  d := 3 - 2 · r;
  while  x < y  do  begin
    CirclePoints(x, y);
    if  d < 0 then d  := d + 4  · x + 6
      else { d := d + 4 · (x - y) + 10;  y := y - 1 };
    x := x + 1
  end;
  if  x = y  then  CirclePoints(x, y)
end;     .i).Bresenham's algorithm:circle;
```

Exercises and programming projects

- ❖ Design and implement an efficient geometric primitive which determines whether two aligned rectangles (i.e. rectangles with sides parallel to the coordinate axes) intersect.
- ❖ Design and implement a geometric primitive which takes a triangle t given by its three vertices and a point p and returns a ternary value: p is inside t, p is on the boundary of t, p is outside t.

```
function inTriangle(t: triangle; p: point): ...;
```

- ❖ Use the functions 'intersect' of in "intersection" and 'inTriangle' above to program a to check whether segment s and triangle t share common points. 'SegmentintersectsTriangle' returns a ternary value: yes, degenerate, no. List all distinct cases of degeneracy that may occur, and show how your code handles them.

```
function SegmentIntersectsTriangle(s: segment; t: triangle): …;
```

- Implement Bresenham's incremental algorithms for drawing digitized straight lines and circles.

- Two circles (x', y', r') and (x'', y'', r'') are given by the coordinates of their center and their radius. Find effective formulas for deciding the three-way question whether (a) the circles intersect as lines, (b) the circles intersect as disks, or (c) neither. Avoid the square-root operation whenever possible.

Part-IV

Complexity of Problems and Algorithms

Fundamental issues of computation

A successful search for better and better algorithms naturally leads to the question "Is there a best algorithm?", whereas an unsuccessful search leads one to ask apprehensively: "Is there *any* algorithm (of a certain type) to solve this problem?" These questions turned out to be difficult and fertile. Historically, the question about the *existence* of an algorithm came first, and led to the concepts of computability and decidability in the 1930s. The question about a "best" algorithm led to the development of complexity theory in the 1960s.

The study of these fundamental issues of computation requires a mathematical arsenal that includes mathematical logic, discrete mathematics, probability theory, and certain parts of analysis, in particular asymptotics. We introduce a few of these topics, mostly by example, and illustrate the use of mathematical techniques of algorithm analysis on the important problem of sorting.

15

Computability and Complexity

Models of computation: the ultimate RISC

Algorithm and *computability* are originally intuitive concepts. They can remain intuitive as long as we only want to show that some specific result can be computed by following a specific algorithm. Almost always an informal explanation suffices to convince someone with the requisite background that a given algorithm computes a specified result. We have illustrated this informal approach throughout Part III. Everything changes if we wish to show that a desired result is *not computable.* The question arises immediately: "What tools are we allowed to use?" Everything is computable with the help of an oracle that knows the answers to all questions. The attempt to prove negative results about the nonexistence of certain algorithms forces us to agree on a rigorous definition of *algorithm*.

The question "What can be computed by an algorithm, and what cannot?" was studied intensively during the 1930s by Emil Post (1897–1954), Alan Turing (1912-1954), Alonzo Church (1903), and other logicians. They defined various formal models of computation, such as production systems, Turing machines, and recursive functions, to capture the intuitive concept of "computation by the application of precise rules". All these different formal models of computation turned out to be equivalent. This fact greatly strengthens Church's thesis that the intuitive concept of algorithm is formalized correctly by any one of these mathematical systems.

We will not define any of these standard models of computation. They all share the trait that they were designed to be conceptually simple: their primitive operations are chosen to be as weak as possible, as long as they retain their property of being universal computing systems in the sense that they can simulate any computation performed on any other machine. it usually comes as a surprise to novices that the set of primitives of a universal computing machine can be so simple as long as these machines possess two essential ingredients: *unbounded memory* and *unbounded time.*

Most simulations of a powerful computer on a simple one share three characteristics: it is straightforward in principle, it involves laborious coding in practice, and it explodes the space and time requirements of a computation. The weakness of the primitives, desirable from a theoretical point of view, has the consequence that as simple an operation as integer addition becomes an exercise in programming.

The model of computation used most often in algorithm analysis is significantly more powerful than a Turing machine in two respects: (1) its memory is not a tape, but an array, and (2) in one primitive operation it can deal with numbers of arbitrary size. This model of computation is called *random access machine*, abbreviated as RAM. A RAM is essentially a *random access memory*, also abbreviated as RAM, of unbounded capacity, as suggested in Exhibit 15.1. The memory consists of an infinite array of memory cells, addressed 0, 1, 2, ... Each cell can hold a number, say an integer, of arbitrary size, as the arrow pointing to the right suggests.

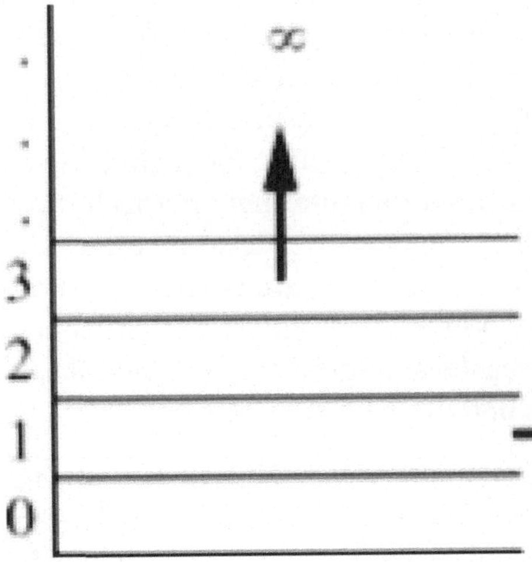

Exhibit 15.1: RAM - unbounded address space, unbounded cell size.

A RAM has an arithmetic unit and is driven by a program. The meaning of the word *random* is that any memory cell can be accessed in unit time (as opposed to a tape memory, say, where access time depends on distance). A further crucial assumption in the RAM model is that an arithmetic operation (+, –, ·, /) also takes unit time, regardless of the size of the numbers involved. This assumption is unrealistic in a computation where numbers may grow very large, but often is a useful assumption. As is the case with all models, the responsibility for using them properly lies with the user. To give the reader the flavor of a model of computation, we define a RAM whose architecture is rather similar to real computers, but is unrealistically simple.

The ultimate RISC

RISC stands for *Reduced* instruction *Set Computer*, a machine that has only a few types of instructions built into the hardware. What is the minimum number of instructions a computer needs to be universal? in theory, one.

Consider a stored-program computer of the "von Neumann type" where data and program are stored in the same memory (John von Neumann, 1903–1957). Let the random access memory (RAM) be "doubly infinite": There is a *countable* infinity of memory cells addressed 0, 1, ... , each of which can hold an integer of arbitrary size, or an instruction. We assume that the constant 1 is hardwired into memory cell 1; from 1 any other integer can be constructed. There is a single type of "three-address instruction" which we call "subtract, test and jump", abbreviated as

STJ x, y, z

where x, y, and z are addresses. its semantics is equivalent to

```
STJ  x, y, z   ⇔   x := x - y;  if  x ≤ 0  then  goto z;
```

x, y, and z refer to cells Cx, Cy, and Cz. The contents of Cx and Cy are treated as data (an integer); the contents of Cz, as an instruction (Exhibit 15.2).

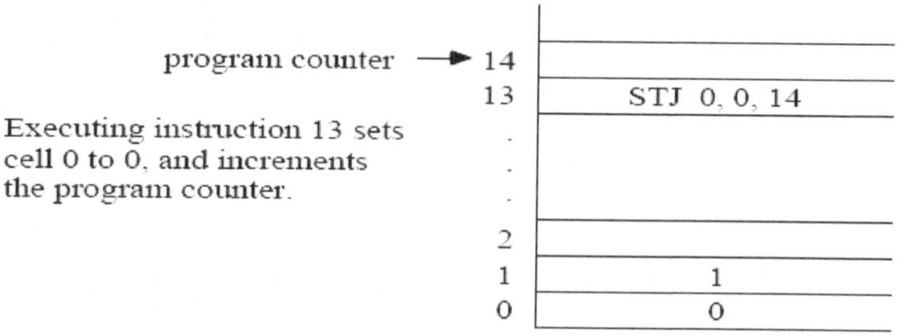

Exhibit 15.2: Stored program computer: data and instructions share the memory.

Since this RISC has just one type of instruction, we waste no space on an op-code field. An instruction contains three addresses, each of which is an unbounded integer. in theory, fortunately, three unbounded integers can be packed into the same space required for a single unbounded integer. in the following exercise, this simple idea leads to a well-known technique introduced into mathematical logic by Kurt Gödel (1906 – 1978).

Exercise: Gödel numbering

- ❖ Motel infinity has a countable infinity of rooms numbered 0, 1, 2, ... Every room is occupied, so the sign claims "No Vacancy". Convince the manager that there is room for one more person.

- ❖ Assume that a memory cell in our RISC stores an integer as a sign bit followed by a sequence d0, d_1, d_2, ... of decimal digits, least significant first. Devise a scheme for storing three addresses in one cell.

- ❖ Show how a sequence of positive integers i1, i2, ... , in of arbitrary length n can be encoded in a single natural number j: Given j, the sequence can be uniquely reconstructed. Gödel's solution:

$$j = 2^{i_1} \cdot 3^{i_2} \cdot 5^{i_3} \cdot \ldots \cdot (\text{n-th prime})^{i_n}.$$

Basic program fragments

This computer is best understood by considering program fragments for simple tasks. These fragments implement simple operations, such as setting a variable to a given constant, or the assignment operator, that are given as primitives in most programming languages. Programming these fragments naturally leads us to introduce basic concepts of assembly language, in particular symbolic and relative addressing.

Set the content of cell

0 to 0: STJ 0, 0, .+1

Whatever the current content of cell 0, subtract it from itself to obtain the integer 0. This instruction resides at some address in memory, which we abbreviate as '.', read as "the current value of the program counter". '.+1' is the next address, so regardless of the outcome of the test, control flows to the next instruction in memory.

a := b, where a and b are symbolic addresses. Use a temporary variable t:

STJ t, t, .+1 *{ t := 0 }*

STJ t, b, .+1 *{ t := –b }*

STJ a, a, .+1 *{ a := 0 }*

STJ a, t, .+1 *{ a := –t, so now a = b }*

Exercise: a program library

- ❖ Write RISC programs for a:= b + c, a := b · c, a := b div c, a := b mod c, a := |b|, a : = min(b, c), a := gcd(b, c).

- ❖ Show how this RISC can compute with rational numbers represented by a pair [a, b] of integers denoting numerator and denominator.

- ❖ (Advanced) Show that this RISC is universal, in the sense that it can simulate any computation done by any other computer.

The exercise of building up a RISC program library for elementary functions provides the same experience as the equivalent exercise for Turing machines, but leads to the goal much faster, since the primitive STJ is much more powerful than the primitives of a Turing machine.

The purpose of this section is to introduce the idea that conceptually simple models of computation are as powerful, in theory, as much more complex models, such as a high-level programming language. The next two sections demonstrate results of an opposite nature: Universal computers, in the sense we have just introduced, are subject to striking limitations, even if we remove any limit on the memory and time they may use. We prove the existence of non-computable functions and show that the "halting problem" is undecidable.

The theory of computability was developed in the 1930s, and greatly expanded in the 1950s and 1960s. its basic ideas have become part of the foundation that any computer scientist is expected to know. Computability theory is not directly useful. it is based on the concept "computable in principle" but offers no concept of a "feasible computation". Feasibility, rather than "possible in principle", is the touchstone of computer science. Since the 1960s, a theory of the complexity of computation is being developed, with the goal of partitioning the range of computability into complexity classes according to time and space requirements. This theory is still in full development and breaking new ground, in particular in the area of concurrent computation. We have used some of its concepts throughout Part III and continue to illustrate these ideas with simple examples and surprising results.

Almost nothing is computable

Consider as a model of computation any programming language, with the fictitious feature that it is implemented on a machine with infinite memory and no operational time limits. Nevertheless we reach the conclusion that "almost nothing is computable". This follows simply from the observation that there are fewer programs than problems to be solved (functions to be computed). Both the number of programs and the number of functions are infinite, but the latter is an infinity of higher cardinality.

A programming language L is defined over an alphabet $A= \{a_1, a_2, \dots, a_k\}$ of k characters. The set of programs in L is a subset of the set A∗ of all strings over A. A∗ is *countable*, and so is its subset L, as it is in one-to-one correspondence with the natural numbers under the following mapping:

- ❖ Generate all strings in A∗ in order of increasing length and, in case of equal length, in lexicographic order.
- ❖ Erase all strings that do not represent a program according to the syntax rules of L.

❖ Enumerate the remaining strings in the originally given order.

Among all programs in L we consider only those which compute a (partial) function from the set N = {1, 2, 3, ...} of natural numbers into N. This can be recognized by their heading; for example,

```
function f(x: N): N;
```

As this is a subset of L, there exist only countably many such programs.

However, there are uncountably many functions f: N → N, as Georg Cantor (1845–1918) proved by his famous diagonalization argument. it starts by assuming the opposite, that the set {f | f: N → N} is countable, then derives a contradiction. If there were only a countable number of such functions, we could enumerate all of them according to the following scheme:

	1	2	3	4	...
f_1	$f_1(1)$	$f_1(2)$	$f_1(3)$	$f_1(4)$	
f_2	$f_2(1)$	$f_2(2)$	$f_2(3)$	$f_2(4)$	
f_3	$f_3(1)$	$f_3(2)$	$f_3(3)$	$f_3(4)$	
f_4	$f_4(1)$	$f_4(2)$	$f_4(3)$	$f_4(4)$	
⋮					

Construct a function g: N → N, g(i) = fi(i) + 1, which is obtained by adding 1 to the diagonal elements in the scheme above. Hence g is different from each f_i, at least for the argument i: g(i) ≠ fi(i). Therefore, our assumption that we have enumerated all functions f: N → N is wrong. Since there exists only a countable infinity of programs, but an uncountable infinity of functions, almost all functions are non-computable.

The halting problem is undecidable

If we could predict, for any program P executed on any data set D, whether P terminates or not (i.e. whether it will get into an infinite loop), we would have an interesting and useful technique. If this prediction were based on rules that prescribe exactly how the pair (P, D) is to be tested, we could write a program H for it. A fundamental result of computability theory states that under reasonable assumptions about the model of computation, such a *halting program* H cannot exist.

Consider a programming language L that contains the constructs we will use: mainly recursive procedures and procedure parameters. Consider all procedures P in L that have no parameters, a property that can be recognized from the heading

```
procedure P;
```

This simplifies the problem by avoiding any data dependency of termination.

Assume that there exists a program H in L that takes as argument any parameterless procedure P in L and decides whether P halts or loops (i.e. runs indefinitely):

$$\mathbf{H(P)} = \begin{cases} \mathbf{true} & \mathbf{if\ P\ halts,} \\ \mathbf{false} & \mathbf{if\ P\ loops.} \end{cases}$$

Consider the behavior of the following parameterless procedure X:

```
procedure X;
begin  while  H(X)  do;  end;
```

Consider the reference of X to itself; this trick corresponds to the diagonalization in the previous example.

Consider further the loop which is infinite if H(X) returns true (i.e. exactly when X should halt) and terminates if H(X) returns false (i.e. exactly when X should run forever). This trick corresponds to the change of the diagonal $g(i) = f_i(i) + 1$. We obtain:

```
while  H(X)  do;
```

By definition of X:

$$\mathbf{H(X)} = \begin{cases} \mathbf{true} & \mathbf{if\ X\ halts,} \\ \mathbf{false} & \mathbf{if\ X\ loops.} \end{cases}$$

By construction of X:

$$\mathbf{H(X)} = \begin{cases} \mathbf{true} & \mathbf{if\ X\ loops,} \\ \mathbf{false} & \mathbf{if\ X\ halts.} \end{cases}$$

The fiendishly crafted program X traps H in a web of contradictions. We blame the weakest link in the chain of reasoning that leads to this contradiction, namely the unsupported assumption of the existence of a halting program H. This proves that the halting problem is undecidable.

Computable, yet unknown

In the preceding two sections we have illustrated the limitations of computability: clearly stated questions, such as the halting problem, are undecidable. This means that the halting question cannot be answered, in *general*, by any computation no matter how extensive in time and space. There are, of course, lots of individual halting questions that can be answered, asserting that a particular program running on a particular data set terminates, or fails to do so. To illuminate this key concept of theoretical computer science further, the following examples will highlight a different type of practical limitation of computability.

Computable or decidable is a concept that naturally involves *one* algorithm and a *denumerably infinite* set of problems, indexed by a parameter, say n. Is there a uniform procedure that will solve any one problem in the infinite set? For example, the "question" (really a denumerable infinity of questions) "Can a given integer n > 2 be expressed as the sum of two primes?" is decidable because there exists the algorithm 's2p' that will answer any single instance of this question:

```
procedure s2p(n: integer): boolean;
  { for n>2, s2p(n) returns true if n is the sum of two primes,
    false otherwise }

  function p(k: integer): integer;
    { for k>0, p(k) returns the k-th prime: p(1) = 2, p(2) = 3, p(3)
= 5, … }
  end;

begin
  for all i, j such that p(i) < n and p(j )< n do
    if  p(i) + p(j)  = n  then  return(true);
  return(false);
end;  { s2p }
```

So the general question "Is any given integer the sum of two primes?" is solved readily by a simple program. A *single* related question, however, is much harder: "Is every even integer >2 the sum of two primes?" Let's try:

4 = 2 + 2, 6 = 3 + 3, 8 = 5 + 3, 10 = 7 + 3 = 5 + 5, 12 = 7 + 5,

14 = 11 + 3 = 7 + 7, 16 = 13 + 3 = 11 + 5, 18 = 13 + 5 = 11 + 7,

20 = 17 + 3 = 13 + 7, 22 = 19 + 3 = 17 + 5 = 11 + 11,

24 = 19 + 5 = 17 + 7 = 13 + 11, 26 = 23 + 3 = 21 + 5 = 19 + 7 = 13 + 13,

28 = 23 + 5 = 17 + 11, 30 = 23 + 7 = 19 + 11 = 17 + 13,

32 = 29 + 3 = 19 + 13, 34 = 31 + 3 = 29 + 5 = 23 + 11 = 17 + 17,

36 = 33 + 3 = 31 + 5 = 29 + 7 = 23 + 13 = 19 + 17.

A bit of experimentation suggests that the number of distinct representations as a sum of two primes increases as the target integer grows. Christian Goldbach (1690–1764) had the good fortune of stating the plausible conjecture "yes" to a problem so hard that it has defied proof or counterexample for three centuries.

One might ask: Is the Goldbach conjecture decidable? The straight answer is that the concept of decidability does not apply to a single yes/no question such as Goldbach's conjecture. Asking for an algorithm that tells us whether the conjecture is true or false is meaninglessly trivial. Of course, there is such an algorithm! If the Goldbach conjecture is true, the algorithm that says 'yes' decides. If the conjecture is false, the algorithm that says 'no' will do the job. The problem that we *don't know* which algorithm is the right one is quite compatible with saying that *one of those two* is the right algorithm. If we package two trivial algorithms into one, we get the following trivial algorithm for deciding Goldbach's conjecture:

```
function GoldbachOracle(): boolean:
begin  return(GoldbachIsRight)  end;
```

Notice that 'GoldbachOracle' is a function without arguments, and 'GoldbachIsRight' is a boolean constant, either true or false. Occasionally, the star$_k$ triviality of the argument above is concealed so cleverly under technical jargon as to sound profound. Watch out to see through the following plot.

Let us call an even integer > 2 that is *not* a sum of two primes a *counterexample.* None have been found as yet, but we can certainly reason about them, whether they exist or not. Define the

```
function G(k: cardinal): boolean;
```

as follows:

$$\mathbf{G(k)} = \begin{cases} \textbf{true} & \textbf{if the number of counterexamples is } \leq \mathbf{k}, \\ \textbf{false} & \textbf{otherwise}. \end{cases}$$

Goldbach's conjecture is equivalent to G(0) = true. The (implausible) rival conjecture that there is exactly one counterexample is equivalent to G(0) = false, G(1) = true. Although we do not know the value of G(k) for any single k, the definition of G tells us a lot about this artificial function, namely:

if G(i) = true for any i, then G(k) = true for all k > i.

With such a strong monotonicity property, how can G look?

- ❖ If Goldbach is right, then G is a constant: G(k) = true for all k.
- ❖ If there are a finite number i of exceptions, then G is a step function:
- ❖ G(k) = false for k < i, G(k) = true for k ≥ i.
- ❖ If there is an infinite number of exceptions, then G is again a constant:

G(k) = false for all k.

Each of the infinitely many functions listed above is obviously computable. Hence G is computable. The value of G(0) determines truth or falsity of Goldbach's conjecture. Does that help us settle this time-honored mathematical puzzle? Obviously not. All we have done is to rephrase the honest statement with which we started this section, "The answer is yes or no, but I don't know which" by the circuitous "The answer can be obtained by evaluating a computable function, but I don't know which one".

Multiplication of complex numbers

Let us turn our attention from noncomputable functions and undecidable problems to very simple functions that are obviously computable, and ask about their complexity: How many primitive operations must be executed in evaluating a specific function? As an example, consider arithmetic operations on real numbers to be primitive, and consider the product z of two complex numbers x and y:

$x = x_1 + i \cdot x_2$ and $y = y_1 + i \cdot y_2$,

$x \cdot y = z = z_1 + i \cdot z_2$.

The complex product is defined in terms of operations on real numbers as follows:

$z_1 = x_1 \cdot y_1 - x_2 \cdot y_2$,

$z_2 = x_1 \cdot y_2 + x_2 \cdot y_1$.

it appears that one complex multiplication requires four real multiplications and two real additions/subtractions. Surprisingly, it turns out that multiplications can be traded for additions. We first compute three intermediate variables using one multiplication for each, and then obtain z by additions and subtractions:

$p_1 = (x_1 + x_2) \cdot (y_1 + y_2)$,

$p_2 = x_1 \cdot y_1$,

$p_3 = x_2 \cdot y_2$,

$z_1 = p_2 - p_3$, $z_2 = p_1 - p_2 - p_3$.

This evaluation of the complex product requires only 3 real multiplications, but 5 real additions / subtractions. This trade of one multiplication for three additions may not look like a good deal in practice, because many computers have arithmetic chips with fast multiplication circuitry. in theory, however, the trade is clearly favorable. The cost of an addition grows linearly in the number of digits, whereas the cost of a multiplication using the standard method grows quadratically. The key idea behind this algorithm is that "linear combinations of k products of sums can generate more than k products of simple terms". Let us exploit this idea in a context where it makes a real difference.

Complexity of matrix multiplication

The *complexity of an algorithm* is given by its time and space requirements. Time is usually measured by the number of operations executed, space by the number of variables needed at any one time (for input, intermediate results, and output). For a given algorithm it is often easy to count the number of operations performed in the worst and in the best case; it is usually difficult to determine the average number of operations performed (i.e. averaged over all possible input data). Practical algorithms often have time complexities of the order O(log n), $O(n^2)$, O(n · log n), $O(n^2)$, and space complexity of the order O(n), where n measures the size of the input data.

The *complexity of a problem* is defined as the minimal complexity of all algorithms that solve this problem. It is almost always difficult to determine the complexity of a problem, since all possible algorithms must be considered, including those yet unknown. This may lead to surprising results that disprove obvious assumptions.

The complexity of an algorithm is an upper bound for the complexity of the problem solved by this algorithm. An algorithm is a witness for the assertion: You need at most this many operations to solve this problem. A specific algorithm never provides a *lower bound* on the complexity of a problem— it cannot extinguish the hope for a more efficient algorithm. Occasionally, algorithm designers engage in races lasting decades that result in (theoretically) faster and faster algorithms for solving a given problem. Volker Strassen started such a race with his 1969 paper "Gaussian Elimination Is Not Optimal" [Str 69], where he showed that matrix multiplication requires fewer operations than had commonly been assumed necessary. The race has not yet ended.

The obvious way to multiply two n × n matrices uses three nested loops, each of which is iterated n times, as we saw in a transitive hull algorithm in the chapter, "Matrices and graphs: transitive closure". The fact that the obvious algorithm for matrix multiplication is of time complexity Θ(n3), however, does not imply that the matrix multiplication problem is of the same complexity.

Strassen's matrix multiplication

The standard algorithm for multiplying two n × n matrices needs n3 scalar multiplications and n2 · (n – 1) additions; for the case of 2 × 2 matrices, eight multiplications and four additions. Seven scalar multiplications suffice if we accept 18 additions/subtractions.

$$\begin{bmatrix} r_{11} & r_{12} \\ r_{21} & r_{22} \end{bmatrix} = \begin{bmatrix} a_{11} & a_{12} \\ a_{21} & a_{22} \end{bmatrix} * \begin{bmatrix} b_{11} & b_{12} \\ b_{21} & b_{22} \end{bmatrix}.$$

Evaluate seven expressions, each of which is a product of sums:

$p_1 = (a_{11} + a_{22}) \cdot (B_{11} + b_{22}),$

$p_2 = (a_{21} + a_{22}) \cdot B_{11}$

$p_3 = a_{11} \cdot (B_{12} - b_{22})$

$p_4 = a_{22} \cdot (-B_{11} + b_{21})$ $p_5 = (a_{11} + a_{12}) \cdot b_{22}$

$p_6 = (-a_{11} + a_{21}) \cdot (B_{11} + B_{12})$ $p_7 = (a_{12} - a_{22}) \cdot (b_{21} + b_{22})$.

The elements of the product matrix are computed as follows:

$r_{11} = p_1 + p_4 - p_5 + p_7$,

$r_{12} = p_3 + p_5$,

$r_{21} = p_2 + p_4$, $r_{22} = p_1 - p_2 + p_3 + p_6$.

This algorithm does not rely on the commutativity of scalar multiplication. Hence it can be generalized to n × n matrices using the divide-and-conquer principle. For reasons of simplicity consider n to be a power of 2 (i.e. $n = 2^k$); for other values of n, imagine padding the matrices with rows and columns of zeros up to the next power of 2. An n × n matrix is partitioned into four n/2 × n/2 matrices:

$$\begin{bmatrix} R_{11} & R_{12} \\ R_{21} & R_{22} \end{bmatrix} = \begin{bmatrix} A_{11} & A_{12} \\ A_{21} & A_{22} \end{bmatrix} * \begin{bmatrix} B_{11} & B_{12} \\ B_{21} & B_{22} \end{bmatrix}.$$

The product of two n × n matrices by Strassen's method requires seven (not eight) multiplications and 18 additions/subtractions of n/2 × n/2 matrices. For large n, the Work required for the 18 additions is negligible compared to the Work required for even a single multiplication (why?); thus we have saved one multiplication out of eight, asymptotically at no cost.

Each n/2 × n/2 matrix is again partitioned recursively into four n/4 × n/4 matrices; after log2 n partitioning steps we arrive at 1 × 1 matrices for which matrix multiplication is the primitive scalar multiplication. Let T(n) denote the number of scalar arithmetic operations used by Strassen's method for multiplying two n × n matrices. For n > 1, T(n) obeys the recursive equation

$$T(n) = 7 \cdot T\left(\frac{n}{2}\right) + 18 \cdot \left(\frac{n}{2}\right)^2.$$

If we are only interested in the leading term of the solution, the constants 7 and 2 justify omitting the quadratic term, thus obtaining

$$T(n) = 7 \cdot T\left(\frac{n}{2}\right) = 7 \cdot 7 \cdot T\left(\frac{n}{4}\right) = 7 \cdot 7 \cdot 7 \cdot T\left(\frac{n}{8}\right) = \cdots$$
$$= 7^{\log_2 n} \cdot T(1) = n^{\log_2 7} \cdot T(1) \approx n^{2.81} \cdot T(1).$$

Thus the number of primitive operations required to multiply two n × n matrices using Strassen's method is proportional to $n^{2.81}$, a statement that we abbreviate as "Strassen's matrix multiplication takes time $\Theta(n^{2.81})$".

Does this asymptotic improvement lead to a more efficient program in practice?

Probably not, as the ratio grows too slowly to be of practical importance: For n ≈ 1000, for example, we have $5\sqrt{1024} = 4$ (remember: 210 = 1024). A factor of 4 is not to be disdained, but there are many ways to win or lose a factor of 4. Trading an algorithm with simple code, such as straightforward matrix multiplication, for another that requires more elaborate bookkeeping, such as Strassen's, can easily result in a fourfold increase of the constant factor that measures the time it takes to execute the body of the innermost loop.

$$\frac{n^3}{n^{2.81}} \approx n^{02} = \sqrt[5]{n}$$

Exercises

- ❖ Prove that the set of all ordered pairs of integers is countably infinite.
- ❖ A *recursive function* is defined by a finite set of rules that specify the function in terms of variables, nonnegative integer constants, increment ('+1'), the function itself, or an expression built from these by composition of functions. As an example, consider *Ackermann's function* defined as $A(n) = A_n(n)$ for $n \geq 1$, where Ak(n) is determined by,

  where $A_k(n)$ is determined by

  $A_k(1) = 2$ for $k \geq 1$

  $A_1(n) = A_1(n-1) + 2$ for $n \geq 2$

  $A_k(n) = A_{k-1}(A_k(n-1))$ for $k \geq 2$
- ❖ Calculate A(1) , A(2) , A(3), A(4).
- ❖ Prove that

  $A_k(2) = 4$ for $k \geq 1$,

  $A_1(n) = 2 \cdot n$ for $n \geq 1$,

  $A_2(n) = 2n$ for $n \geq 1$,

  $A_3(n) = 2^{A_3(n-1)}$ for $n \geq 2$.
- ❖ Define the inverse of Ackermann's function as

  $\alpha(n) = \min\{m: A(m) \geq n\}$.

Show that $\alpha(n) \leq 3$ for $n \leq 16$, that $\alpha(n) \leq 4$ for n at most a "tower" of 65536 2's, and that $\alpha(n) \to \infty$ as $n \to \infty$.

- Complete Strassen's algorithm by showing how to multiply n × n matrices when n is not an exact power of 2.
- Assume that you can multiply 3 × 3 matrices using k multiplications. What is the largest k that will lead to an asymptotic improvement over Strassen's algorithm?
- A permutation matrix P is an n × n matrix that has exactly one '1' in each row and each column; all other entries are '0'. A permutation matrix can be represented by an array

```
var a: array[1 .. n] of integer;
```

  as follows: a[i] = j if the i-th row of P contains a '1' in the j-th column.

- Prove that the product of two permutation matrices is again a permutation matrix.
- Design an algorithm that multiplies in time $\Theta(n)$ two permutation matrices given in the array representation above, and stores the result in this same array representation.

16

The Mathematics of Algorithm Analysis

Growth rates and orders of magnitude

To understand a specific algorithm, it is useful to ask and answer the following questions, usually in this order: What is the problem to be solved? What is the main idea on which this algorithm is based? Why is it correct? How efficient is it?

The variety of problems is vast, and so is the variety of "main ideas" that lead one to design an algorithm and establish its correctness. True, there are general algorithmic principles or schemas which are problem-independent, but these rarely suffice: interesting algorithms typically exploit specific features of a problem, so there is no unified approach to understanding the logic of algorithms. Remar$_k$ably, there *is* a unified approach to the efficiency analysis of algorithms, where efficiency is measured by a program's time and storage requirements. This is remar$_k$able because there is great variety in (1) sets of input data and (2) environments (computers, operating systems, programming languages, coding techniques), and these differences have a great influence on the run time and storage consumed by a program. These two types of differences are overcome as follows.

Different sets of input data: worst-case and average performance

The most important characteristic of a set of data is its size, measured in terms of any unit convenient to the problem at hand. This is typically the number of primitive objects in the data, such as bits, bytes, integers, or any monotonic function thereof, such as the magnitude of an integer. *Examples:* For sorting, the number n of elements is natural; for square matrices, the number n of rows and columns is convenient; It is a monotonic function (square root) of the actual size n2 of the data. An algorithm may well behave very differently on different data sets of equal size n—among all possible configurations of given size n some will be favorable, others less so. Both the *worst-case* data set of size n and the *average* over all data sets of size n provide well-defined and important measures of efficiency. *Example:* When sorting data sets about whose order nothing is known, average performance is well characterized by averaging run time over all n! permutations of the n elements.

Different environments: focus on growth rate and ignore constants

The Work performed by an algorithm is expressed as a function of the problem size, typically measured by size n of the input data. By focusing on the growth rate of this function but ignoring specific constants, we succeed in losing a lot of detail information that changes wildly from one computing environment to another, while retaining some essential information that is remar$_k$ably invariant when moving a computation from a micro- to a supercomputer, from machine language to Pascal, from amateur to professional programmer. The definition of general measures for the complexity of problems and for the efficiency of algorithms is a major achievement of computer science. It is based on the notions of *asymptotic time and space complexity*. Asymptotics renounces exact measurement but states how the Work grows as the problem size increases. This information often suffices to distinguish efficient algorithms from inefficient ones. The asymptotic behavior of an algorithm is described by the O(), Ω(), Θ(), and o() notations. To determine the amount of Work to be performed by an algorithm we count operations that take constant time (independently of n) and data objects that require constant storage space. The time required by an addition, comparison, or exchange of two numbers is typically independent of how many numbers we are processing; so is the storage requirement for a number.

Assume that the time required by four algorithms A1, A2, A3, and A4 is log2n, n, n · log2n, and n2, respectively. The following table shows that for sizes of data sets that frequently occur in practice, from n ≈ 10 3 to 106, the difference in growth rate translates into large numerical differences:

n	$A_1 = \log_2 n$	$A_2 = n$	$A_3 = n \cdot \log_2 n$	$A_4 = n_2$
$2^5 = 32$	5	$2^5 = 3^2$	$5 \cdot 2^5 = 160$	$2^{10} \approx 10^3$
$2^{10} = 1024$	10	$2^{10} \approx 10^3$	$10 \cdot 2^{10} \approx 10^4$	$2^{20} \approx 10^6$
$2^{20} \approx 10_6$	20	$2^{20} \approx 1^{06}$	$20 \cdot 2^{20} \approx 2 \cdot 10^7$	$2^{40} \approx 10^{12}$

For a specific algorithm these functions are to be multiplied by a constant factor proportional to the time it takes to execute the body of the innermost loop. When comparing different algorithms that solve the same problem, it may well happen that one innermost loop is 10 times faster or slower than another. it is rare that this difference approaches a factor of 100. Thus for n ≈ 1000 an algorithm with time complexity Θ(n · log n) will almost always be much more efficient than an algorithm with time complexity Θ(n2). For small n, say n = 32, an algorithm of time complexity Θ(n2) may be more efficient than one of complexity Θ(n · log n) (e.g. if its constant is 10 times smaller).

When we wish to predict exactly how many seconds and bytes a program needs, asymptotic analysis is still useful but is only a small part of the Work. We now have to go back over the formulas and keep track of all the constant factors discarded in cavalier fashion by the O() notation. We have to assign numbers to the time consumed by scores of primitive O(1) operations. it may be sufficient to estimate the time consuming primitives, such as floating-point operations; or it may be necessary to include those that are hidden by a high-level programming language and answer questions such as: How long does an array access a[i, j] take? A procedure call? incrementing the index i in a loop "for i := 0 to n"?

Asymptotics

Asymptotics is a technique used to estimate and compare the growth behavior of functions. Consider the function

$$\mathbf{f(x)} = \frac{\mathbf{1+x^2}}{\mathbf{x}}.$$

f(x) is said to behave like x for $x \to \infty$ and like 1/x for $x \to 0$. The motivation for such a statement is that both x and 1/x are intuitively simpler, more easily understood functions than f(x). A complicated function is unlike any simpler one across its entire domain, but it usually behaves like a simpler one as x approaches some particular value. Thus all asymptotic statements include the qualifier $x \to x0$. For the purpose of algorithm analysis we are interested in the behavior of functions for large values of their argument, and all our definitions below assume $x \to \infty$.

The asymptotic behavior of functions is described by the O(), Ω(), Θ (), and o() notations, as in $f(x) \in O(g(x))$. Each of these notations assigns to a given function g the *set of all functions* that are related to g in a well-defined way. intuitively, O(), Ω(), Θ(), and o() are used to compare the growth of functions, as ≤, ≥, =, and < are used to compare numbers. O(g) is the set of all functions that are ≤ g in a precise technical sense that corresponds to the intuitive notion "grows no faster than g". The definition involves some technicalities signaled by the preamble ∃∃c > 0,∃> $\exists x_0 \in X, \forall x \geq x_0$. it says that we ignore constant factors and initial behavior and are interested only in a function's behavior from some point on. N_0 is the set of nonnegative integers, R_0 the set of nonnegative reals. In the following definitions X stands for either N0 or R0. Let $g: X \to X$.

Definition of O(), "big oh":

$$O(g) := \{f: X \to X \mid \exists\, c > 0, \exists\, x_0 \in X, \forall\, x \geq x_0 : f(x) \leq c \cdot g(x)\}$$

We say that $f \in O(g)$, or that f grows at most as fast as g(x) for $x \to \infty$. Definition of Ω(), "omega":

$\Omega(g) := \{f: X \to X \mid \exists\, c > 0\; x_0 \in X, \forall\, x \geq x_0: f(x) \geq c \cdot g(x)\}$.

We say that $f \in O(g)$, or that f grows at least as fast as g(x) for $x \to \infty$. Definition of Θ(), "theta":

$\Theta(g) := O(g) \cap \Omega(g)$.

We say that $f \in \Theta(g)$, or that f has the same growth rate as g(x) for $x \to \infty$. Definition of o(), "small oh":

$$o(g) := \{f: X \to X \mid \lim_{x \to \infty} \frac{f(x)}{g(x)} = 0\}.$$

We say that $f \in o(g)$, or that f grows slower than g(x) for $x \to \infty$.

Notation: Most of the literature uses = in place of our ∈, such as in $x = O(x^2)$. If you do so, just remember that this = has none of the standard properties of an equality relation—it is neither commutative nor transitive. Thus $O(x^2) = x$ is not used, and from $x = O(x^2)$ and $x^2 = O(x^2)$ it does not follow that $x = x^2$. The key to avoiding confusion is the insight that O() is not a function but a set of functions.

Summation formulas

$\log_2$ denotes the logarithm to the base 2, ln the natural logarithm to the base e.

$$\sum_{i=1}^{n} i = \frac{n \cdot (n+1)}{2} = \frac{n^2 + n}{2}$$

$$\sum_{i=1}^{n} i^2 = \frac{n \cdot (n+1) \cdot (2n+1)}{6} = \frac{n^3}{3} + \frac{n^2}{2} + \frac{n}{6}$$

$$\sum_{i=1}^{n} i^3 = \frac{n^2 \cdot (n+1)^2}{4} = \frac{n^4}{4} + \frac{n^3}{2} + \frac{n^2}{4}$$

The asymptotic behavior of a sum can be derived by comparing the sum to an integral that can be evaluated in closed form. Let f(x) be a monotonically increasing, integrable function. Then

$$\int_a^b f(x)\, dx$$

is bounded below and above by sums (Exhibit 16.1):

$$\sum_{i=1}^{n} f(x_{i-1}) \cdot (x_i - x_{i-1}) \leq \int_{x_0}^{x_n} f(x)\, dx \leq \sum_{i=1}^{n} f(x_i) \cdot (x_i - x_{i-1}).$$

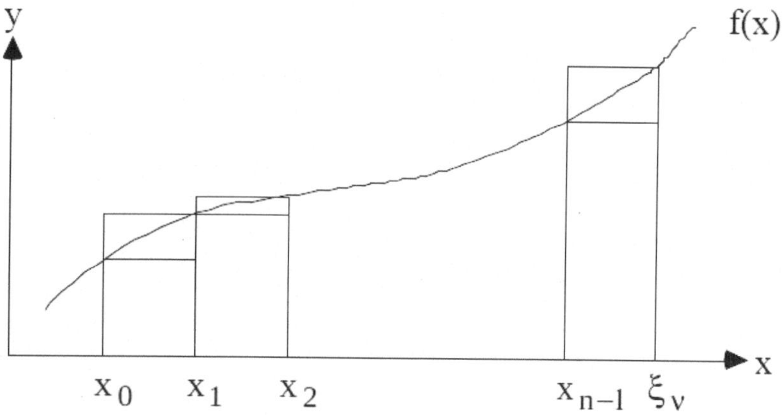

Exhibit 16.1: Bounding a definite integral by lower and upper sums.

Letting $x_i = i + 1$, this inequality becomes

$$\sum_{i=1}^{n} f(i) \le \int_{1}^{n+1} f(x)dx \le \sum_{i=1}^{n} f(i+1),$$

so

$$\int_{1}^{n+1} f(x)\,dx - f(n+1) + f(1) \le \sum_{i=1}^{n} f(i) \le \int_{1}^{n+1} f(x)\,dx. \qquad (*)$$

Example

By substituting

$$f(x) = x^k \quad \text{and} \quad \int x^k\,dx = \frac{x^{k+1}}{k+1}$$

with $k > 0$ in $(*)$ we obtain

$$\frac{(n+1)^{k+1}}{k+1} - \frac{1}{k+1} - (n+1)^k + 1 \le \sum_{i=1}^{n} i^k \le \frac{(n+1)^{k+1}}{k+1} - \frac{1}{k+1},$$

and therefore

$$\forall k > 0: \sum_{i=1}^{n} i^k = \frac{n^{k+1}}{k+1} + g(n) \quad \text{with} \quad g(n) \in O(n^k).$$

Example

By substituting

$f(x)=\ln x$ and $\int \ln x\, dx = x\cdot \ln x - x$

in (**) we obtain

$$(n+1)\cdot\ln(n+1)-n-\ln(n+1)\leq\sum_{i=1}^{n}\ln i\leq(n+1)\text{cdotln}(n+1)-n,$$

and therefore

$$\sum_{i=1}^{n}\log_2 i=(n+1)\cdot\log_2(n+1)-\frac{n}{\ln 2}+g(n)\text{ with }g(n)\in O(\log n)$$

Example

By substituting

$$\mathbf{f(x)=\ln x \quad and \quad \int \ln x\, dx = x\cdot \ln x - x}$$

in (**) we obtain

$$\sum_{i=1}^{n} i\cdot\log_2 i=\frac{(n+1)^2}{2}\cdot\log_2(n+1)-\frac{(n+1)^2}{4\cdot\ln 2}+g(n)$$

with $g(n) \in O(n \cdot \log n)$.

Recurrence relations

A *homogeneous linear* recurrence relation with constant coefficients is of the form

$x_n = a_1 \cdot x_{n-1} + a_2 \cdot x_{n-2} + \ldots + a_k \cdot x_{n-k}$

where the coefficients ai are independent of n and $x_1, x_2, \ldots, x_{n-1}$ are specified. There is a general technique for solving linear recurrence relations with constant coefficients - that is, for determining x_n as a function of n. We will demonstrate this technique for the Fibonacci sequence which is defined by the recurrence

$x_n = x_{n-1} + x_{n-2}, x_0 = 0, x_1 = 1.$

We seek a solution of the form

$x_n = c \cdot r^n$

with constants c and r to be determined. Substituting this into the Fibonacci recurrence relation yields

$c \cdot r^n = c \cdot r^{n-1} + c \cdot r^{n-2}$

or

$c \cdot r^{n-2} \cdot (r^2 - r - 1) = 0.$

This equation is satisfied if either $c = 0$ or $r = 0$ or $r^2 - r - 1 = 0$. We obtain the trivial solution $x_n = 0$ for all n if $c = 0$ or $r = 0$. More interestingly, $r^2 - r - 1 = 0$ for

$$r_1 = \frac{1-\sqrt{5}}{2} \quad \text{and} \quad r_2 = \frac{1+\sqrt{5}}{2}.$$

The sum of two solutions of a homogeneous linear recurrence relation is obviously also a solution, and it can be shown that any linear combination of solutions is again a solution. Therefore, the most general solution of the Fibonacci recurrence has the form

$$x_n = c_1 \cdot \left(\frac{1+\sqrt{5}}{2}\right)^n + c_2 \cdot \left(\frac{1-\sqrt{5}}{2}\right)^n$$

where c_1 and c_2 are determined as solutions of the linear equations derived from the initial conditions:

$$x_0 = 0 \quad \Rightarrow \quad c_1 + c_2 = 0$$

$$x_1 = 1 \quad \Rightarrow \quad c_1 \cdot \frac{1+\sqrt{5}}{2} + c_2 \cdot \frac{1-\sqrt{5}}{2} = 1$$

which yield

$$c_1 = \frac{1}{\sqrt{5}} \quad \text{and} \quad c_2 = -\frac{1}{\sqrt{5}}.$$

the complete solution for the Fibonacci recurrence relation is therefore

$$x_n = \frac{1}{\sqrt{5}} \cdot \left(\frac{1+\sqrt{5}}{2}\right)^n - \frac{1}{\sqrt{5}} \cdot \left(\frac{1-\sqrt{5}}{2}\right)^n.$$

Recurrence relations that are not linear with constant coefficients have no general solution techniques comparable to the one discussed above. General recurrence relations are solved (or their solutions are approximated or bounded) by trial-and-error techniques. If the trial and error is guided by some general technique, it will yield at least a good estimate of the asymptotic behavior of the solution of most recurrence relations.

Example

Consider the recurrence relation

$$x_n = \frac{2}{n}\sum_{i=0}^{n-1} x_i + a \cdot n + b \qquad (*)$$

with a > 0 and b > 0, which appears often in the average-case analysis of algorithms and data structures. When we know from the interpretation of this recurrence that its solution is monotonically nondecreasing, a systematic trial- and-error process leads to the asymptotic behavior of the solution. The simplest possible try is a constant, $x_n = c$.

Substituting this into (**) leads to

$$c \ ?=? \ 2 \cdot c + a \cdot n + b,$$

so $x_n = c$ is not a solution. Since the left-hand side xn is smaller than an average of previous values on the right-hand side, the solution of this recurrence relation must grow faster than c. Next, we try a linear function $x_n = c \cdot n$:

$$\begin{aligned} c \cdot n \ &?=? \ \frac{2}{n}\sum_{i=0}^{n-1} c \cdot i + a \cdot n + b \\ &?=? \ (c+a) \cdot n - c + b. \end{aligned}$$

At this stage of the analysis it suffices to focus on the leading terms of each side: $c \cdot n$ on the left and $(c + a) \cdot n$ on the right. The assumption a > 0 makes the right side larger than the left, and we conclude that a linear function also grows too slowly to be a solution of the recurrence relation. A new attempt with a function that grows yet faster, $x_n = c \cdot n^2$, leads to

$$\begin{aligned} c \cdot n^2 \ &?=? \ \frac{2}{n}\sum_{i=0}^{n-1} c \cdot i^2 + a \cdot n + b \\ &?=? \ \frac{2}{3} \cdot c \cdot n^2 + (a - c) \cdot n + \frac{c}{3} + b. \end{aligned}$$

Comparing the leading terms on both sides, we find that the left side is now larger than the right, and conclude that a quadratic function grows too fast. Having bounded the growth rate of the solution from below and above, we try functions whose growth rate lies between that of a linear and a quadratic function, such as $x_n = c \cdot n^{1.5}$. A more sophisticated approach considers a family of functions of the form $x_n = c \cdot n^{1+e}$ for any ε > 0: All of them grow too fast. This suggests $x_n = c \cdot n \cdot \log_2 n$, which gives

$$c \cdot n \cdot \log_2 n \quad ?=? \quad \frac{2}{n} \sum_{i=0}^{n-1} c \cdot i \cdot \log_2 i + a \cdot n + b$$

$$?=? \quad \frac{2 \cdot c}{n} \left(\frac{n^2}{2} \cdot \log_2 n - \frac{n^2}{4 \cdot \ln 2} + g(n) \right) + a \cdot n + b$$

$$?=? \quad c \cdot n \cdot \log_2 n + \left(a - \frac{c}{2 \cdot \ln 2} \right) \cdot n + h(n)$$

with $g(n) \in O(n \cdot \log n)$ and $h(n) \in O(\log n)$. To match the linear terms on each side, we must choose c such that

$$a - \frac{c}{2 \cdot \ln 2} = 0$$

or $c = a \cdot \ln 4 \approx 1.386 \cdot a$. Hence we now know that the solution to the recurrence relation (∗) has the form

$$x_n = (\ln 4) \cdot a \cdot n \cdot \log_2 n + g(n) \quad \text{with} \quad g(n) \in O(n).$$

Asymptotic performance of divide-and-conquer algorithms

We illustrate the power of the techniques developed in previous sections by analyzing the asymptotic performance not of a specific algorithm, but rather, of an entire class of divide-and-conquer algorithms. in "Divide and conquer recursion" we presented the following schema for divide-and-conquer algorithms that partition the set of data into two parts:

```
A(D):  if  simple(D) then return(A0(D))
                     else 1. divide: partition D into D1 and D2;
                          2. conquer: R1 := A(D1);  R2 := A(D2);
                          3. combine: return(merge(R1, R2));
```

Assume further that the data set D can always be partitioned into two halves, d_1 and d_2, at every level of recursion. Two comments are appropriate:

- For repeated halving to be possible it is not necessary that the size n of the data set D be a power of 2, $n = 2^k$. it is not important that D be partitioned into two exact halves—approximate halves will do. Imagine padding any data set D whose size is not a power of 2 with dummy elements, up to the next power of 2. Dummies can always be found that do not disturb the real computation: for example, by replicating elements or by appending sentinels. Padding is usually just a conceptual trick that may help in understanding the process, but need not necessarily generate any additional data.
- Whether or not the divide step is guaranteed to partition D into two approximate halves, on the other hand, depends critically on the problem and on the data structures used. Example: Binary search in an ordered array partitions D into

halves by probing the element at the midpoint; the same idea is impractical in a linked list because the midpoint is not directly accessible.

Under our assumption of halving, the time complexity T(n) of algorithm A applied to data D of size n satisfies the recurrence relation

$$T(n) = 2 \cdot T\left(\frac{n}{2}\right) + f(n)$$

where f(n) is the sum of the partitioning or splitting time and the "stitching time" required to merge two solutions of size n / 2 into a solution of size n. Repeated substitution yields

$$\begin{aligned} T(n) &= 4 \cdot T\left(\frac{n}{4}\right) + f(n) + 2 \cdot f\left(\frac{n}{2}\right) \\ &= 8 \cdot T\left(\frac{n}{8}\right) + f(n) + 2 \cdot f\left(\frac{n}{2}\right) + 4 \cdot f\left(\frac{n}{4}\right) \\ &\cdots \\ &= n \cdot T(1) + \sum_{k=0}^{(\log_2 n)-1} 2^k \cdot f\left(\frac{n}{2^k}\right). \end{aligned}$$

The term $n \cdot T(1)$ expresses the fact that every data item gets looked at, the second sums up the splitting and stitching time. Three typical cases occur:

- Constant time splitting and merging f(n) = c yields

  $T(n) = (T(1) + c) \cdot n$.

  Example: Find the maximum of n numbers.

- Linear time splitting and merging

  $f(n) = a \cdot n + b$ yields $T(n) = a \cdot n \cdot \log2\ n + (T(1) + b) \cdot n$.

  Examples: Mergesort, quicksort.

- splitting and merging: $n \in o(f(n))$ yields and therefore rarely leads to interesting algorithms.

  $T(n) = n \cdot T(1) + O(f(n) \cdot \log n)$

Permutations

Inversions

Let (a_k: $1 \le k \le n$) be a permutation of the integers 1.. n. A pair (a_i, a_j), $1 \le I < j \le n$, is called an *inversion* iff $a_i > a_j$. What is the average number of inversions in a permutation? Consider all permutations in pairs; that is, with any permutation A:

$a_1 = x_1; a_2 = x_2; \ldots ; a_n = x_n$

consider its inverse A', which contains the elements of A in inverse order:

$a_1 = x_n; a_2 = x_{n-1}; \ldots ; a_n = x_1.$

in one of these two permutations x_i and x_j are in the correct order, in the other, they form an inversion. Since there are n·(n – 1) / 2 pairs of elements (x_i, x_j) with $1 \le i < j \le n$ there are, on average, inversions.

$$\mathrm{inv}_{average} = \frac{1}{2} \cdot n \cdot \frac{n-1}{2} = \frac{n^2 - n}{4}$$

Average distance

Let $(a_k: 1 \le k \le n)$ be a permutation of the natural numbers from 1 to n. The distance of the element a_i from its correct position is $|a_i - i|$. The total distance of all elements from their correct positions is

$$TD((a_k : 1 \le k \le n)) = \sum_{i=1}^{n} |a_i - i|.$$

Therefore, the average total distance (i.e. the average over all n! permutations) is

$$TD_{average} = \frac{1}{n!} \sum_{\substack{\text{all permutations} \\ (a_k : 1 \le k \le n)}} TD((a_k : 1 \le k \le n))$$

$$= \frac{1}{n!} \sum_{\substack{\text{all permutations} \\ (a_k : 1 \le k \le n)}} \sum_{i=1}^{n} |a_i - i|$$

$$= \frac{1}{n!} \sum_{i=1}^{n} \sum_{\substack{\text{all permutations} \\ (a_k : 1 \le k \le n)}} |a_i - i|.$$

Let $1 \le i \le n$ and $1 \le j \le n$. Consider all permutations for which ai is equal to j. Since there are (n – 1)! such permutations, we obtain

$$\sum_{\substack{\text{all permutations} \\ (a_k : 1 \le k \le n)}} |a_i - i| = (n-1)! \sum_{j=1}^{n} |j - i|.$$

Therefore,

$$TD_{average} = \frac{1}{n!} \sum_{i=1}^{n} \left((n-1)! \sum_{j=1}^{n} |j-i| \right)$$

$$= \frac{1}{n} \sum_{i=1}^{n} \left(\sum_{j=1}^{i-1} (i-j) + \sum_{j=i+1}^{n} (j-i) \right)$$

$$= \frac{n^2-1}{3}.$$

the average distance of an element ai from its correct position is therefore

$$\frac{1}{n} \cdot \frac{n^2-1}{3} = \frac{n}{3} - \frac{1}{3 \cdot n}.$$

Trees

Trees are ubiquitous in discrete mathematics and computer science, and this section summarizes some of the basic concepts, terminology, and results. Although trees come in different versions, in the context of algorithms and data structures, "tree" almost always means an ordered rooted tree. An *ordered rooted tree* is either empty or it consists of a node, called a *root,* and a sequence of k ordered subtrees T1, T2, ... , Tk (Exhibit 16.2). The nodes of an ordered tree that have only empty subtrees are called leaves or external nodes, the other nodes are called in*ternal nodes* (Exhibit 16.3). The roots of the subtrees attached to a node are its *children*; and this node is their *parent.*

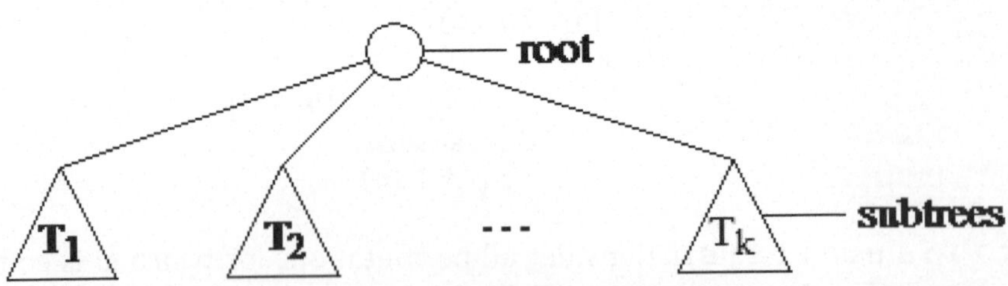

Exhibit 16.2: Recursive definition of a rooted, ordered tree.

The *level* of a node is defined recursively. The root of a tree is at level 0. The children of a node at level t are at level t + 1. The level of a node is the length of the path from the root of the tree to this node. The *height* of a tree is defined as the maximum level of all leaves. The *path length* of a tree is the sum of the levels of all its nodes (Exhibit 16.3).

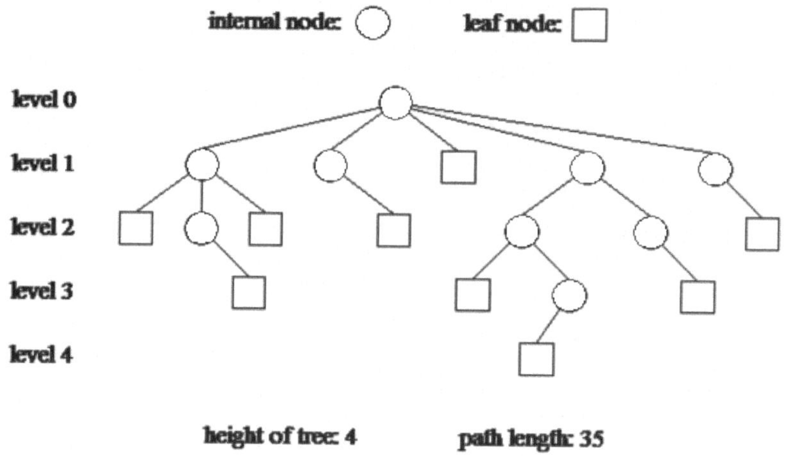

Exhibit 16.3: A tree of height = 4 and path length = 35.

A *binary tree* is an ordered tree whose nodes have at most two children. A 0-2 binary tree is a tree in which every node has zero or two children but not one. A 0-2 tree with n leaves has exactly n – 1 internal nodes. A binary tree of height h is called *complete* (completely balanced) if it has $2^{h+1} - 1$ nodes (Exhibit 16.4. A binary tree of height h is called *almost complete if* all its leaves are on levels h – 1 and h, and all leaves on level h are as far left as possible (Exhibit 16.4).

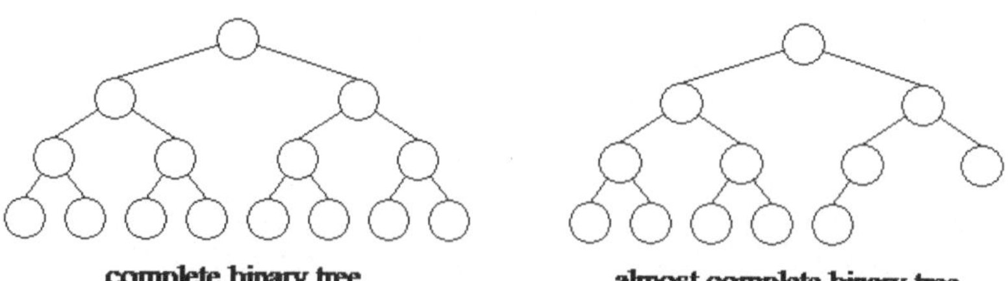

Exhibit 16.4: Examples of well-balanced binary trees.

Exercises

- Suppose that we are comparing implementations of two algorithms on the same machine. For inputs of size n, the first algorithm runs in 9 · n2 steps, while the second algorithm runs in 81 · n · log2 n steps. Assuming that the steps in both algorithms take the same time, for which values of n does the first algorithm beat the second algorithm?
- What is the smallest value of n such that an algorithm whose running time is 256 · n2 runs faster than an algorithm whose running time is 2n on the same machine?
- For each of the following functions fi(n), determine a function g(n) such that fi(n) ∈ Θ(g(n)). The function g(n) should be as simple as possible.

$f_1(n) = 0.001 \cdot n^7 + n^2 + 2 \cdot n$

$f_2(n) = n \cdot \log n + \log n + 1234 \cdot n$

$f_3(n) = 5 \cdot n \cdot \log n + n^2 \cdot \log n + n^2$

$f_4(n) = 5 \cdot n \cdot \log n + n^3 + n^2 \cdot \log n$

- Prove formally that $1024 \cdot n^2 + 5 \cdot n \in \Theta(n^2)$.
- Give an asymptotically tight bound for the following summation:

$$\sum_{i=1}^{n} i^k \cdot \log_2 i \quad \text{for } k > 0.$$

- Find the most general solutions to the following recurrence relations.

(a) $x_n = x_{n-1} - \frac{1}{4} \cdot x_{n-2}$ (b) $x_n = x_{n-1} - \frac{1}{4} \cdot x_{n-2} + 2^{-n}$

- Solve the recurrence $T(\sqrt{n}) = 2 \cdot T() + \log2\ n$. *Hint:* Make a change of variables $m = \log2\ n$.
- Compute the number of inversions and the total distance for the permutation (3 1 2 4).

17

Sorting and its Complexity

What is sorting? How difficult is it?

The problem

Assume that S is a set of n elements $x_1, x_2, \dots, x_n$ drawn from a domain X, on which a total order ≤ is defined (i.e. a relation that satisfies the following axioms):

```
≤ is reflexive (i.e ∀x ∈ X:  x ≤ x)
≤ is antisymmetric (i.e ∀x, y ∈ X:  x ≤ y  ∧  y ≤ x ⇒ x = y)
≤ is transitive (i.e ∀x, y, z ∈ X:  x ≤ y  ∧  y ≤ z  ⇒  x ≤ z)
≤ is total (i.e. ∀x, y ∈ X  ⇒ x ≤ y  ∨  y ≤ x)
```

Sorting is the process of generating a sequence

$$x_{i_1}, x_{i_2}, \dots, x_{i_n}$$

such that (i1, i2, ... , in) is a permutation of the integers from 1 to n and, holds. Phrased abstractly, sorting is the problem of finding a specific permutation (or one among a few permutations, when distinct elements may have equal values) out of n! possible permutations of the n given elements. Usually, the set S of elements to be sorted will be given in a data structure; in this case, the elements of S are ordered implicitly by this data structure, but not necessarily according to the desired order ≤. Typical sorting problems assume that S is given in an array or in a sequential file (magnetic tape), and the result is to be generated in the same structure. We characterize elements by their position in the structure (e.g. A[i] in the array A or by the value of a pointer in a sequential file). The access operations provided by the underlying data structure determine what sorting algorithms are possible.

$$\forall k, 1 \leq k \leq n-1: \ x_{i_k} \leq x_{i_{k+1}}$$

Algorithms

Most sorting algorithms are refinements of the following idea:

```
while ∃(i, j):  i < j  ∧  A[i] > A[j]  do  A[i] :=: A[j];
```

where :=: denotes the exchange operator. Even sorting algorithms that do not explicitly exchange pairs of elements, or do not use an array as the underlying data structure, can usually be thought of as conforming to the schema above. An insertion sort, for example, takes one element at a time and inserts it in its proper place among those already sorted. To find the correct place of insertion, we can think of a ripple effect whereby the new element successively displaces (exchanges position with) all those larger than itself.

As the schema above shows, two types of operations are needed in order to sort:

- collecting information about the order of the given elements
- ordering the elements (e.g. by exchanging a pair)

When designing an efficient algorithm we seek to economize the number of operations of both types: We try to avoid collecting redundant information, and we hope to move an element as few times as possible. The nondeterministic algorithm given above lets us perform any one of a number of exchanges at a given time, regardless of their usefulness. For example, in sorting the sequence

$x_1 = 5, x_2 = 2, x_3 = 3, x_4 = 4, x_5 = 1$

the nondeterministic algorithm permits any of seven exchanges

x_1 :=: x_i for $2 \leq i \leq 5$ and x_j :=: x_5 for $2 \leq j \leq 4$.

We might have reached the state shown above by following an exotic sorting technique that sorts "from the middle toward both ends", and we might know at this time that the single exchange x_1 :=: x_5 will complete the sort.

The nondeterministic algorithm gives us no handle to express and use this knowledge.

The attempt to economize Work forces us to depart from nondeterminacy and to impose a control structure that carefully sequences the operations to be performed so as to make maximal use of the information gained so far. The resulting algorithms will be more complex and difficult to understand. It is useful to remember, though, that sorting is basically a simple problem with a simple solution and that all the acrobatics in this chapter are due to our quest for efficiency.

Intrinsic complexity

There are obvious limits to how much we can economize. in the absence of any previously acquired information, it is clear that each element must be inspected and, in general, moved at least once. Thus we cannot hope to get away with fewer than $\Omega(n)$ primitive operations. There are less obvious limits, we mention two of them here.

- If information is collected by asking binary questions only (any question that may receive one of two answers (e.g. a yes/no question, or a comparison of two elements that yields either $\leq$ or >), then at least n · log2 n questions are necessary in general, as will be proved in the section "A lower bound Ωn · logn". Thus in this model of computation, sorting requires time $\Theta(n \cdot \log n)$.
- In addition to collecting information, one must rearrange the elements. in the section "Permutation" in chapter 16, we have shown that in a permutation the average distance of an element from its correct position is approximately n/3. Therefore elements have to move an average distance of approximately n/3 elements to end up at their destination. Depending on the access operations of the underlying storage structure, an element can be moved to its correct position in a single step of average length n/3, or in n/3 steps of average length 1. If elements are rearranged by exchanging adjacent elements only, then on average Θ(n2) moving operations are required. Therefore, short steps are insufficient to obtain an efficient $\Theta(n \cdot \log n)$ sorting algorithm.

Practical aspects of sorting

Records instead of elements. We discuss sorting assuming only that the elements to be sorted are drawn from a totally ordered domain. in practice these elements are just the keys of records that contain additional data associated with the key: for example,

```
type recordtype =  record
        key: keytype;   { totally ordered by ≤ }
        data: anytype
    end;
```

We use the relational operators =, <, ≤ to compare keys, but in a given programming language, say Pascal, these may be undefined on values of type keytype. in general, they must be replaced by procedures: for example, when comparing strings with respect to the lexicographic order.

If the key field is only a small part of a large record, the exchange operation :=:, interpreted literally, becomes an unnecessarily costly copy operation. This can be avoided by leaving the record (or just its data field) in place, and only moving a small surrogate record consisting of a key and a pointer to its associated record.

Sort generators. On many systems, particularly in the world of commercial data processing, you may never need to write a sorting program, even though sorting is a frequently executed operation. Sorting is taken care of by a sort generator, a program akin to a compiler; it selects a suitable sorting algorithm from its repertoire and tailors it to the problem at hand, depending on parameters such as the number of elements to be sorted, the resources available, the key type, or the length of the records.

Partially sorted sequences. The algorithms we discuss ignore any order that may exist in the sequence to be sorted. Many applications call for sorting files that are *almost sorted*, for example, the case where a sorted *master file* is updated with an unsorted *transaction file*. Some algorithms take advantage of any order present in the input data; their time complexity varies from O(n) for almost sorted files to O(n · log n) for randomly ordered files.

Types of sorting algorithms

Two important classes of incremental sorting algorithms create order by processing each element in turn and placing it in its correct position. These classes, *insertion sorts* and *selection sorts*, are best understood as maintaining two disjoint, mutually exhaustive structures called 'sorted' and 'unsorted'.

```
    Initialize: 'sorted' := Ø;   'unsorted' := {x1, x2, … , xn};
    Loop:   for i := 1 to n do
                move an element from 'unsorted' to its correct place in
'sorted';
```

The following illustrations show 'sorted' and 'unsorted' sharing an array[1 .. n]. in this case the boundary between 'sorted' and 'unsorted' is represented by an index i that increases as more elements become ordered. The important distinction between the two types of sorting algorithms emerges from the question: in which of the two structures is most of the Work done? insertion sorts remove the first or most easily accessible element from 'unsorted' and search through 'sorted' to find its proper place. Selection sorts search through 'unsorted' to find the next element to be appended to 'sorted'.

Insertion sort

The i-th step inserts the i-th element into the sorted sequence of the first (i – 1) elements Exhibit 17.1).

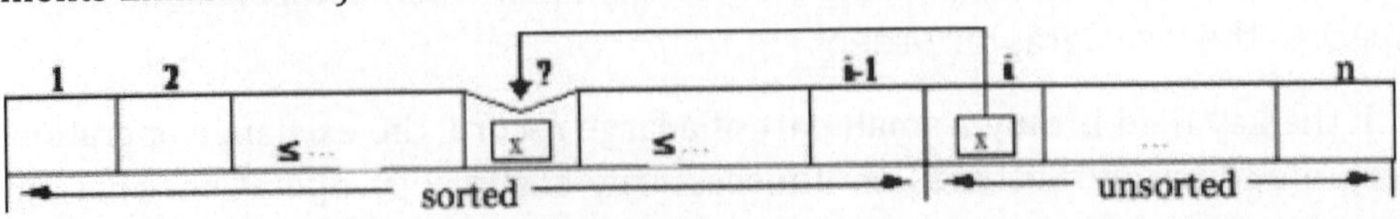

Exhibit 17.1: insertion sorts move an easily accessed element to its correct place.

Selection sort

The i-th step selects the smallest among the n – i + 1 elements not yet sorted, and moves it to the i-th position (Exhibit 17.2).

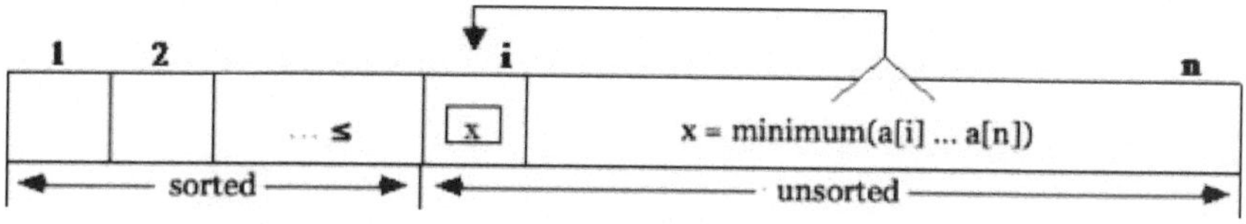

Exhibit 17.2: Selection sorts search for the correct element to move to an easily accessed place.

Insertion and selection sorts repeatedly search through a large part of the entire data to find the proper place of insertion or the proper element to be moved. Efficient search requires random access, hence these sorting techniques are used primarily for in*ternal sorti*ng in central memory.

Merge sort

Merge sorts process (sub)sequences of elements in unidirectional order and thus are well suited for *external* sorting on secondary storage media that provide sequential access only, such as magnetic tapes; or random access to large blocks of data, such as disks. Merge sorts are also efficient for internal sorting. The basic idea is to merge two sorted sequences of elements, called *runs,* into one longer sorted sequence. We read each of the input runs, and write the output run, starting with small elements and ending with the large ones. We keep comparing the smallest of the remaining elements on each input run, and append the smaller of the two to the output run, until both input runs are exhausted (Exhibit 17.3).

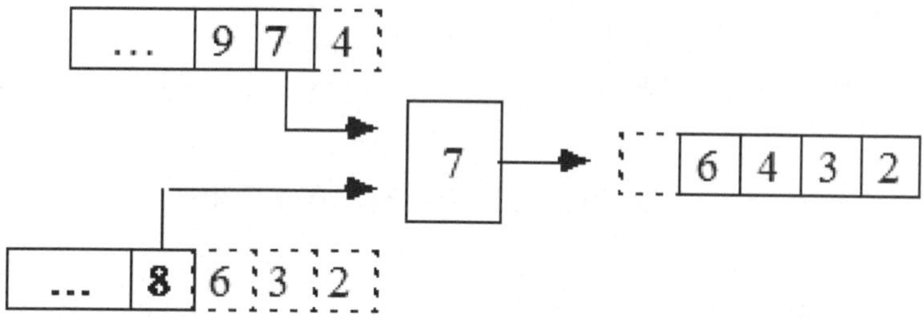

Exhibit 17.3: Merge sorts exploit order already present.

The processor shown at left in Exhibit 17.4 reads two tapes, A and B. Tape A contains runs 1 and 2; tape B contains runs 3 and 4. The processor merges runs 1 and 3 into the single run 1 & 3 on tape C, and runs 2 and 4 into the single run 2 & 4 on tape D. in a second merge step, the processor shown at the right reads tapes C and D and merges the two runs 1 & 3 and 2 & 4 into one run, 1 & 3 & 2 & 4.

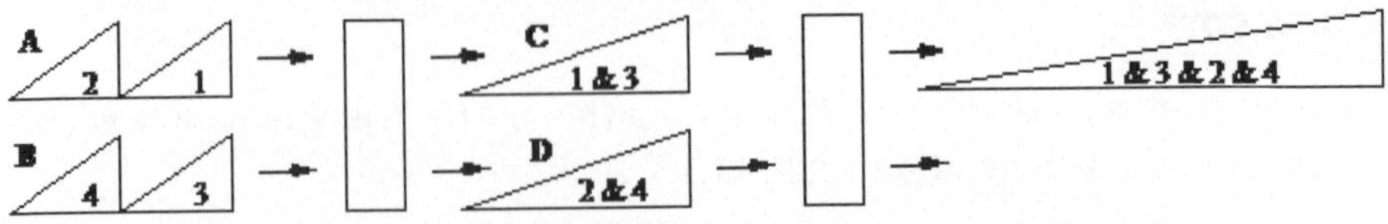

Exhibit 17.4: Two merge steps in sequence.

Distribution sort

Distribution sorts process the *representation* of an element as a value in a radix number system and use primitive arithmetic operations such as "extract the k-th digit". These sorts do not compare elements directly. They introduce a different model of computation than the sorts based on comparisons, exchanges, insertions, and deletions that we have considered thus far. As an example, consider numbers with at most three digits in radix 4 representation. in a first step these numbers are distributed among four queues according to their least significant digit, and the queues are concatenated in increasing order. The process is repeated for the middle digit, and finally for the leftmost, most significant digit, as shown in Exhibit 17.5

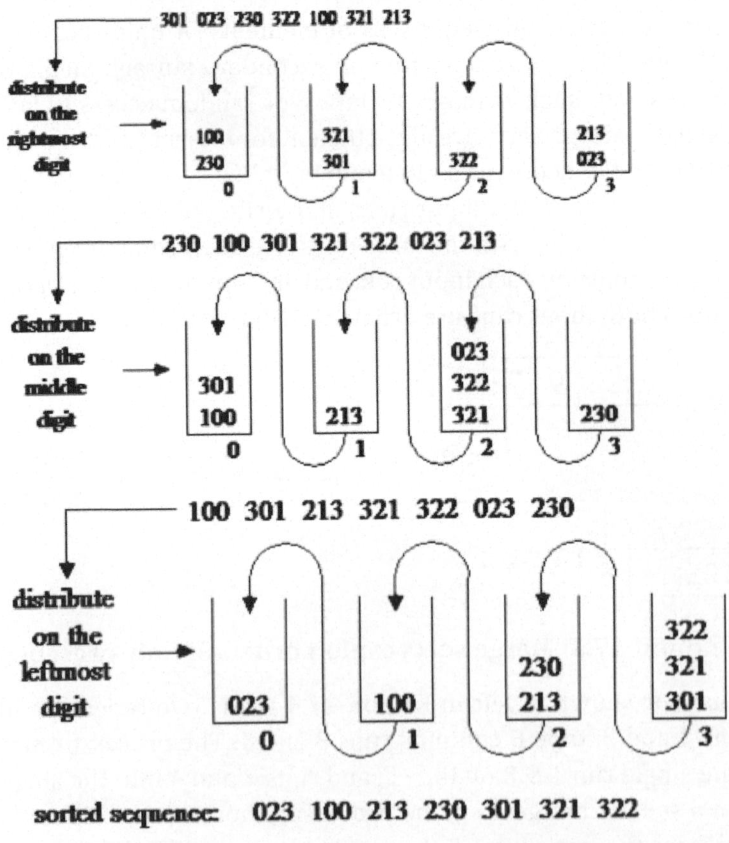

Exhibit 17.5 Distribution sorts use the radix representation of keys to organize elements in buckets

We have now seen the basic ideas on which all sorting algorithms are built. it is more important to understand these ideas than to know dozens of algorithms based on them. To appreciate the intricacy of sorting, you must understand some algorithms in detail: we begin with simple ones that turn out to be inefficient.

Simple sorting algorithms that Work in time Θ(n2)

If you invent your own sorting technique without prior study of the literature, you will probably "discover" a well-known inefficient algorithm that Works in time O(n2), requires time Θ(n2) in the worst case, and thus is of time complexity Ω(n2). Your algorithm might be similar to one described below.

Consider in-*place algorithms* that Work on an array declared as

```
var  A: array[1 .. n] of elt;
```

and place the elements in ascending order. Assume that the comparison operators are defined on values of type elt. Let c_{best}, $c_{average}$, and c_{worst} denote the number of comparisons, and e_{best}, $e_{average}$, and e_{worst} the number of exchange operations performed in the best, average, and worst case, respectively. Let $inv_{average}$ denote the average number of inversions in a permutation.

Insertion sort (Exhibit 17.6)

Let −∞ denote a constant ≤ any key value. The smallest value in the domain often serves as a sentinel −∞.

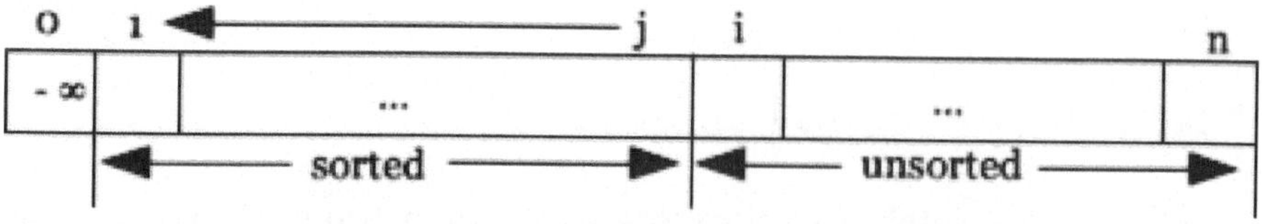

Exhibit 17.6: Straight insertion propagates a ripple-effect across the sorted part of the array.

```
A[0] := -∞;
for i := 2 to n do  begin
   j := i;
   while  A[j] < A[j - 1]  do  { A[j] :=: A[j - 1]; { exchange }
j := j - 1 }
end;
```

$$c_{best} = n - 1$$

$$c_{average} = inv_{average} + (n-1) = \frac{n^2 + 3 \cdot n - 4}{4}$$

$$c_{worst} = \sum_{i=2}^{n} i = \frac{n^2 + n - 2}{2}$$

$$e_{best} = 0$$

$$e_{average} = inv_{average} = \frac{n^2 - n}{4}$$

$$e_{worst} = \sum_{i=2}^{n} (i-1) = \frac{n^2 - n}{2}$$

This straight insertion sort is an Θ(n) algorithm in the best case and an Θ(n2) algorithm in the average and worst cases. in the program above, the point of insertion is found by a linear search interleaved with exchanges. A binary search is possible but does not improve the time complexity in the average and worst cases, since the actual insertion still requires a linear-time ripple of exchanges.

Selection sort (Exhibit 17.7)

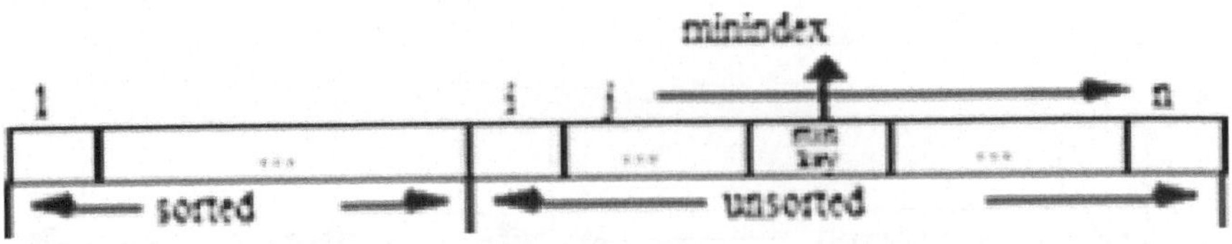

Exhibit 17.7: Straight selection scans the unsorted part of the array.

```
for i := 1 to n - 1 do  begin
  minindex := i;  minkey := A[i];
  for j := i + 1 to n do
    if  A[j] < minkey  then  { minkey := A[j];  minindex := j }
  A[i] :=: A[minindex]  { exchange }
end;
```

$$c_{best} = c_{average} = c_{worst} = \sum_{i=1}^{n-1} (n-i) = \frac{n^2 - n}{2}$$

$$e_{best} = e_{average} = e_{worst} = n - 1$$

The sum in the formula for the number of comparisons reflects the structure of the two nested for loops. The body of the inner loop is executed the same number of times for each of the three cases. Thus this straight selection sort is of time complexity Θ(n2).

A lower bound $\Omega(n \cdot \log n)$

A straightforward counting argument yields a lower bound on the time complexity of any sorting algorithm that collects information about the ordering of the elements by asking only binary questions. A binary question has a two-valued answer: yes or no, true or false. A comparison of two elements, $x \leq y$, is the most obvious example, but the following theorem holds for binary questions in general.

Theorem: Any sorting algorithm that collects information by asking binary questions only executes at least binary questions both in the worst case, and averaged over all n! permutations. Thus the average and worst-case time complexity of such an algorithm is $\Omega(n \cdot \log n)$.

$$n \cdot \log_2 (n+1) - \frac{n}{\ln 2}$$

Proof: A sorting algorithm of the type considered here can be represented by a *binary decision tree*. Each internal node in such a tree represents a binary question, and each leaf corresponds to a result of the decision process. The decision tree must distinguish each of the n! possible permutations of the input data from all the others; and thus must have at least n! leaves, one for each permutation.

Example: The decision tree shown in Exhibit 17.8 collects the information necessary to sort three elements, x, y and z, by comparisons between two elements.

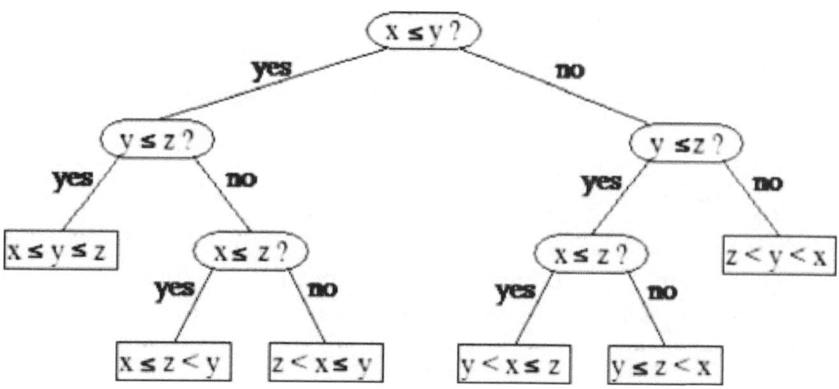

Exhibit 17.8 The decision tree shows the possible n! Outcomes when sorting n elements.

The average number of binary questions needed by a sorting algorithm is equal to the average depth of the leaves of this decision tree. The lemma following this theorem will show that in a binary tree with k leaves the average depth of the leaves is at least log2k. Therefore, the average depth of the leaves corresponding to the n! permutations is at least log2n!. Since

$$\log_2 n! = \log_2\left(\prod_{i=1}^{n} i\right) = \sum_{i=1}^{n} \log_2 i$$

$$\geq (n+1)\cdot\log_2(n+1) - \frac{n}{\ln 2} - \log_2(n+1) = n\cdot\log_2(n+1) - \frac{n}{\ln 2}$$

it follows that on average at least.binary questions are needed, that is, the time complexity of each such sorting algorithm is $\Omega(n \cdot \log n)$ in the average, and therefore also in the worst case.

$$n * \log_2(n+1) - \frac{n}{\ln_2}$$

Lemma: in a binary tree with k leaves the average depth of the leaves is at least $\log_2 k$.

Proof: Suppose that the lemma is not true, and let T be the counterexample with the smallest number of nodes. T cannot consist of a single node because the lemma is true for such a tree. If the root r of T has only one child, the subtree T' rooted at this child would contain the k leaves of T that have an even smaller average depth in T' than in T. Since T was the counterexample with the smallest number of nodes, such a T' cannot exist. Therefore, the root r of T must have two children, and there must be $k_L > 0$ leaves in the left subtree and $k_R > 0$ leaves in the right subtree of r ($k_L + k_R = k$). Since T was chosen minimal, the k_L leaves in the left subtree must have an average depth of at least $\log_2 k_L$, and the k_R leaves in the right subtree must have an average depth of at least $\log_2 k_R$. Therefore, the average depth of all k leaves in T must be at least

$$\frac{k_L}{k_L + k_R}\cdot\log_2 k_L + \frac{k_R}{k_L + k_R}\cdot\log_2 k_R + 1. \qquad (*)$$

It is easy to see that (∗) assumes its minimum value if $k_L = k_R$. Since ∗()has the value $\log_2 k$ if $k_L = k_R = k/2$ we have found a contradiction to our assumption that the lemma is false.

Quicksort

Quicksort (C. A. R. Hoare, 1962) [Hoa 62] combines the powerful algorithmic principle of divide-and- conquer with an efficient way of moving elements using few exchanges. The *divide phase* partitions the array into two disjoint parts: the "small" elements on the left and the "large" elements on the right. The *conquer phase* sorts each part separately. Thanks to the Work of the divide phase, the *merge phase* requires no Work at all to combine two partial solutions. Quicksort's efficiency depends crucially on the expectation that the divide phase cuts two sizable subarrays rather than merely slicing off an element at either end of the array (Exhibit 17.9).

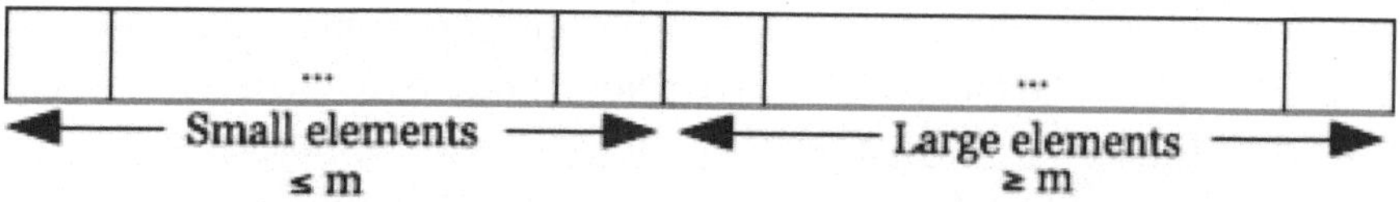

Exhibit 17.9: Quicksort partitions the array into the "small" elements on the left and the "large" elements on the right.

We chose an arbitrary *threshold* value m to define "small" as ≤ m, and "large" as ≥ m, thus ensuring that any "small element" ≤ any "large element". We partition an arbitrary subarray A[L .. R] to be sorted by executing a left- to-right scan (incrementing an index i) "concurrently" with a right-to-left scan (decrementing j) (Exhibit 17.10). The left-to-right scan pauses at the first element A[i] ≥ m, and the right-to-left scan pauses at the first element A[j] ≤ m. When both scans have paused, we exchange A[i] and A[j] and resume the scans. The partition is complete when the two scans have crossed over with j < i. Thereafter, quicksort is called recursively for A[L .. j] and A[i .. R], unless one or both of these subarrays consists of a single element and thus is trivially sorted. *Example* of partitioning (m = 16):

```
25  23   3 16   4    7    29   6
i                              j
6   23   3 16   4    7    29   25
    i                j
6   7    3 16   4    23   29   25
           i    j
6   7    3 4    16   23   29   25
           j    i
```

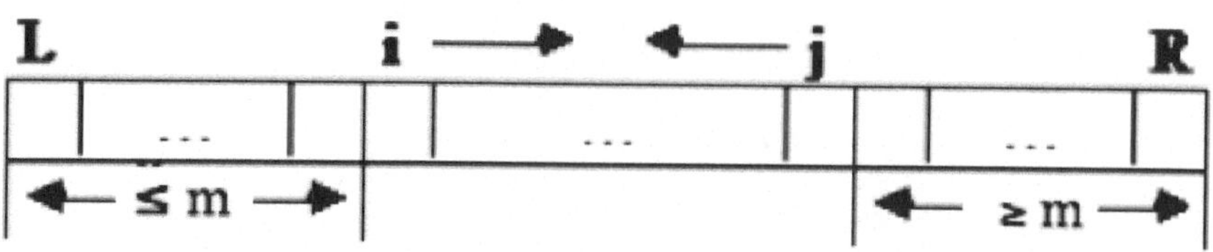

Exhibit 17.10: Scanning the array concurrently from left to right and from right to left.

Although the threshold value m appeared arbitrary in the description above, it must meet criteria of correctness and efficiency. *Correctness:* if either the set of elements ≤ m or the set of elements ≥ m is empty, quicksort fails to terminate. Thus we require that min(xi) ≤ m ≤ max(xi). *Efficiency* requires that m be close to the median.

How do we find the median of n elements? The obvious answer is to sort the elements and pick the middle one, but this leads to a chicken-and-egg problem when trying to sort in the first place. There exist sophisticated algorithms that determine the exact median of n elements in time O(n) in the worst case [BFPRT

72]. The multiplicative constant might be large, but from a theoretical point of view this does not matter. The elements are partitioned into two equal-sized halves, and quicksort runs in time O(n · log n) even in the worst case. From a practical point of view, however, it is not worthwhile to spend much effort in finding the exact median when there are much cheaper ways of finding an acceptable approximation. The following techniques have all been used to pick a threshold m as a "guess at the median":

- An array element in a fixed position such as A[(L + R) div 2]. *Warning:* stay away from either end, A[L] or A[R], as these thresholds lead to poor performance if the elements are partially sorted.
- An array element in a random position: a simple technique that yields good results.
- The median of three or five array elements in fixed or random positions.
- The average between the smallest and largest element. This requires a separate scan of the entire array in the beginning; thereafter, the average for each subarray can be calculated during the previous partitioning process.

The recursive procedure 'rqs' is a possible implementation of quicksort. The function 'guessmedian' must yield a threshold that lies on or between the smallest and largest of the elements to be sorted. If an array element is used as the threshold, the procedure 'rqs' should be changed in such a way that after finishing the partitioning process this element is in its final position between the left and right parts of the array.

```
procedure rqs (L, R: 1 .. n);   { sorts A[L], … , A[R] }
var  i, j: 0 .. n + 1;

  procedure partition;
  var  m: elt;
  begin  { partition }
    m := guessmedian (L, R);
    { min(A[L], … , A[R]) ≤ m ≤ max(A[L], … , A[R]) }
    i := L;   j := R;
    repeat
      { A[L], … , A[i - 1] ≤ m ≤ A[j + 1], … , A[R] }
      while  A[i] < m  do  i := i + 1;
      { A[L], … , A[i - 1] ≤ m ≤ A[i] }
      while  m < A[j]  do  j := j - 1;
      { A[j] ≤ m ≤ A[j + 1], … , A[R] }
      if  i ≤ j  then  begin
        A[i] :=: A[j];   { exchange }
        { i ≤ j ⇒ A[i] ≤ m ≤ A[j] }
        i := i + 1;   j := j - 1
        { A[L], … , A[i - 1] ≤ m ≤ A[j + 1], … , A[R] }
      end
      else
        { i > j ⇒ i = j + 1 ⇒ exit }
      end
    until  i > j
  end;  { partition }

begin  { rqs }
  partition;
  if L < j then  rqs(L, j);
  if i < R then  rqs(i, R)
end;  { rqs }
```

An initial call 'rqs(1, n)' with n > 1 guarantees that L < R holds for each recursive call.

An iterative implementation of quicksort is given by the following procedure, 'iqs', which sorts the whole array A[1 .. n]. The boundaries of the subarrays to be sorted are maintained on a stack.

```
procedure iqs;
const  stacklength = ... ;
type  stackelement = record  L, R: 1 .. n  end;
var  i, j, L, R, s: 0 .. n;
   stack: array[1 .. stacklength] of stackelement;

   procedure partition;   { same as in rqs }
   end;   { partition }

begin   { iqs }
   s := 1;   stack[1].L := 1;   stack[1].R := n;
   repeat
      L := stack[s].L;   R := stack[s].R;   s := s - 1;

        repeat
          partition;
          if  j - L < R - i  then  begin
             if  i <  R then  { s := s + 1;  stack[s].L := i;
stack[s].R := R };
             R := j
          end
          else  begin
             if  L < j  then  { s := s + 1;  stack[s].L := L;
stack[s].R := j };
             L := i
          end
        until  L ≥ R
     until  s = 0
  end;  { iqs }
```

After partitioning, 'iqs' pushes the bounds of the larger part onto the stack, thus making sure that part will be sorted later, and sorts the smaller part first. Thus the length of the stack is bounded by log2n.

For very small arrays, the overhead of managing a stack makes quicksort less efficient than simpler $O(n^2)$ algorithms, such as an insertion sort. A practically efficient implementation of quicksort might switch to another sorting technique for subarrays of size up to 10 or 20. [Sed 78] is a comprehensive discussion of how to optimize quicksort.

Analysis for three cases: best, "typical", and worst

Consider a quicksort algorithm that chooses a guessed median that differs from any of the elements to be sorted and thus partitions the array into two parts, one with k elements, the other with n – k elements. The Work q(n) required to sort n elements satisfies the recurrence relation

$$q(n) = q(k) + q(n-k) + a \cdot n + b. \qquad (*)$$

The constant b measures the cost of calling quicksort for the array to be sorted. The term a · n covers the cost of partitioning, and the terms q(k) and q(n – k) correspond to the Work involved in quicksorting the two subarrays. Most quicksort algorithms partition the array into three parts: the "small" left part, the single array element used to guess the median, and the "large" right part. Their Work is expressed by the equation

$$q(n) = q(k) + q(n-k-1) + a \cdot n + b.$$

We analyze equation (*); it is close enough to the second equation to have the same asymptotic solution. Quicksort's behavior in the best and worst cases are easy to analyze, but the average over all permutations is not. Therefore, we analyze another average which we call the *typical case.*

Quicksort's *best-case behavior* is obtained if we guess the correct median that partitions the array into two equal-sized subarrays. For simplicity's sake the following calculation assumes that n is a power of 2, but this assumption does not affect the solution. Then (*) can be rewritten as

$$q(n) = 2 \cdot q\left(\frac{n}{2}\right) + a \cdot n + b.$$

We use this recurrence equation to calculate

$$q\left(\frac{n}{2}\right) = 2 \cdot q\left(\frac{n}{4}\right) + a \cdot \frac{n}{2} + b$$

and substitute on the right-hand side to obtain

$$q(n) = 2 \cdot \left(2 \cdot q\left(\frac{n}{4}\right) + a \cdot \frac{n}{2} + b\right) + a \cdot n + b = 4 \cdot q\left(\frac{n}{4}\right) + 2 \cdot a \cdot n + 3 \cdot b.$$

Repeated substitution yields

$$\begin{aligned} q(n) &= n \cdot q(1) + a \cdot n \cdot \log_2 n + b \cdot (n-1) \\ &= a \cdot n \cdot \log_2 n + g(n) \quad \text{with} \quad g(n) \in O(n). \end{aligned}$$

The constant q(1), which measures quicksort's Work on a trivially sorted array of length 1, and b, the cost of a single procedure call, do not affect the dominant term n · log2n. The constant factor a in the dominant term can be estimated by analyzing the code of the procedure 'partition'. When these details do not matter, we summarize: Quicksort's time complexity in the best case is Θ(n · log n).

Quicksort's *worst-case* behavior occurs when one of the two subarrays consists of a single element after each partitioning. in this case equation (∗) becomes

$$q(n) = q(n-1) + q(1) + a \cdot n + b.$$

We use this recurrence equation to calculate

$$q(n-1) = q(n-2) + q(1) + a \cdot (n-1) + b$$

and substitute on the right-hand side to obtain

$$q(n) = q(n-2) + 2 \cdot q(1) + a \cdot n + a \cdot (n-1) + 2 \cdot b.$$

Repeated substitution yields

$$q(n) = n \cdot q(1) + a \cdot \frac{n^2 + n - 2}{2} + b \cdot (n-1).$$

Therefore the time complexity of quicksort in the worst case is Θ(n2).

For the analysis of quicksort's *typical behavior* we make the plausible assumption that the array is equally likely to get partitioned between any two of its elements: For all k, 1 ≤ k < n, the probability that the array A is partitioned into the subarrays A[1 .. k] and A[k + 1 .. n] is 1 / (n – 1). Then the average Work to be performed by quicksort is expressed by the recurrence relation

$$q(n) = \frac{1}{n-1} \sum_{k=1}^{n-1} (q(k) + q(n-k)) + a \cdot n + b$$

$$= \frac{2}{n-1} \sum_{k=1}^{n-1} q(k) + a \cdot n + b.$$

This recurrence relation approximates the recurrence relation discussed in chapter 16 well enough to have the same solution

$$q(n) = (\ln 4) \cdot a \cdot n \cdot \log_2 n + g(n) \quad \text{with} \quad g(n) \in O(n).$$

Since ln 4 ≈ 1.386, quicksort's asymptotic behavior in the typical case is only about 40% worse than in the best case, and remains in Θ(n · log n). [Sed 77] is a thorough analysis of quicksort.

Merging and merge sorts

The in*ternal* sorting algorithms presented so far require *direct access* to each element. This is reflected in our analyses by treating an array access A[i], or each exchange A[i] :=: A[j], as a primitive operation whose cost is constant (independent of n). This assumption is not valid for elements stored on secondary storage devices such as magnetic tapes or disks. A better assumption that mirrors the realities of *external* sorting is that the elements to be sorted are stored as a *sequential file* f.

The file is accessed through a file pointer which, at any given time, provides direct access to a single element. Accessing other elements requires repositioning of the file pointer. Sequential files may permit the pointer to advance in one direction only, as in the case of Pascal files, or to move backward and forward. In either case, our theoretical model assumes that the time required for repositioning the pointer is proportional to the distance traveled. This assumption obviously favors algorithms that process (compare, exchange) pairs of adjacent elements, and penalizes algorithms such as quicksort that access elements in random positions.

The following external sorting algorithm is based on the merge sort principle. To make optimal use of the available main memory, the algorithm first creates initial runs; a *run* is a sorted subsequence of elements f_i, f_{i+1}, … , fj stored consecutively in file f, $f_k \leq f_{k+1}$ for all k with $i \leq k \leq j - 1$. Assume that a buffer of capacity m elements is available in main memory to create init*ial runs* of length m (perhaps less for the last run). in processing the r-th run, r = 0, 1, … , we read the m elements $f_{r \cdot m+1}$, $f_{r \cdot m+2}$, … , $f_{r \cdot m+m}$ into memory, sort them internally, and write the sorted sequence to a modified file f, which may or may not reside in the same physical storage area as the original file f. This new file f is partially sorted into runs: $f_k \leq f_{k+1}$ for all k with $r \cdot m + 1 \leq k < r \cdot m + m$.

At this point we need two files, g and h, in addition to the file f, which contains the initial runs. In a *copy phase* we distribute the initial runs by copying half of them to g, the other half to h. in the subsequent *merge phase* each run of g is merged with exactly one run of h, and the resulting new run of double length is written onto f (Exhibit 17.11). After the first cycle, consisting of a copy phase followed by a merge phase, f contains half as many runs as it did before. After $\lceil \log_2(n / m) \rceil$ cycles f contains one single run, which is the sorted sequence of all elements.

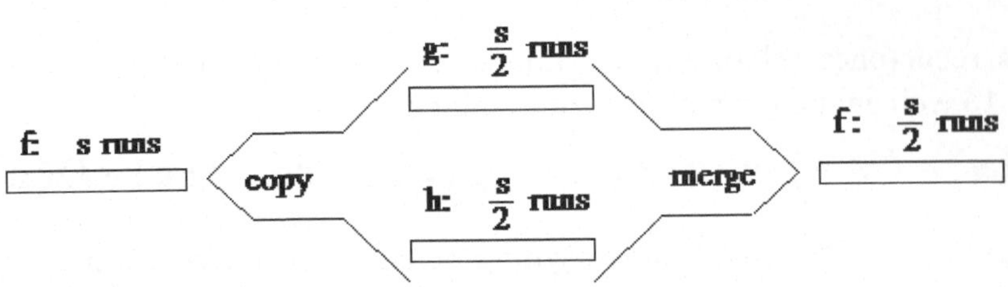

Exhibit 17.11: Each copy-merge cycle halves the number of runs and doubles their lengths.

Exercise: a merge sort in main memory

Consider the following procedure that sorts the array A:

```
const  n = ... ;
var  A: array[1 .. n] of integer;
...
procedure sort (L, R: 1 .. n);
  var  m: 1 .. n;

  procedure combine;
  var B: array [1 .. n] of integer;
        i, j, k: 1 .. n;
  begin  { combine }
    i := L;  j := m + 1;
    for k := L to R do
      if  (i > m) cor ((j ≤ R) cand (A[j] < A[i]))  then
        { B[k] := A[j];  j := j + 1 }
      else
        { B[k] := A[i];  i := i + 1 } ;
    for k := L to R do  A[k] := B[k]
  end;  { combine }

begin  { sort}
  if  L < R  then
    { m := (L + R) div 2;  sort(L, m);  sort(m + 1, R);  combine }
end;  { sort }
```

The relational operators 'cand' and 'cor' are conditional! The procedure is initially called by

```
sort(1,n);
```

- Draw a picture to show how 'sort' Works on an array of eight elements.
- Write down a recurrence relation to describe the Work done in sorting n elements.
- Determine the asymptotic time complexity by solving this recurrence relation.
- Assume that 'sort' is called for m subarrays of equal size, not just for two. How does the asymptotic time complexity change?

Solution

- 'sort' depends on the algorithmic principle of divide and conquer. After dividing an array into a left and a right subarray whose numbers of elements differ by at most one, 'sort' calls itself recursively on these two subarrays. After these two calls are finished, the procedure 'combine' merges the two sorted subarrays A[L .. m] and A[m + 1 .. R] together in B. Finally, B is copied to A. An example is shown in Exhibit 17.12.

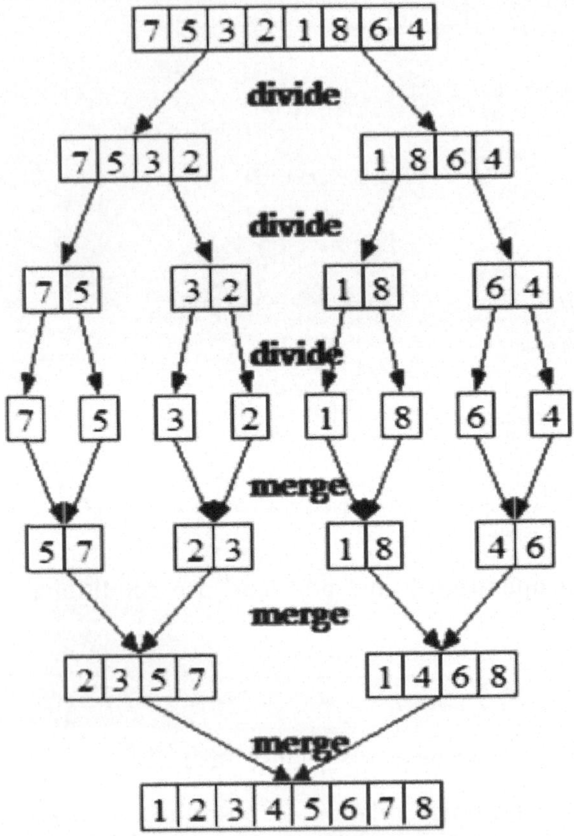

Exhibit 17.12: Sorting an array by using a divide-and-conquer scheme.

❖ The Work w(n) performed while sorting n elements satisfies

$$\mathbf{w(n)=2\cdot w\left(\frac{n}{2}\right)+a\cdot n+b.} \qquad (*)$$

The first term describes the cost of the two recursive calls of 'sort', the term a · n is the cost of merging the two sorted subarrays, and the constant b is the cost of calling 'sort' for the array.

❖ If is substituted in (**), we obtain

$$\mathbf{w\left(\frac{n}{2}\right)=2\cdot w\left(\frac{n}{4}\right)+a\cdot\frac{n}{2}+b}$$

$$\mathbf{w(n)=4\cdot w\left(\frac{n}{4}\right)+2\cdot a\cdot n+3\cdot b.}$$

Continuing this substitution process results in

$$\begin{aligned} w(n) &= n \cdot w(1) + a \cdot n \cdot \log_2 n + b \cdot (n-1) \\ &= a \cdot n \cdot \log_2 n + g(n) \quad \text{with} \quad g(n) \in O(n). \end{aligned}$$

since w(1) is constant the time complexity of 'sort' is $\Theta(n \cdot \log n)$.

- If 'sort' is called recursively for m subarrays of equal size, the cost w'(n) is solving this recursive equation shows that the time complexity does not change [i.e. it is $\Theta(n \cdot \log n)$].

$$w'(n) = m \cdot w'\left(\frac{n}{m}\right) + a \cdot m \cdot n + b.$$

Is it possible to sort in linear time?

The lower bound $\Omega(n \cdot \log n)$ has been derived for sorting algorithms that gather information about the ordering of the elements by binary questions and nothing else. This lower bound need not apply in other situations.

Example 1: sorting a permutation of the integers from 1 to n

If we know that the elements to be sorted are a permutation of the integers 1 .. n, it is possible to sort in time $\Theta(n)$ by storing element i in the array element with index i.

Example 2: sorting elements from a finite domain

Assume that the elements to be sorted are samples from a finite domain W = 1 .. w. Then it is possible to sort in time $\Theta(n)$ if gaps between the elements are allowed (Exhibit 17.13). The gaps can be closed in time $\Theta(w)$.

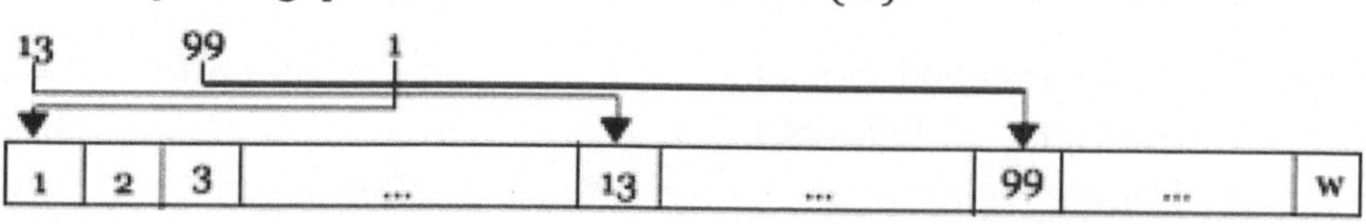

Exhibit 17.13: Sorting elements from a finite domain in linear time.

Do these examples contradict the lower bound $\Omega(n \cdot \log n)$? No, because in these examples the information about the ordering of elements is obtained by asking questions more powerful than binary questions: namely, n- valued questions in Example 1 and w-valued questions in Example 2.

A k-valued question is equivalent to log2k binary questions. When this "exchange rate" is taken into consideration, the theoretical time complexities of the two sorting techniques above are $\Theta(n \cdot \log n)$ and $\Theta(n \cdot \log w)$, respectively, thus conforming to the lower bound in the section "A lower bound $\Omega(n \cdot \log n)$".

Sorting algorithms that sort in linear time (expected linear time, but not in the worst case) are described in the literature under the terms *bucket sort*, distribution sort, and *radix sort*.

Sorting netWorks

The sorting algorithms above are designed to run on a sequential machine in which all operations, such as comparisons and exchanges, are performed one at a time with a single processor. If algorithms are to be efficient, they need to be rethought when the ground rules for their execution change: when the theoretician uses another model of computation, or when they are executed on a computer with a different architecture. This is particularly true of the many different types of multiprocessor architectures that have been built or conceived. When many processors are available to share the Workload, questions of how to distribute the Work among them, how to synchronize their operation, and how to transport data, prevail. it is not our intention to discuss sorting on general- purpose parallel machines. We wish to illustrate the point that algorithms must be redesigned when the model of computation changes. For this purpose a discussion of special-purpose sorting netWorks suffices. The "processors" in a sorting netWork are merely *comparators*: Their only function is to compare the values on two input wires and switch them onto two output wires such that the smaller is on top, the larger at the bottom (Exhibit 17.14).

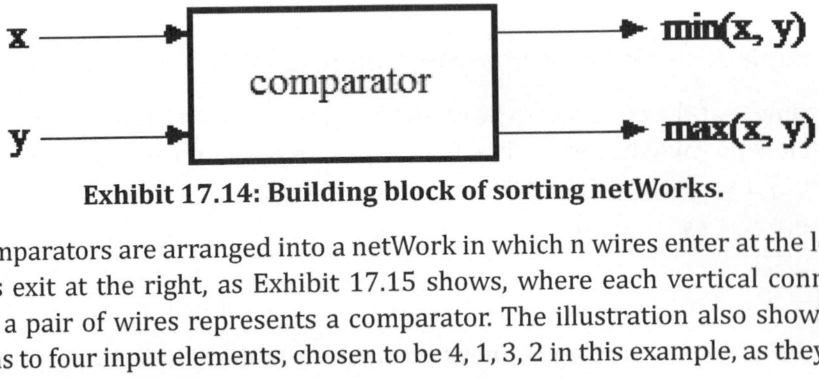

Exhibit 17.14: Building block of sorting netWorks.

Comparators are arranged into a netWork in which n wires enter at the left and n wires exit at the right, as Exhibit 17.15 shows, where each vertical connection joining a pair of wires represents a comparator. The illustration also shows what happens to four input elements, chosen to be 4, 1, 3, 2 in this example, as they travel from left to right through the netWork.

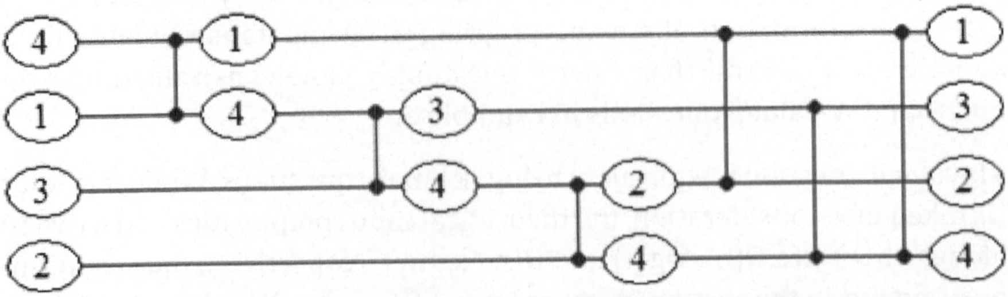

Exhibit 17.15: A comparator netWork that fails to sort. The output of each comparator performing an exchange is shown in the ovals.

A netWork of comparators is a *sorting net*Work if it sorts every input configuration. We consider an input configuration to consist of distinct elements, so that without loss of generality we may regard it as one of the n! permutations of the sequence (1, 2, ... , n). A netWork that sorts a duplicate-free configuration will also sort a configuration containing duplicates.

The comparator netWork above correctly sorts many of its 4! = 24 input configurations, but it fails on the sequence (4, 1, 3, 2). Hence it is not a sorting netWork. it is evident that a netWork with a sufficient number of comparators in the right places will sort correctly, but as the example above shows, it is not immediately evident what number suffices or how the comparators should be placed. The netWork in Exhibit 17.16 shows that five comparators, arranged judiciously, suffice to sort four elements.

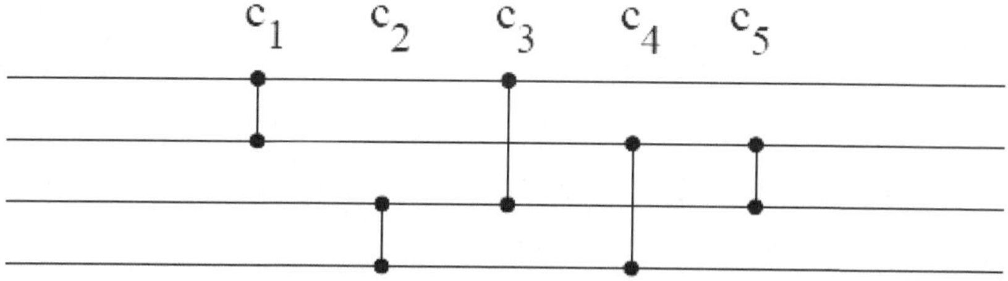

Exhibit 17.16: Five comparators suffice to sort four elements.

How can we tell if a given netWork sorts successfully? Exhaustive testing is feasible for small netWorks such as the one above, where we can trace the flow of all 4! = 24 input configurations. NetWorks with a regular structure usually admit a simpler correctness proof. For this example, we observe that c_1, c_2, and c_3 place the smallest element on the top wire. Similarly, c_1, c_2, and c_4 place the largest on the bottom wire. This leaves the middle two elements on the middle two wires, which c_5 then puts into place.

What design principles might lead us to create large sorting netWorks guaranteed to be correct? Sorting algorithms designed for a sequential machine cannot, in general, be mapped directly into netWork notation, because the netWork is a more restricted model of computation: Whereas most sequential sorting algorithms make comparisons based on the outcome of previous comparisons, a sorting netWork makes the same comparisons for all input configurations. The same fundamental algorithm design principles useful when designing sequential algorithms also apply to parallel algorithms.

Divide-and-conquer. Place two sorting netWorks for n wires next to each other, and combine them into a sorting netWork for 2 · n wires by appending a *merge net*Work to merge their outputs. in sequential computation merging is simple because we can choose the most useful comparison depending on the outcome of previous comparisons. The rigid structure of comparator netWorks makes merging netWorks harder to design.

*Incremental algorithm.*We place an n-th wire next to a sorting netWork with n – 1 wires, and either precede or follow the netWork by a "ladder" of comparators that tie the extra wire into the existing netWork, as shown in the following figures. This leads to designs that mirror the straight insertion and selection algorithms in the section "Simple sorting algorithms that Work in time $\Theta(n^2)$

Insertion sort. With the top n – 1 elements sorted, the element on the bottom wire trickles into its correct place.

Induction yields the expanded diagram on the right in Exhibit 17.17.

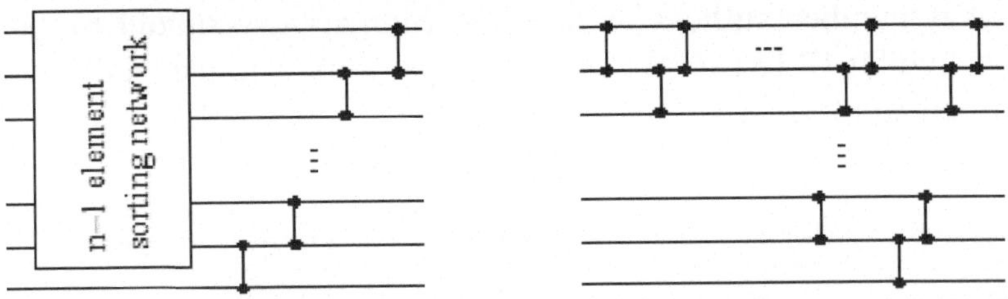

Exhibit 17.17: insertion sort leads by induction to the sorting netWork on the right.

Selection sort. The maximum element first trickles down to the bottom, then the remaining elements are sorted.

The expanded diagram is on the right in Exhibit 17.18.

Exhibit 17.18: Selection sort leads by induction to the sorting netWork on the right.

Comparators can be shifted along their pair of wires so as to reduce the number of stages, provided that the topology of the netWork remains unchanged. This compression reduces both insertion and selection sort to the triangular netWork shown in Exhibit 17.19. Thus we see that the distinction between insertion and selection was more a distinction of sequential order of operations rather than one of data flow.

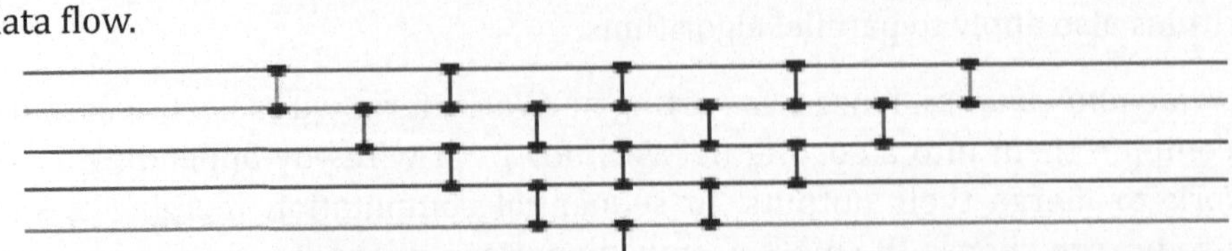

Exhibit 17.19: Shifting comparators reduces the number of stages.

Any number of comparators that are aligned vertically require only a single unit of time. The compressed triangular netWork has $O(n^2)$ comparators, but its time complexity is $2 \cdot n - 1 \in O(n)$. There are netWorks with better asymptotic behavior, but they are rather exotic [Knu 73b].

Exercises and programming projects

- Implement insertion sort, selection sort, merge sort, and quicksort and animate the sorting process for each of these algorithms: for example, as shown in the snapshots in "Algorithm animation". Compare the number of comparisons and exchange operations needed by the algorithms for different input configurations.

- What is the smallest possible depth of a leaf in a decision tree for a sorting algorithm?

- Show that $2 \cdot n - 1$ comparisons are necessary in the worst case to merge two sorted arrays containing n elements each.

- The most obvious method of systematically interchanging the out-of-order pairs of elements in an array var A: array[1 .. n] of elt; is to scan adjacent pairs of elements from bottom to top (imagine that the array is drawn vertically, with A[1] at the top and A[n] at the bottom) repeatedly, interchanging those found out of order:

```
for  i := 1  to  n - 1  do
      for  j := n  downto  i + 1  do
        if  A[j - 1] > A[j]  then  A[j - 1] :=: A[j];
```

This technique is known as *bubble sort,* since smaller elements "bubble up" to the top.

- Explain by words, figures, and an example how bubble sort Works. Show that this algorithm sorts correctly.

- Determine the exact number of comparisons and exchange operations that are performed by bubble sort in the best, average, and worst case.

- What is the worst-case time complexity of this algorithm?

- A sorting algorithm is called *stable* if it preserves the original order of equal elements. Which of the sorting algorithms discussed in this chapter is stable?

- Assume that quicksort chooses the threshold m as the first element of the sequence to bc sorted. Sliow that the running time of such a quicksort algorithm is Θ(n2) when the input array is sorted in nonincreasing or nondecreasing order.

- Find a worst-case input configuration for a quicksort algorithm that chooses the threshold m as the median of the first, middle, and last elements of the sequence to be sorted.

❖ Array A contains m and array B contains n different integers which are not necessarily ordered:

```
const  m = ... ;    { length of array A }
       n = ... ;    { length of array B }

var  A: array[1 .. m] of integer;
     B: array[1 .. n] of integer;
```

A *duplicate* is an integer that is contained in both A and B. *Problem:* How many duplicates are there in A and B?

❖ Determine the time complexity of the brute-force algorithm that compares each integer contained in one array to all integers in the other array.

❖ Write a more efficient

```
function duplicates: integer;
```

Your solution may rearrange the integers in the arrays.

❖ What is the worst-case time complexity of your improved algorithm?

Part-V

Data Structures

The tools of bookkeeping

When thinking of algorithms we emphasize a dynamic sequence of actions: "Take this and do that, then that, then ... ." in human experience, "take" is usually a straightforward operation, whereas "do" means Work. in programming, on the other hand, there are lots of interesting examples where "do" is nothing more complex than incrementing a counter or setting a bit; but "take" triggers a long, sophisticated search. Why do we need fancy data structures at all? Why can't we just spread out the data on a desk top? Everyday experience does not prepare us to appreciate the importance of data structure—it takes programming experience to see that algorithms are nothing without data structures. The algorithms presented so far were carefully chosen to require only the simplest of data structures: static arrays. The geometric algorithms of Part VI, on the other hand, and lots of other useful algorithms, depend on sophisticated data structures for their efficiency.

The key insight in understanding data structures is the recognition that an algorithm in execution is, at all times, in some state, chosen from a potentially huge state space. The state records such vital information as what steps have already been taken with what results, and what remains to be done. Data structures are the bookkeepers that record all this state information in a tidy manner so that any part can be accessed and updated efficiently. The remarkable fact is that there are a relatively small number of standard data structures that turn out to be useful in the most varied types of algorithms and problems, and constitute essential knowledge for any programmer.

The literature on data structures. Whereas one can present some algorithms without emphasizing data structures, as we did in Part III, it appears pointless to discuss data structures without some of the typical algorithms that use them; at the very least, access and update algorithms form a necessary part of any data structure. Accordingly, a new data structure is typically published in the context of a particular new algorithm. Only later, as one notices its general applicability, it may find its way into textbooks. The data structures that have become standard today can be found in many books, such as [AHU 83], [CLR 90], [GB 91], [HS 82], [Knu 73a], [Knu 73b], [Meh 84a], [Meh 84c], [RND 77], [Sam 90a], [Sam 90b], [Tar 83], and [Wir 86].

18

What is a Data Structure?

Data structures old and new

The discipline of data structures, as a systematic body of knowledge, is truly a creation of computer science. The question of how best to organize data was a lot simpler to answer in the days before the existence of computers: the organization had to be simple, because there was no automatic device that could have processed an elaborate data structure, and there is no human being with enough patience to do it. Consider two examples.

- Manual files and catalogs, as used in business offices and libraries, exhibit several distinct organizing principles, such as sequential and hierarchical order and cross-references. From today's point of view, however, manual files are not well-defined data structures. For good reasons, people did not rigorously define those aspects that we consider essential when characterizing a data structure: what constraints are imposed on the data, both on the structure and its content; what operations the data structure must support; what constraints these operations must satisfy. As a consequence, searching and updating a manual file is not typically a process that can be automated: it requires common sense, and perhaps even expert training, as is the case for a library catalog.
- In manual computing (with pencil and paper or a nonprogrammable calculator) the *algorithm* is the focus of attention, not the data structure. Most frequently, the person computing writes data (input, intermediate results, output) in any convenient place within his field of vision, hoping to find them again when he needs them. Occasionally, to facilitate highly repetitive computations (such as income tax declarations), someone designs a form to prompt the user, one operation at a time, to write each data item into a specific field. Such a form specifies both an algorithm and a data structure with considerable formality. Compared to the general-purpose data structures we study in this chapter, however, such forms are highly special purpose.

Edge-notched cards are perhaps the most sophisticated data structures ever designed for manual use. Let us illustrate them with the example of a database of English words organized so as to help in solving crossword puzzles. We write one word per card and index it according to which vowels it contains and which ones it does not contain. Across the top row of the card we punch 10 holes labeled A, E, I, O, U, ~A, ~E, ~I, ~O, ~U. When a word, say ABACA, exhibits a given vowel, such as A, we cut a notch above the hole for A; when it does not, such as E, we cut a notch above the hole for ~E (pronounced "not E"). Exhibit 18.1 shows the encoding of the words BEAUTIFUL, EXETER, OMAHA, OMEGA. For example, we search for words that contain at least one E, but no U, by sticking two needles through the pack of cards at the holes E and ~U. EXETER and OMEGA will drop out. in principle it is easy to make this sample database more powerful by including additional attributes, such as "A occurs exactly once", "A occurs exactly twice", "A occurs as the first letter in the word", and so on. in practice, a few dozen attributes and thousands of cards will stretch this mechanical implementation of a multikey data structure to its limits of feasibility.

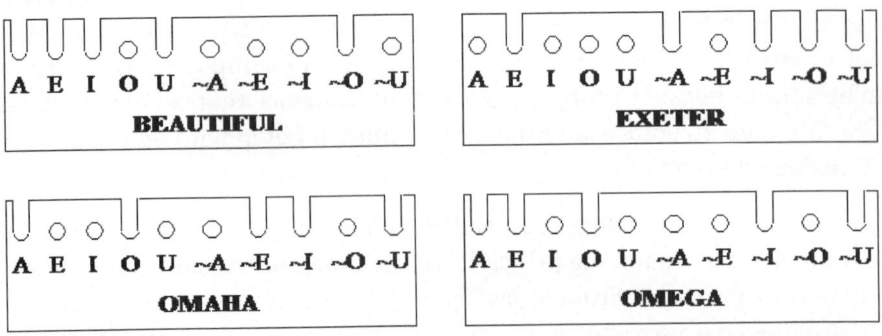

Exhibit 18.1: Encoding of different words in edge-notched cards.

In contrast to data structures suitable for manual processing, those developed for automatic data processing can be complex. Complexity is not a goal in itself, of course, but it may be an unavoidable consequence of the search for efficiency. Efficiency, as measured by processing time and memory space required, is the primary concern of the discipline of data structures. Other criteria, such as simplicity of the code, play a role, but the first question to be asked when evaluating a data structure that supports a specified set of operations is typically: How much time and space does it require?

In contrast to the typical situation of manual computing (consideration of the algorithm comes first, data gets organized only as needed), programmed computing typically proceeds in the opposite direction: First we define the organization of the data rigorously, and from this the structure of the algorithm follows. Thus algorithm design is often driven by data structure design.

The range of data structures studied

We present generally useful data structures along with the corresponding query, update, and maintenance algorithms; and we develop concepts and techniques designed to organize a vast body of knowledge into a coherent whole. Let us elaborate on both of these goals.

"Generally useful" refers to data structures that occur naturally in many applications. They are relatively simple from the point of view of the operations they support—tables and queues of various types are typical examples. These basic data structures are the building blocks from which an applications programmer may construct more elaborate structures tailored to her particular application. Although our collection of specific data structures is rather small, it covers the great majority of techniques an applications programmer is likely to need.

We develop a unified scheme for understanding many data structures as special cases of general concepts. This includes:

- The separation of abstract data types, which specify only functional properties, from data structures, which also involve aspects of implementation
- The classification of all data structures into three major types: implicit data structures, lists, and address computation
- A rough assessment of the performance of data structures based on the asymptotic analysis of time and memory requirements

The simplest and most common assumption about the elements to be stored in a data structure is that they belong to a domain on which a total order ≤ is defined. *Examples:* integers ordered by magnitude, a character set with its alphabetic order, character strings of bounded length ordered lexicographically. We assume that each element in a domain requires as much storage as any other element in that domain; in other words, that a data structure manages memory fragments of fixed size. Data objects of greatly variable size or length, such as fragments of text, are typically not considered to be "elements"; instead, they are broken into constituent pieces of fixed size, each of which becomes an element of the data structure.

The elements stored in a data structure are often processed according to the order ≤ defined on their domain. The topic of *sorting*, which we surveyed in "Sorting and its complexity", is closely related to the study of data structures: indeed, several sorting algorithms appear "for free" in "List structures", because every structure that implements the abstract data type *dictionary* leads to a sorting algorithm by successive insertion of elements, followed by a traversal.

Performance criteria and measures

The design of data structures is dominated by considerations of efficiency, specifically with respect to time and memory. But efficiency is a multifaceted quality not easily defined and measured. As a scientific discipline, the study of data structures is not directly concerned with the number of microseconds, machine cycles, or bytes required by a specific program processing a given set of data on a particular system. it is concerned with general statements from which an expert practitioner can predict concrete outcomes for a specific processing task. Thus, measuring run times and memory usage is not the typical way to evaluate data structures. We need concepts and notations for expressing the performance of an algorithm independently of machine speed, memory size, programming language, and operating system, and a host of other details that vary from run to run.

The solution to this problem emerged over the past two decades as the discipline of computational complexity was developed. in this theory, algorithms are "executed" on some "mathematical machine", carefully designed to be as simple as possible to reflect the bare essentials of a problem. The machine makes available certain *primitive operations*, and we measure "time" by counting how many of those are executed. For a given algorithm and all the data sets it accepts as input, we analyze the number of primitive operations executed as a function of the size of the data. We are often interested in the *worst case*, that is, a data set of given size that causes the algorithm to run as long as possible, and the *average case*, the run time averaged over all data sets of a given size.

Among the many different mathematical machines that have been defined in the theory of computation, data structures are evaluated almost exclusively with respect to a theoretical *random access mach*ine (RAM). A RAM is essentially a memory with as many locations as needed, each of which can hold a data element, such as an integer, or a real number; and a processing unit that can read from any one or two locations, operate on their content, and write the result back into a third location, all in one time unit. This model is rather close to actual sequential computers, except that it incorporates no bounds on the memory size—either in terms of the number of locations or the size of the content of this location. it implies, for example, that a multiplication of two very large numbers requires no more time than $2 \cdot 3$ does. This assumption is unrealistic for certain problems, but is an excellent one for most program runs that fit in central memory and do not require variable-precision arithmetic or variable- length data elements. The point is that the programmer has to understand the model and its assumptions, and bears responsibility for applying it judiciously.

In this model, time and memory requirements are expressed as functions of input data size, and thus comparing the performance of two data structures is reduced to comparing functions. *Asymptotics* has proven to be just the right tool for this comparison: sharp enough to distinguish different growth rates, blunt enough to ignore constant factors that differ from machine to machine.

As an example of the concise descriptions made possible by asymptotic operation counts, the following table evaluates several implementations for the abstract data type 'dictionary'. The four operations 'find', 'insert', 'delete', and 'next' (with respect to the order ≤) exhibit different asymptotic time requirements for the different implementations. The student should be able to explain and derive this table after studying this part of the book.

	Ordered array	Linear list	Balanced tree	Hash table
find	O(log n)	O(n)	O(log n)	O(1)[a]
next	O(1)	O(1)	O(log n)	O(n)
insert	O(n)	O(n)	O(log n)	O(1)[a]
delete	O(n)	O(n)	O(log n)	O(1)[b]

[a]On the average, but not necessarily in the worst case

[b]Deletions are possible but may degrade performance

Exercise

❖ Describe the manual data structures that have been developed to organize libraries (e.g. catalogs that allow users to get access to the literature in their field of interest, or circulation records, which keep track of who has borrowed what book). Give examples of queries that can be answered by these data structures.

19

Abstract Data Types

Concepts: What and why?

A data structure organizes the data to be processed in such a way that the relations among the data elements are reflected and the operations to be performed on the data are supported. *How* these goals can be achieved efficiently is the central issue in data structures and a major concern of this book. In this chapter, however, we ask not how but *what?* in particular, we ask: what is the exact functional behavior a data structure must exhibit to be called a stack, a queue, or a dictionary or table?

There are several reasons for seeking a formal functional specification for common data structures. The primary motivation is increased generality through abstraction; specifically, to separate input/output behavior from implementation, so that the implementation can be changed without affecting any program that uses a particular data type. This goal led to the earlier introduction of the concept of *type* in programming languages: the type *real* is implemented differently on different machines, but usually a program using reals does not require modification when run on another machine. A secondary motivation is the ability to prove general theorems about all data structures that exhibit certain properties, thus avoiding the need to verify the theorem in each instance. This goal is akin to the one that sparked the development of algebra: from the axioms that define a field, we prove theorems that hold equally true for real or complex numbers as well as quaternions.

The primary motivation can be further explained by calling on an analogy between data and programs. All programming languages support the concept of *procedural abstraction*: operations or algorithms are isolated in procedures, thus making it easy to replace or change them without affecting other parts of the program. Other program parts do not know how a certain operation is realized; they know only how to call the corresponding procedure and what effect the procedure call will have. Modern programming languages increasingly support the analogous concept of *data abstraction* or *data encapsulation*: the organization of data is encapsulated (e.g. in a module or a package) so that it is possible to change the data structure without having to change the whole program.

The secondary motivation for formal specification of data types remains an unrealized goal: although abstract data types are an active topic for theoretical research, it is difficult today to make the case that any theorem of use to programmers has been proved.

An *abstract data type* consists of a domain from which the data elements are drawn, and a set of operations. The specification of an abstract data type must identify the domain and define each of the operations. Identifying and describing the domain is generally straightforward. The definition of each operation consists of a syntactic and a semantic part. The *syntactic part*, which corresponds to a procedure heading, specifies the operation's name and the type of each operand. We present the syntax of operations in mathematical function notation, specifying its domain and range. The *semantic part* attaches a meaning to each operation: what values it produces or what effect it has on its environment. We specify the semantics of abstract data types algebraically by axioms from which other properties may be deduced. This formal approach has the advantage that the operations are defined rigorously for any domain with the required properties. A formal description, however, does not always appeal to intuition, and often forces us to specify details that we might prefer to ignore. When every detail matters, on the other hand, a formal specification is superior to a precise specification in natural language; the latter tends to become cumbersome and difficult to understand, as it often takes many words to avoid ambiguity.

In this chapter we consider the abstract data types: stack, first-in-first-out queue, priority queue, and dictionary. For each of these data types, there is an ideal, unbounded version, and several versions that reflect the realities of finite machines. From a theoretical point of view we only need the ideal data types, but from a practical point of view, that doesn't tell the whole story: in order to capture the different properties a programmer intuitively associates with the vague concept "stack", for example, we are forced into specifying different types of stacks. in addition to the ideal *unbounded stack*, we specify a *fixed-length stack* which mirrors the behavior of an array implementation, and a *variable-length stack* which mirrors the behavior of a list implementation. Similar distinctions apply to the other data types, but we only specify their unbounded versions.

Let X denote the domain from which the data elements are drawn. Stacks and fifo queues make no assumptions about X; priority queues and dictionaries require that a total order $\leq$ be defined on X. Let $X*$denote the set of all finite sequences over X.

Stack

A *stack* is also called a *last*-in-*first-out queue*, or *lifo queue*. A brief informal description of the abstract data type stack (more specifically, unbounded stack, in contrast to the versions introduced later) might merely state that the following operations are defined on it:

create Create a new, empty stack.

empty Return true if the stack is empty.

push insert a new element.

top Return the element most recently inserted, if the stack is not empty.

pop Remove the element most recently inserted, if the stack is not empty.

Exhibit 19.1 helps to clarify the meaning of these words.

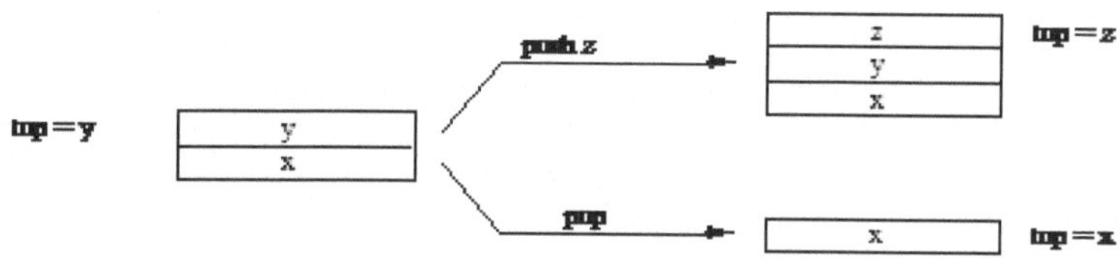

Exhibit 19.1: Elements are inserted at and removed from the top of the stack.

A definition that uses conventional mathematical notation to capture the intention of the description above might define the operations by explicitly showing their effect on the contents of a stack. Let S = X∗ be the set of

possible states of a stack, let s = x_1 x_2 ... x_k ∈ S be an arbitrary stack state with k elements, and let λdenote the empty state of the stack, corresponding to the null string ∈ X*. Let 'cat' denote string concatenation. Define the functions

```
create: → S
empty: S → {true, false}
push: S × X → S
top: S - {λ} → X
pop: S - {λ} → S
```

as follows:

```
∀s ∈ S,∀x, y ∈ X:
create - λ
empty(λ) = true
s ≠ λ ⇒ empty(s) = false
push(s, y) = s cat y = x1 x2 ... xk y
s ≠ λ top(s) = xk
s ≠ pop(s) = x1 x2 ... xk-1
```

This definition refers explicitly to the contents of the stack. If we prefer to hide the contents and refer only to operations and their results, we are led to another style of formal definition of abstract data types that expresses the semantics of the operations by relating them to each other rather than to the explicitly listed contents of a data structure. This is the commonly used approach to define abstract data types, and we follow it for the rest of this chapter.

Let S be a set and $s_0 \in S$ a distinguished state. s_0 denotes the empty stack, and S is the set of stack states that can be obtained from the empty stack by performing finite sequences of 'push' and 'pop' operations. The following functions represent stack operations:

```
create: → S
empty: S → {true, false}
push: S  X → S
top: S - {s0} → X
pop: S - {s0} → S
```

```
The semantics of the stack operations is specified by the following
axioms:
 ∀s ∈ S, ∀x ∈ X:
(1)  create = s0
(2)  empty(s0) = true
(3)  empty(push(s, x)) = false
(4)  top(push(s, x)) = x
(5)  pop(push(s, x)) = s

These axioms can be described in natural language as follows:
(1)  'create' produces a stack in the distinguished state.
(2)  The distinguished state is empty.
(3)  A stack is not empty after an element has been inserted.
(4)  The element most recently inserted is on top of the stack.
(5)  'pop' is the inverse of 'push'.
```

Notice that 'create' plays a different role from the other stack operations: it is merely a mechanism for causing a stack to come into existence, and could have been omitted by postulating the existence of a stack in st ate s0. in any implementation, however, there is always some code that corresponds to 'create'. *Technical note:* we could identify 'create' with s0, but we choose to make a distinction between the act of creating a new empty stack and the empty state that results from this creation; the latter may recur during normal operation of the stack.

Reduced sequences

Any $s \in S$ is obtained from the empty stack s0 by performing a finite sequence of 'push' and 'pop' operations. By axiom (5) this sequence can be reduced to a sequence that transforms s 0 into s and consists of 'push' operations only.

Example

```
s  =  pop(push(pop(push(push(s0, x), y)), z))
   =  pop(push(push(s0, x), z))
   =  push(s0, x)
```

An implementation of a stack may provide the following procedures:

```
procedure create(var s: stack);
function empty(s: stack): boolean;
procedure push(var s: stack; x: elt);
function top(s: stack): elt;
procedure pop(var s: stack);
```

Any program that uses this data type is restricted to calling these five procedures for creating and operating on stacks; it is not allowed to use information about the underlying implementation. The procedures may only be called within the constraints of the specification; for example, 'top' and 'pop' may be called only if the stack is not empty:

```
if  not empty(s)  then  pop(s);
```

The specification above assumes that a stack can grow without a bound; it defines an abstract data type called *unbounded stack*. However, any implementation imposes some bound on the size (depth) of a stack: the size of the underlying array in an array imple→d reflect such→ limitations. The following *fixed-length stack* describes an implementation as an array of fixed size m, which limits the maximal stack depth.

Fixed-length stack

```
create:→ S
empty: S → {true, false}
full: S → {true, false}
push: {s ∈ S: not full(s)} × X → S
top: S - {s0} → X
pop: S - {s0} → S
```

To specify the behavior of the function 'full' we need an internal function depth: S → {0, 1, 2, ... , m} that measures the stack depth, that is, the number of elements currently in the stack. The function 'depth' interacts with the other functions in the following axioms, which specify the stack semantics:

```
∀s ∈ S, ∀x ⊂ X:
create = s0
empty(s) = true
not full(s) ⇒ empty(push(s, x)) = false
depth(s0) = 0
```

```
not empty(s) ⇒ depth(pop(s)) = depth(s) - 1
not full(s) ⇒ depth(push(s, x)) = depth(s) + 1
full(s) = (depth(s) = m)
not full(s) ⇒
   top(push(s, x)) = x
   pop(push(s, x)) = s
```

Variable-length stack

A stack implemented as a list may overflow at unpredictable moments depending on the contents of the entire memory, not just of the stack. We specify this behavior by postulating a function 'space-available'. it has no domain and thus acts as an oracle that chooses its value independently of the state of the stack (if we gave 'space-available' a domain, this would have to be the set of states of the entire memory).

```
create: → S
empty: S → {true, false}
space-available: →{true, false}
push: S × X → S
top: S - {s0} → X
pop: S - {s0} → S

∀s ∈ S, ∀x ∈ X:
create = s0
empty(s0) = true
space-available ⇒
   empty(push(s, x)) = false
   top(push(s, x)) = x
   pop(push(s, x)) = s
```

Implementation

We have seen that abstract data types cannot capture our intuitive, vague concept of a stack in one single model. The rigor enforced by the formal definition makes us aware that there are different types of stacks with different behavior (quite apart from the issue of the domain type X, which specifies what type of elements are to be stored). This clarity is an advantage whenever we attempt to process abstract data types automatically; it may be a disadvantage for human communication, because a rigorous definition may force us to (over)specify details.

The different types of stacks that we have introduced are directly related to different styles of implementation.

The fixed-length stack, for example, describes the following implementation:

```
const  m = … ;   { maximum length of a stack }
type elt = … ;
    stack =record
          a: array[1 .. m] of elt;
          d: 0 .. m;   { current depth of stack }
       end;

procedure create(var s: stack);
begin  s.d := 0  end;

function empty(s: stack): boolean;
begin  return(s.d = 0)  end;

function full(s: stack): boolean;
begin  return(s.d = m)  end;

procedure push(var s: stack; x: elt);   { not to be called if the stack
is full }
  begin  s.d := s.d + 1;  s.a[s.d] := x  end;

function top(s: stack): elt;   { not to be called if the stack is
empty }
  begin  return(s.a[s.d])  end;

procedure pop(var s: stack);   { not to be called if the stack is
empty }
  begin  s.d := s.d - 1  end;
```

Since the function 'depth' is not exported (i.e. not made available to the user of this data type), it need not be provided as a procedure. Instead, we have implemented it as a variable d which also serves as a stack pointer.

Our implementation assumes that the user checks that the stack is not full before calling 'push', and that it is not empty before calling 'top' or 'pop'. We could, of course, write the procedures 'push', 'top', and 'pop' so as to "protect themselves" against illegal calls on a full or an empty stack simply by returning an error message to the calling program. This requires adding a further argument to each of these three procedures and leads to yet other types of stacks which are formally different abstract data types from the ones we have discussed.

First-in-first-out queue

The following operations (Exhibit 19.2) are defined for the abstract data type *fifo queue* (first-in-first-out queue):

```
empty Return true if the queue is empty.
enqueue Insert a new element at the tail end of the queue.
front Return the front element of the queue.
dequeue Remove the front element.
```

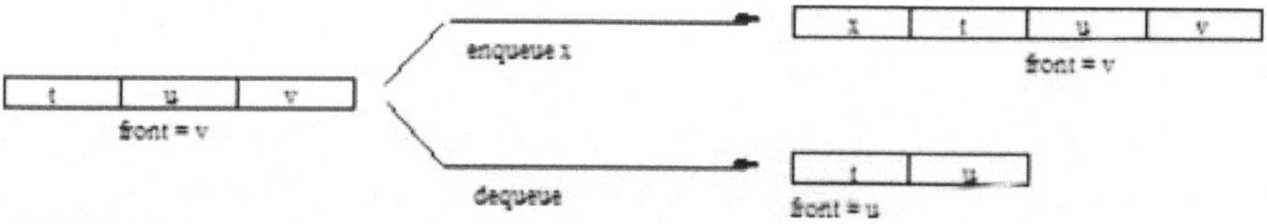

Exhibit 19.2: Elements are inserted at the tail and removed from the head of the fifo queue.

Let F be the set of queue states that can be obtained from the empty queue by performing finite sequences of 'enqueue' and 'dequeue' operations. f0 ∈ F denotes the empty queue. The following functions represent fifo queue operations:

```
create: → F
empty: F → {true, false}
enqueue: F × X → F
front: F - {f0} → X
dequeue: F - {f0} → F
```

```
The semantics of the fifo queue operations is specified by the
following axioms:
  ∀f ∈ F, ∀x ∈ X:
(1)  create = f0
(2)  empty(f0) = true
(3)  empty(enqueue(f, x)) = false
(4)  front(enqueue(f0, x)) = x
(5)  not empty(f) ⇒ front(enqueue(f, x)) = front(f)
(6)  dequeue(enqueue(f0, x)) = f0
(7)  not empty(f) ⇒ dequeue(enqueue(f, x)) = enqueue(dequeue(f), x)
```

Any f ∈ F is obtained from the empty fifo queue f0 by performing a finite sequence of 'enqueue' and 'dequeue' operations. By axioms (6) and (7) this sequence can be reduced to a sequence consisting of 'enqueue' operations only which also transforms f0 into f.

Example

```
f = dequeue(enqueue(dequeue(enqueue(enqueue(f0, x), y)), z))
  = dequeue(enqueue(enqueue(dequeue(enqueue(f0, x)), y), z))
  = dequeue(enqueue(enqueue(f0, y), z))
  = enqueue(dequeue(enqueue(f0, y)), z)
  = enqueue(f0, z)
```

An implementation of a fifo queue may provide the following procedures:

```
procedure create(var f: fifoqueue);
function empty(f: fifoqueue): boolean;
procedure enqueue(var f: fifoqueue; x: elt);
function front(f: fifoqueue): elt;
procedure dequeue(var f: fifoqueue);
```

Priority queue

A priority queue orders the elements according to their *value* rather than their arrival time. Thus we assume that a total order ≤ is defined on the domain X. in the following examples, X is the set of integers; a small integer means high priority. The following operations (Exhibit 19.3) are defined for the abstract data type *priority queue*:

```
- empty  Return true if the queue is empty.
- insert Insert a new element into the queue.
- min  Return the element of highest priority contained in the queue.
- delete Remove the element of highest priority from the queue.
```

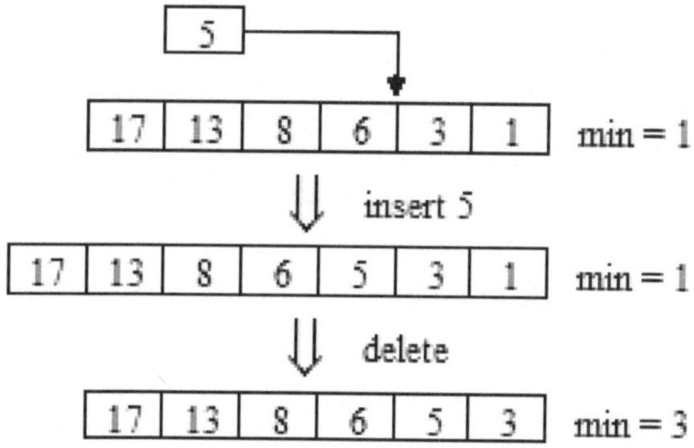

Exhibit 19.3: An element's priority determines its position in a priority queue.

Let P be the set of priority queue states that can be obtained from the empty queue by performing finite sequences of 'insert' and 'delete' operations. The empty priority queue is denoted by p0 ∈ P. The following functions represent priority queue operations:

```
create: → P
empty: P → {true, false}
insert: P × X → P
min: P - {p0} → X
delete: P - {p0} → P
```

The semantics of the priority queue operations is specified by the following axioms. For x, y ∈ X, the function Min(x, y) returns the smaller of the two values.

```
∀p ∈ P, ∀x ∈ X:
(1)  create = p0
(2)  empty(p0) = true
(3)  empty(insert(p, x)) = false
(4)  min(insert(p0, x)) = x
(5)  not empty(p) ⇒ min(insert(p, x)) = MIN(x, min(p))
(6)  delete(insert(p0, x)) = p0
(7) not empty(p)⇒
```

$$\text{delete (insert(p,x))} = \begin{cases} \text{pifx} \leq \text{min(p)} \\ \text{insert(delete(p), x)else} \end{cases}$$

Any p ∈ P is obtained from the empty queue p0 by a finite sequence of 'insert' and 'delete' operations. By axioms and (7) any such sequence can be reduced to a shorter one that also transforms p 0 into p and consists of 'insert' operations only.

Example

```
Assume that x < z, y < z.
  p = delete(insert(delete(insert(insert(p0, x), z)), y))
    = delete(insert(insert(delete(insert(p0, x)), z), y))
    = delete(insert(insert(p0, z), y))
    = insert(p0, z)
```

An implementation of a priority queue may provide the following procedures:

```
procedure create(var p: priorityqueue);
function empty(p: priorityqueue): boolean;
procedure insert(var p: priorityqueue; x: elt);
function min(p: priorityqueue): elt;
procedure delete(var p: priorityqueue);
```

Dictionary

Whereas stacks and fifo queues are designed to retrieve and process elements depending on their order of arrival, a dictionary (or table) is designed to process elements exclusively by their value (name). A priority queue is a hybrid: insertion is done according to value, as in a dictionary, and deletion according to position, as in a fifo queue.

The simplest type of *dictionary* supports the following operations:
- member Return true if a given element is contained in the dictionary.
- insert Insert a new element into the dictionary.
- delete Remove a given element from the dictionary.

Let D be the set of dictionary states that can be obtained from the empty dictionary by performing finite sequences of 'insert' and 'delete' operations. d0 ∈ D denotes the empty dictionary. Then the operations can be represented by functions as follows:

```
create: → D
insert: D × X → D
member: D × X → {true, false}
delete: D × X → D
```

The semantics of the dictionary operations is specified by the following axioms:

```
∀d ∈ D,∀x, y ∈ X:
(1)  create = d0
(2)  member(d0, x) = false
(3)  member(insert(d, x), x) = true
(4)  x ≠ y ⇒ member(insert(d, y), x) = member(d, x)
(5)  delete(d0, x) = d0
(6)  delete(insert(d, x), x) = delete(d, x)
(7)  x ≠ y ⇒ delete(insert(d, x), y) = insert(delete(d, y), x)
```

Any d ∈ D is obtained from the empty dictionary d0 by a finite sequence of 'insert' and 'delete' operations. By axioms (6) and (7) any such sequence can be reduced to a shorter one that also transforms d0 into d and consists of 'insert' operations only.

Example

```
d  = delete(insert(insert(insert(d0, x), y), z), y)
   = insert(delete(insert(insert(d0, x), y), y), z)
   = insert(delete(insert(d0, x), y), z)
   = insert(insert(delete(d0, y), x), z)
   = insert(insert(d0, x), z)
```

This specification allows duplicates to be inserted. However, axiom (6) guarantees that all duplicates are removed if a delete operation is performed. To prevent duplicates, the following axiom is added to the specification above:

```
(8)  member(d, x)  ⇒ insert(d, x)  = d
In this case axiom (6) can be weakened to
(6') not member(d, x)  ⇒ delete(insert(d, x), x)  = d
```

An implementation of a dictionary may provide the following procedures:

```
procedure create(var d: dictionary);
function member(d: dictionary; x: elt): boolean;
procedure insert(var d: dictionary; x: elt);
procedure delete(var d: dictionary; x: elt);
```

In actual programming practice, a dictionary usually supports the additional operations 'find', 'predecessor', and 'successor'. 'find' is similar to 'member' but in addition to a true/false answer, provides a pointer to the element found. Both 'predecessor' and 'successor' take a pointer to an element e as an argument, and return a pointer to the element in the dictionary that immediately precedes or follows e, according to the order ≤. Repeated call of 'successor' thus processes the dictionary in sequential order.

Exercise: extending the abstract data type 'dictionary'

We have defined a dictionary as supporting the three operations 'member', 'insert' and 'delete'. But a dictionary, or table, usually supports additional operations based on a total ordering ≤ defined on its domain X. Let us add two operations that take an argument x ∈ X and deliver its two neighboring elements in the table:

```
succ(x) Return the successor of x in the table.
pred(x) Return the predecessor of x in the table.
```

The successor of x is defined as the smallest of all the elements in the table which are larger than x, or as +∞ if none exists. The predecessor is defined symmetrically: the largest of all the elements in the table that are smaller than x, or −∞. Present a formal specification to describe the behavior of the table.

Solution

Let T be the set of states of the table, and t0 a special state that denotes the empty table. The functions and axioms are as follows:

```
member: T × X → {true,false}
insert: T × X → T
delete: T × X → T
succ: T × X → X ∪ {+∞}
pred: T × X → X ∪ {-∞}
```

```
∀t ∈ T,∀x, y ∈ X:
member(t0, x) = false
member(insert(t, x), x) = true
x ≠ y ⇒ member(insert(t, y), x) = member(t, x)
delete(t0, x) = t0
delete(insert(t, x), x) = delete(t, x)
x ≠ y ⇒ delete(insert(t, x), y) = insert(delete(t, y), x)
```

```
-∞ < x < +∞
pred(t, x) < x < succ(t, x)
succ(t, x) ≠ +∞ ⇒ member(t, succ(t, x)) = true
pred(t, x) ≠ -∞ ⇒ member(t, pred(t, x)) = true
x < y, member(t, y), y ≠ succ(t, x) ⇒ succ(t, x) < y
x > y, member(t, y), y ≠ pred(t, x) ⇒ y < pred(t, x)
```

Exercise: the abstract data type 'string'

We define the following operations for the abstract data type *string*:

```
- empty  Return true if the string is empty.
- append Append a new element to the tail of the string.
- head Return the head element of the string.
- tail Remove the head element of the given string.
- length Return the length of the string.
- find Return the index of the first occurrence of a value within the
string.
```

Let X = {a, b, ... , z}, and S be the set of string states that can be obtained from the empty string by performing a finite number of 'append' and 'tail' operations. s0 ∈ S denotes the empty string. The operations can be represented by functions as follows:

```
empty: S → {true, false}
append: S × X → S
head: S - {s0} → X
tail: S - {s0} → S
length: S → {0, 1, 2, ... }
find: S × X → {0, 1, 2, ... }
```

Examples:

```
empty('abc') = false;  append('abc', 'd') = 'abcd';  head('abcd') =
'a';
```

```
tail('abcd') = 'bcd';  length('abcd') = 4;  find('abcd', 'b') = 2.
```

Give the axioms that specify the semantics of the abstract data type 'string'.

The function hchop: S × X → S returns the substring of a string s beginning with the first occurrence of a given value. Similarly, tchop: S × X → S returns the substring of s beginning with head(s) and ending with the last occurrence of a given value. Specify the behavior of these operations by additional axioms. *Examples:*

```
hchop('abcdabc','c')='cdabc'
tchop('abcdabc', 'b') = 'abcdab'
```

The function cat: S × S → S returns the concatenation of two sequences. Specify the behavior of 'cat' by additional axioms. *Example:*

```
cat('abcd', 'efg') = 'abcdefg'
```

The function reverse: S → S returns the given sequence in reverse order. Specify the behavior of reverse by additional axioms. *Example:*

```
reverse('abcd') = 'dcba'
```

Solution

❖ Axioms for the six 'string' operations:

```
∀s ∈ S, ∀x, y ∈ X:
empty(s0) = true
empty(append(s, x)) = false
head(append(s0, x)) = x
not empty(s) ⇒ head(s) = head(append(s, x))
tail(append(s0, x)) = s0
not empty(s) ⇒ tail(append(s, x)) = append(tail(s), x)
length(s0) = 0
length(append(s, x)) = length(s) + 1
find(s0, x) = 0
x ≠ y, find(s, x) = 0 ⇒ find(append(s, y), x) = 0
find(s, x) = 0 ⇒ find(append(s, x), x) = length(s) + 1
find(s, x) = d > 0 ⇒ find(append(s, y), x) = d
```

❖ Axioms for 'hchop' and 'tchop':

```
∀s ∈ S, ∀x, y ∈ X:
hchop(s0, x) = s0
not empty(s), head(s) = x ⇒ hchop(s, x) = s
not empty(s), head(s) ≠ x ⇒ hchop(s, x) = hchop(tail(s), x)
tchop(s0, x) = s0
tchop(append(s, x), x) = append(s, x)
x ≠ y ⇒ tchop(append(s, y), x) = tchop(s, x)
```

- Axioms for 'cat':

```
∀s, s' ∈ S:
cat(s, s0) = s
not empty(s') ⇒ cat(s, s') = cat(append(s, head(s')), tail(s'))
```

- Axioms for 'reverse':

```
∀s ∈ S:
reverse(s0) = s0
s ≠ s0 ⇒ reverse(s) = append(reverse(tail(s)), head(s))
```

Exercises

- Implement two stacks iv onε array a[1 .. m] in such a way that neither stack overflows unless the total number of elements in both stacks together is m. The operations 'push', 'top', and 'pop' should run in O(1) time.
- A double-ended queue (deque) can grow and shrink at both ends, left and right, using the procedures 'enqueue-left', 'dequeue-left', 'enqueue-right', and 'dequeue-right'. Present a formal specification to describe the behavior of the abstract data type deque.
- Extend the abstract data type priority queue by the operation next(x), which returns the element in the priority queue having the next lower priority than x.

20

Implicit Data Structures

What is an implicit data structure?

An important aspect of the art of data structure design is the efficient representation of the structural relationships among the data elements to be stored. Data is usually modeled as a graph, with nodes corresponding to data elements and links (directed arcs, or bidirectional edges) corresponding to relationships. Relationships often serve a double purpose. Primarily, they define the semantics of the data and thus allow programs to interpret the data correctly. This aspect of relationships is highlighted in the database field: for example, in the entity-relationship model. Secondarily, relationships provide a means of accessing data, by starting at some element and following an *access path* that leads to other elements of interest. in studying data structures we are mainly concerned with the use of relationships for access to data.

When the structure of the data is irregular, or when the structure is highly dynamic (extensively modified at run time), there is no practical alternative to representing the relationships explicitly. This is the domain of list structures, presented in the chapter on "List structures". When the structure of the data is static and obeys a regular pattern, on the other hand, there are alternatives that compress the structural information. We can often replace many explicit links by a few formulas that tell us where to find the "neighboring" elements. When this approach Works, it saves memory space and often leads to faster programs.

We use the term *implic*it to denote data structures in which the relationships among data elements are given implicitly by formulas and declarations in the program; no additional space is needed for these relationships in the data storage. The best known example is the array. If one looks at the area in which an array is stored, it is impossible to derive, from its contents, any relationships among the elements without the information that the elements belong to an array of a given type.

Data structures always go hand in hand with the corresponding procedures for accessing and operating on the data. This is particularly true for implicit data structures: They simply do not exist independent of their accessing procedures. Separated from its code, an implicit data structure represents at best an unordered set of data. With the right code, it exhibits a rich structure, as is beautifully illustrated by the *heap* at the end of this chapter.

Array storage

A two-dimensional array declared as

```
var  A: array[1 .. m, 1 .. n] of elt;
```

is usually written in a rectangular shape:

A[1, 1]	A[1, 2]	...	A[1, n]
A[2, 1]	A[2, 2]	...	A[2, n]
...	...	...	...
A[m, 1]	A[m, 2]	...	A[m, n]

But it is stored in a linearly addressed memory, typically row by row (as shown below) or column by column (as in Fortran) in consecutive storage cells, starting at base address b. If an element fits into one cell, we have

	address
A[1, 1]	b
A[1, 2]	b + 1
...	...
A[1, n]	b + n - 1
A[2, 1]	b + n
A[2, 2]	b + n + 1
...	...
A[2, n]	b + 2 · n - 1
...	...
A[m, n]	b + m · n - 1

If an element of type 'elt' occupies c storage cells, the address $\alpha(i, j)$ of A[i, j] is

$$\alpha(i,j) = b + c \cdot (n \cdot (i-1) + j - 1).$$

This linear formula generalizes to k-dimensional arrays declared as

```
var  A: array[1 .. m1, 1 .. m2, ... , 1 .. mk] of elt;
```

The address $\alpha(i_1, i_2, \ldots, i_k)$ of element $A[i_1, i_2, \ldots, i_k]$ is

$$\alpha(i_1, i_2, \ldots, i_k) = b + c \cdot (\quad (i_1 - 1) \cdot m_2 \cdot \ldots \cdot m_k \\ + (i_2 - 1) \cdot m_3 \cdot \ldots \cdot m_k \\ + \ldots \\ + (i_{k-1} - 1) \cdot m_k \\ + i_k - 1 \quad).$$

The point is that access to an element A[i, j, ...] invokes evaluation of a (linear) formula α(i, j, ...) that tells us where to find this element. A high-level programming language hides most of the details of address computation, except when we wish to take advantage of any special structure our matrices may have. The following types of *sparse matrices* occur frequently in numerical linear algebra.

Band matrices. An n × n matrix M is called a *band matrix of width 2 · b + 1* (b = 0, 1, ...) if $M_{i,j} = 0$ for all i and j with |i – j| > b. in other words, all nonzero elements are located on the main diagonal and in b adjacent minor diagonals on both sides of the main diagonal. If n is large and b is small, much space is saved by storing M in a two-dimensional array A with n · (2 · b + 1) cells rather than in an array with n_2 cells:

```
type  bandm = array[1 .. n, -b .. b] of elt;
var  A: bandm;
```

Each row A[i, ·] stores the nonzero elements of the corresponding row of M, namely the diagonal element M i,i, the b elements to the left of the diagonal and the b elements to the right of the diagonal

$$M_{i,i-b}, M_{i,i-b+1}, \ldots, M_{i,i-1}$$
$$M_{i,i+1}, M_{i,i+2}, \ldots, M_{i,i+b}.$$

The first and the last b rows of A contain empty cells corresponding to the triangles that stick out from M in Exhibit 20.1. The elements of M are stored in array A such that A[i, j] contains $M_{i,i+j}$ (1 ≤ i ≤ n, –b ≤ j ≤ b). A total of b · (b + 1) cells in the upper left and lower right of A remain unused. it is not worth saving an additional b · (b + 1) cells by packing the band matrix M into an array of minimal size, as the mapping becomes irregular and the formula for calculating the indices of $M_{i,j}$ becomes much more complicated.

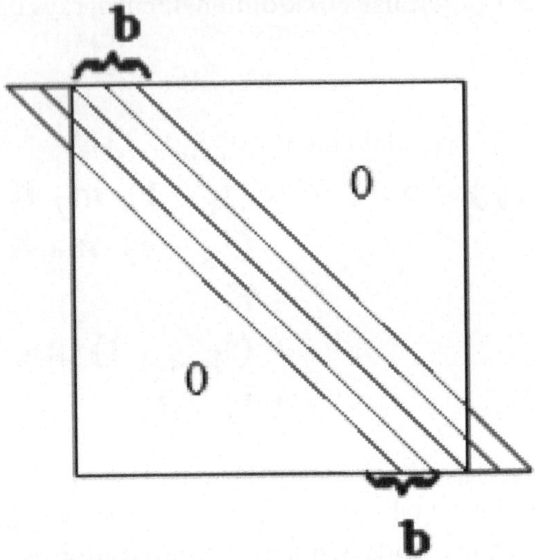

Exhibit 20.1: Extending the diagonals with dummy elements gives the band matrix the shape of a rectangular array.

Exercise: band matrices

- Write a procedure add(p, q: bandm; var r: bandm);

  which adds two band matrices stored in p and q and stores the result in r.

- Write a procedure bmv(p: bandm; v: ... ; var w: ...);

  which multiplies a band matrix stored in p with a vector v of length n and stores the result in w.

Solution

```
(a)   procedure add(p, q: bandm; var r: bandm);
      var  i: 1 .. n;   j: -b .. b;
      begin
        for  i := 1  to  n  do
          for  j := -b  to  b  do
            r[i, j] := p[i, j] + q[i, j]
      end;

(b)    type  vector = array[1 .. n] of real;

      procedure bmv(p: bandm; v: vector; var w: vector);
      var  i: 1 .. n;   j: -b .. b;
      begin
        for  i := 1  to  n  do  begin
          w[i] := 0.0;
          for  j := -b  to  b  do
            if  (i + j ≥ 1) and (i + j ≤ n)  then  w[i] := w[i] + p[i, j]
    v[i + j]
        end
      end;
```

Sparse matrices. A matrix is called *sparse* if it consists mostly of zeros. We have seen that sparse matrices of regular shape can be compressed efficiently using address computation. Irregularly shaped sparse matrices, on the other hand, do not yield gracefully to compression into a smaller array in such a way that access can be based on address computation. instead, the nonzero elements may be stored in an unstructured set of records, where each record contains the pair ((i, j), A[i, j]) consisting of an index tuple (i, j) and the value A[i, j]. Any element that is absent from this set is assumed to be zero. As the position of a data element is stored explicitly as an index pair (i, j), this representation is not an implicit data structure. As a consequence, access to a random element of an irregularly shaped sparse matrix typically requires searching for it, and thus is likely to be slower than the direct access to an element of a matrix of regular shape stored in an implicit data structure.

Exercise: triangular matrices

Let A and B be lower-triangular n × n-matrices; that is, all elements above the diagonal are zero: $A_{i,j} = B_{i,j} = 0$ for $i < j$.

- Prove that the inverse (if it exists) and the matrix product of lower-triangular matrices are again lower-triangular.
- Devise a scheme for storing two lower-triangular matrices A and B in one array C of minimal size. Write a Pascal declaration for C and draw a picture of its contents.
- Write two functions

```
function A(i, j: 1 .. n): real;
function B(i, j: 1 .. n): real;
```

- that access C and return the corresponding matrix elements.
- Write a procedure that computes A := A · B in place: The entries of A in C are replaced by the entries of the product A · B. You may use a (small) constant number of additional variables, independent of the size of A and B.
- Same as (d), but using $A := A^{-1} \cdot B$.

Solution

- The inverse of an n × n-matrix exists iff the determinant of the matrix is non zero. Let A be a lower- triangular matrix for which the inverse matrix B exists, that is,

$$\det(A) = \prod_{i=1}^{n} A_{i,i} \neq 0 \quad \Rightarrow \quad \forall i, 1 \leq i \leq n: \; A_{i,i} \neq 0$$

and

$$\sum_{k=1}^{n} A_{i,k} \cdot B_{k,j} \; = \; \begin{cases} 1 & \text{if } i = j, \\ 0 & \text{if } i \neq j. \end{cases}$$

Let $1 \leq j \leq n$. Then

$$j > 1 \Rightarrow \sum_{k=1}^{n} A_{1,k} \cdot B_{k,j} = A_{1,1} \cdot B_{1,j} = 0 \qquad \Rightarrow B_{1,j} = 0,$$

$$j > 2 \Rightarrow \sum_{k=1}^{n} A_{2,k} \cdot B_{k,j} = A_{2,1} \cdot B_{1,j} + A_{2,2} \cdot B_{2,j} = 0 \Rightarrow B_{2,j} = 0,$$

K

$$j > i \Rightarrow \ldots \qquad \Rightarrow B_{i,j} = 0,$$

and therefore B is a lower-triangular matrix. Let A and B be lower-triangular, C := A · B:

$$C_{i,j} = \sum_{k=1}^{n} A_{i,k} \cdot B_{k,j} = \sum_{k=j}^{i} A_{i,k} \cdot B_{k,j} \cdot$$

If $i < j$, this sum is empty and therefore $C_{i,j} = 0$ (i. e. C is lower-triangular).

❖ A and B can be stored in an array C of size n · (n + 1) as follows (Exhibit 20.2):

```
const  n = … ;
var  C: array [0 .. n, 1 .. n] of real;
```

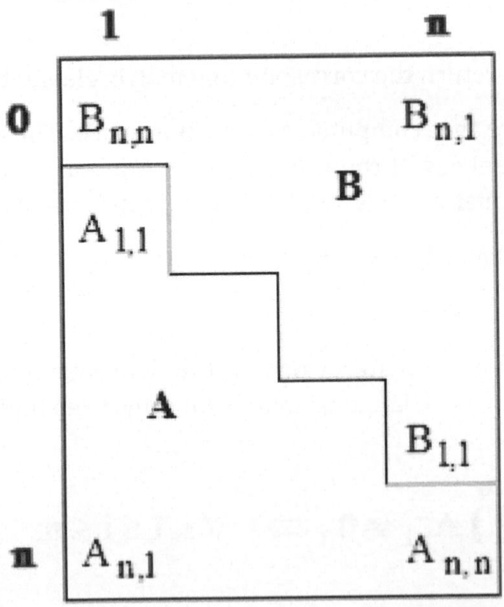

Exhibit 20.2: A staircase separates two triangular matrices

- stored in a rectangular array. (graphic does not match)

```
function A(i, j: 1 .. n): real
begin  if  i < j  then  return(0.0)  else  return(C[i, j])  end;
function B(i, j: 1 .. n): real;
begin  if  i < j  then  return(0.0)  else  return(C[n - i, n + 1 -
j])  end;
```

- Because the new elements of the result matrix C overwrite the old elements of A, it is important to compute them in the right order. Specifically, within every row i of C, elements $C_{i,j}$ must be computed from left to right, that is, in increasing order of j.

```
procedure mult;
var  i, j, k: integer;  x: real;
begin
  for  i := 1  to  n  do
    for  j := 1  to  i  do  begin
      x := 0.0;
      for  k := j  to  i  do  x := x + A(i, k)  · B(k, j);
      C[i, j]  := x
    end
end;
(e)      procedure invertA;
var  i, j, k: integer;  x: real;
begin
    for  i := 1  to  n  do  begin
    for  j := 1  to  i - 1  do  begin
      x := 0.0;
      for  k := j  to i - 1  do  x := x - C[i, k]  · C[k, j];
          C[i, j]  := x / C[i, i]
        end;
          C[i, i]  := 1.0 / C[i, i]
    end
    end;
  procedure AinvertedmultB;
  begin  invertA;  mult  end;
```

Implementation of the fixed-length fifo queue as a circular buffer

A fifo queue is needed in situations where two processes interact in the following way. A process called *producer* generates data for a process called *consumer*. The processes typically Work in bursts: The producer may generate a lot of data while the consumer is busy with something else; thus the data has to be saved temporarily in a buffer, from which the consumer takes it as needed. A keyboard driver and an editor are an example of this producer- consumer interaction. The keyboard driver transfers characters generated by key presses into the buffer, and the editor reads them from the buffer and interprets them (e.g. as control characters or as text to be inserted). it is worth remembering, though, that a buffer helps only if two processes Work at about the same speed over the long run. If the producer is always faster, any buffer will overflow; if the consumer is always faster, no buffer is needed. A buffer can equalize only *temporary* differences in speeds.

With some knowledge about the statistical behavior of producer and consumer one can usually compute a buffer size that is sufficient to absorb producer bursts with high probability, and allocate the buffer statically in an array of fixed size. Among statically allocated buffers, a *circular buffer* is the natural implementation of a fifo queue.

A circular buffer is an array B, considered as a ring in which the first cell B [0] is the successor of the last cell B[m – 1], as shown in Exhibit 20.3. The elements are stored in the buffer in consecutive cells between the two pointers 'in' and 'out': 'in' points to the empty cell into which the next element is to be inserted; 'out' points to the cell containing the next element to be removed. A new element is inserted by storing it in B[in] and advancing 'in' to the next cell. The element in B[out] is removed by advancing 'out' to the next cell.

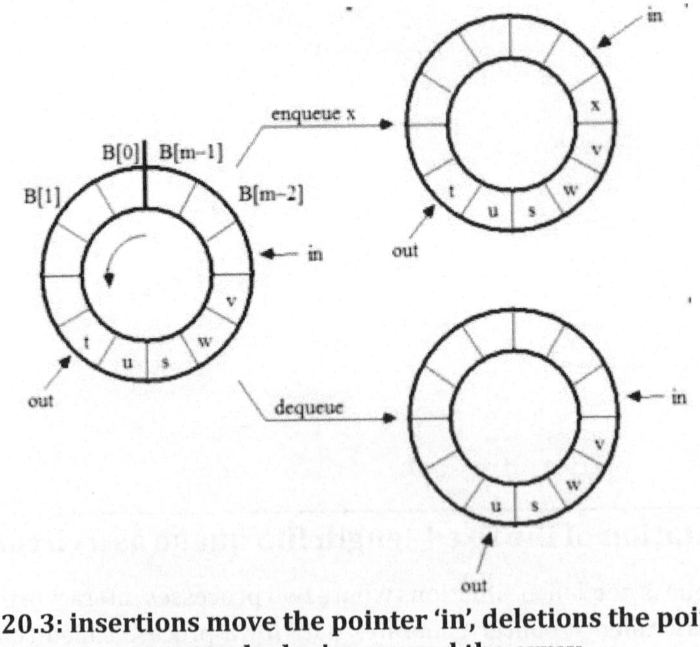

Exhibit 20.3: insertions move the pointer 'in', deletions the pointer 'out' counterclockwise around the array.

Notice that the pointers 'in' and 'out' meet both when the buffer gets full and when it gets empty. Clearly, we must be able to distinguish a full buffer from an empty one, so as to avoid insertion into the former and removal from the latter. At first sight it appears that the pointers 'in' and 'out' are insufficient to determine whether a circular buffer is full or empty. Thus the following implementation uses an additional variable n, which counts how many elements are in the buffer.

```
const  m = … ;   { length of buffer }
type  addr = 0 .. m - 1;   { index range }
var  B: array[addr] of elt;   {storage}
   in, out: addr;   { access to buffer }
   n: 0 .. m;   { number of elements currently in buffer }

procedure create;
begin  in := 0;  out := 0;  n := 0  end;

function empty(): boolean;
begin  return(n = 0)  end;

function full(): boolean;
begin  return(n = m)  end;

procedure enqueue(x: elt);
{ not to be called if the queue is full }
begin  B[in] := x;  in := (in + 1) mod m;  n := n + 1  end;

function front(): elt;
{ not to be called if the queue is empty }
begin  return(B[out])  end;

procedure dequeue;
{ not to be called if the queue is empty }
begin  out := (out + 1) mod m;  n := n - 1  end;
```

The producer uses only 'enqueue' and 'full', as it deletes no elements from the circular buffer. The consumer uses only 'front', 'dequeue', and 'empty', as it inserts no elements.

The state of the circular buffer is described by its contents and the values of 'in', 'out', and n. Since 'in' is changed only within 'enqueue', only the producer needs write-access to 'in'. Since 'out' is changed only by 'dequeue', only the consumer needs write-access to 'out'. The variable n, however, is changed by both processes and thus is a *shared variable* to which both processes have write-access (Exhibit 20.4 (a)).

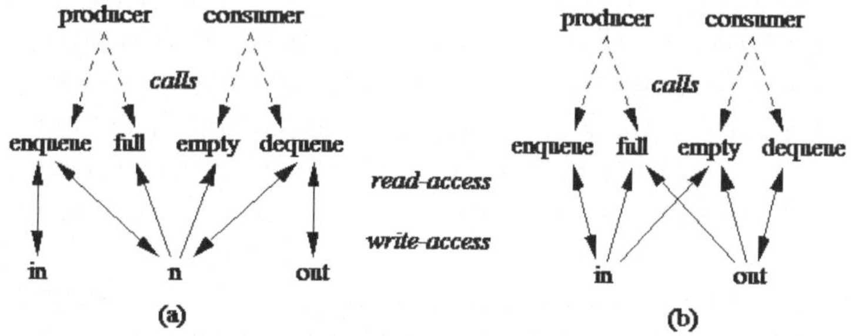

Exhibit 20.4:

- Producer and consumer both have write-access to shared variable n.
- The producer has read/write-access to 'in' and read-only-access to 'out', the consumer has read/write-access to 'out' and read-only-access to 'in'.

In a concurrent programming environment where several processes execute independently, access to shared variables must be synchronized. Synchronization is overhead to be avoided if possible. The shared variable n becomes superfluous (Exhibit 20.4 (b)) if we use the time-honored trick of leaving at least one cell free as a *sentinel*. This ensures that 'empty' and 'full', when expressed in terms of 'in' and 'out', can be distinguished. Specifically, we define 'empty' as in = out, and 'full' as (in + 1) mod m = out. This leads to an elegant and more efficient implementation of the *fixed-length fifo queue* by a circular buffer:

```
const  m = … ;   { length of buffer }
type addr = 0 .. m - 1;   { index range }
  fifoqueue  = record
      B: array[addr] of elt;   { storage }
      in, out: addr   { access to buffer }
    end;

procedure create(var f: fifoqueue);
begin  f.in := 0;   f.out := 0  end;

function empty(f: fifoqueue): boolean;
begin  return(f.in = f.out)  end;

function full(f: fifoqueue): boolean;
begin  return((f.in + 1) mod m = f.out)  end;

procedure enqueue(var f: fifoqueue; x: elt);
{ not to be called if the queue is full }
begin  f.B[f.in] := x;  f.in := ( f.in + 1) mod m  end;

function front(f: fifoqueue): elt;
{ not to be called if the queue is empty }
begin  return(f.B[f.out])  end;

procedure dequeue(f: fifoqueue);
{ not to be called if the queue is empty }
begin  f.out := (f.out + 1) mod m  end;
```

Implementation of the fixed-length priority queue as a heap

A *fixed-length priority queue* can be realized by a circular buffer, with elements stored in the cells between 'in' and 'out', and ordered according to their priority such that 'out' points to the element with highest priority (Exhibit 20.5). in this implementation, the operations 'min' and 'delete' have time complexity O(1), since 'out' points directly to the element with the highest priority. But insertion requires finding the correct cell corresponding to the priority of the element to be inserted, and shifting other elements in the buffer to make space. Binary search could achieve the former task in time O(log n), but the latter requires time O(n).

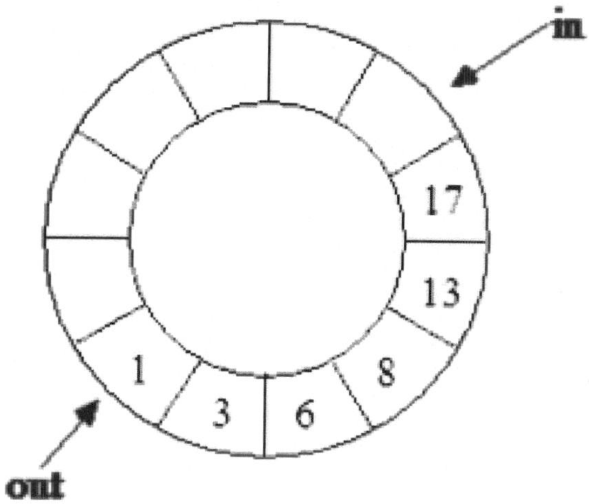

Exhibit 20.5: Implementing a fixed-length priority queue by a circular buffer.

Shifting elements to make space for a new element costs O(n) time.

Implementing a priority queue as a linear list, with elements ordered according to their priority, does not speed up insertion: Finding the correct position of insertion still requires time O(n) (Exhibit 20.6).

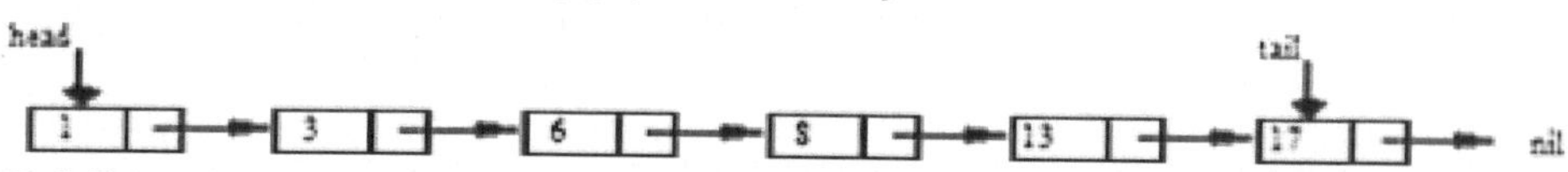

Exhibit 20.6: Implementing a fixed-length priority queue by a linear list. Finding the correct position for a new element costs O(n) time.

The *heap* is an elegant and efficient data structure for implementing a priority queue. it allows the operation 'min' to be performed in time O(1) and allows both 'insert' and 'delete' to be performed in worst-case time O(log n). A heap is a binary tree that:

- obeys a structural property
- obeys an order property
- is embedded in an array in a certain way

Structure: The binary tree is as balanced as possible; all leaves are at two adjacent levels, and the nodes at the bottom level are located as far to the left as possible (Exhibit 20.7).

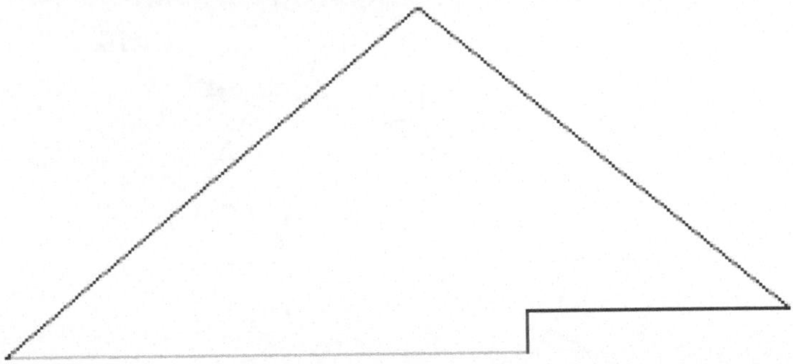

Exhibit 20.7: A heap has the structure of an almost complete binary tree.

Order: The element assigned to any node is ≤ the elements assigned to any children this node may have (Exhibit 20.8).

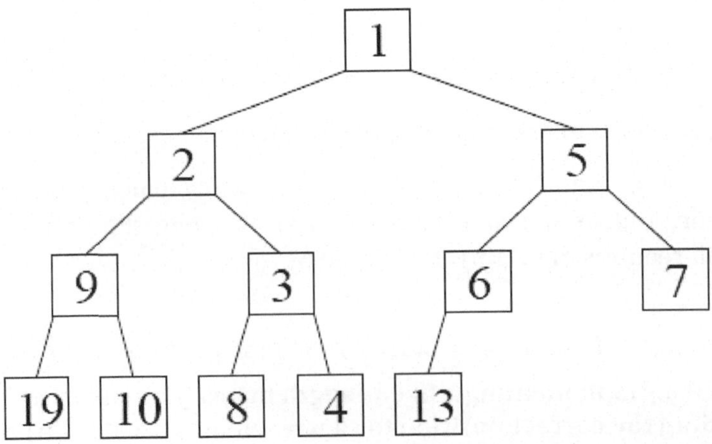

Exhibit 20.8: The order property implies that the smallest element is stored at the root.

The order property implies that the smallest element (the one with top priority) is stored in the root. The 'min' operation returns its value in time O(1), but the most obvious way to delete this element leaves a hole, which takes time to fill. How can the tree be reorganized so as to retain the structural and the order property? The structural condition requires the removal of the rightmost node on the lowest level. The element stored there–13 in our example–is used (temporarily) to fill the vacuum in the root. The root may now violate the order condition, but the latter can be restored by sifting 13 down the tree according to its weight (Exhibit 20.9). If the order condition is violated at any node, the element in this node is exchanged with the smaller of the elements stored in its children; in our example, 13 is exchanged with 2. This *sift-down process* continues until the element finds its proper level, at the latest when it lands in a leaf.

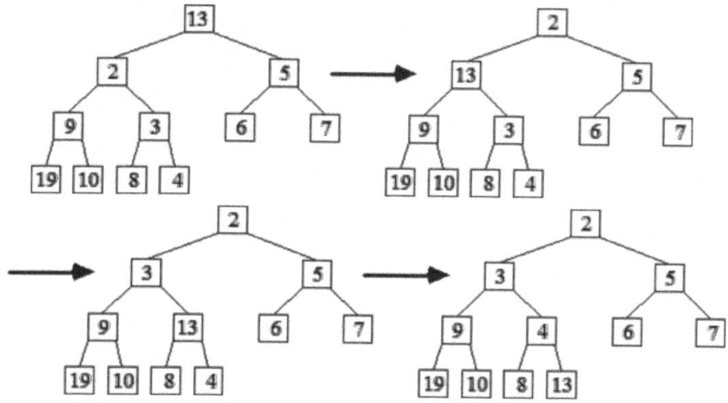

Exhibit 20.9: Rebuilding the order property of the tree in Exhibit 20.8 after 1 has been removed and 13 has been moved to the root.

Insertion is handled analogously. The structural condition requires that a new node is created on the bottom level at the leftmost empty slot. The new element - 0 in our example - is temporarily stored in this node (Exhibit 20.10). If the parent node now violates the order condition, we restore it by floating the new element upward according to its weight. If the new element is smaller than the one stored in its parent node, these two elements - in our example 0 and 6 - are exchanged. This *sift-up process* continues until the element finds its proper level, at the latest when it surfaces at the root.

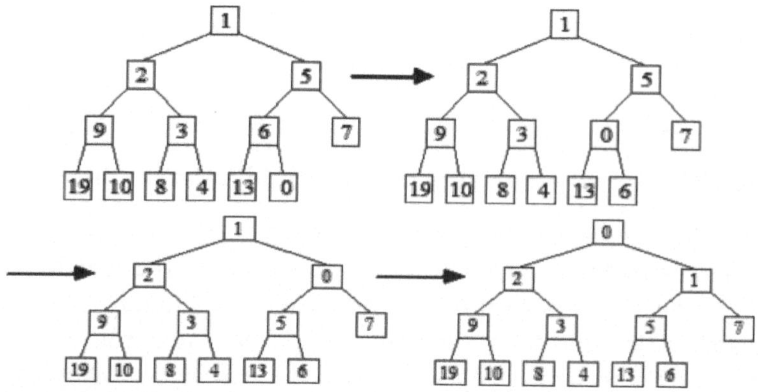

Exhibit 20.10: Rebuilding the order property of the tree in Exhibit 20.8 after 0 has been inserted in a new rightmost node on the lowest level.

The number of steps executed during the sift-up process and the sift-down process is at most equal to the height of the tree. The structural condition implies that this height is [$\log_2$ n]. Thus both 'insert' and 'delete' in a heap Work in time O(log n).

A binary tree can be implemented in many different ways, but the special class of trees that meets the structural condition stated above has a particularly efficient array implementation. A *heap* is a binary tree that satisfies the structural and the order condition and is embedded in a linear array in such a way that the children of a node with index i have indices 2 · i and 2 · i + 1 (Exhibit 20.11). Thus the parent of a node with index j has index j div 2. Any subtree of a heap is also a heap, although it may not be stored contiguously. The order property for the heap implies that the elements stored at indices 2 · i and 2 · i + 1 are ≥ the element stored at index i. This order is called the *heap order*.

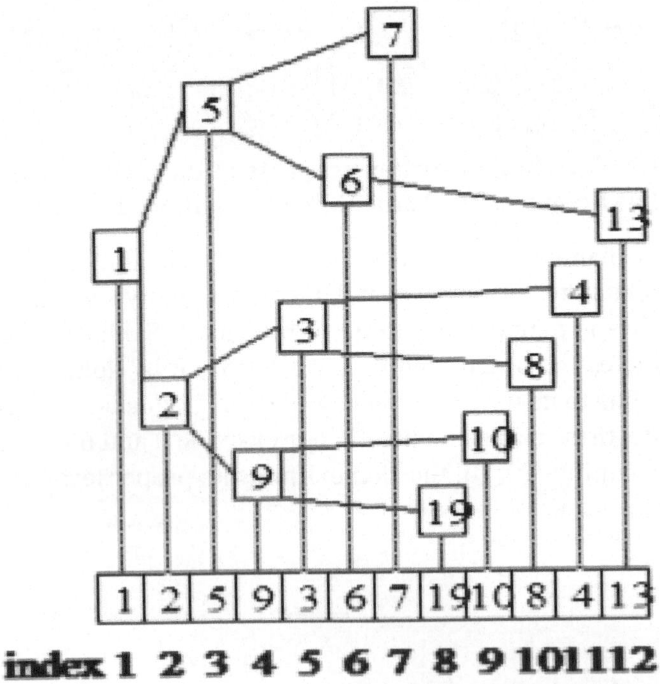

Exhibit 20.11: Embedding the tree of Exhibit 20.8 in a linear array.

The procedure 'restore' is a useful tool for managing a heap. it creates a heap out of a binary tree embedded in a linear array h that satisfies the structural condition, provided that the two subtrees of the root node are already heaps. Procedure 'restore' is applied to subtrees of the entire heap whose nodes are stored between the indices L and R and whose tree structure is defined by the formulas 2 · i and 2 · i + 1.

```
const  m = ... ;  { length of heap }
type  addr = 1 .. m;
var  h: array[addr] of elt;

procedure restore(L, R: addr);
var  i, j: addr;
begin
  i := L;
  while  i ≤ (R div 2)  do  begin
    if  (2 · i < R) cand (h[2 · i + 1] < h[2 · i])  then  j := 2 · i +
1  else  j := 2 · i;
    if  h[j] < h[i]  then  { h[i] :=: h[j];  i := j }  else  i := R
  end
end;
```

Since 'restore' operates along a single path from the root to a leaf in a tree with at most R – L nodes, it Works in time O(log (R – L)).

Creating a heap

An array h can be turned into a heap as follows: for i := n div 2 down to 1 do restore(i, n);

This is more efficient than repeated insertion of a single element into an existing heap. Since the for loop is executed n div 2 times, and n – i ≤ n, the time complexity for creating a heap with n elements is O(n · log n). A more careful analysis shows that the time complexity for creating a heap is O(n).

Heap implementation of the fixed-length priority queue

```
const  m = ... ;   { maximum length of heap }
type addr = 1 .. m;
  priorityqueue = record
      h: array[addr] of elt;   { heap storage }
      n: 0 .. m   { current number of elements }
    end;

procedure restore(var h: array[addr] of elt; L, R: addr);
begin  ...  end;

procedure create(var p: priorityqueue);
begin  p.n := 0  end;

function empty(p: priorityqueue): boolean;
begin  return(p.n = 0)  end;

function full(p: priorityqueue): boolean;
begin  return(p.n = m)  end;

procedure insert(var p: priorityqueue; x: elt);
{ not to be called if the queue is full }
var  i: 1 .. m;
begin
  p.n := p.n + 1;  p.h[p.n] := x;  i := p.n;
  while  (i > 1) cand (p.h[i] < p.h[i div 2])  do
    { p.h[i] :=: p.h[i div 2];  i := i div 2 }
end;

function min(p: priorityqueue): elt;
{ not to be called if the queue is empty }
begin  return(p.h[1])  end;

procedure delete(var p: priorityqueue);
{ not to be called if the queue is empty }
begin  p.h[1] := p.h[p.n];  p.n := p.n - 1;  restore(p.h, 1, p.n)
end;
```

Heapsort

The heap is the core of an elegant O(n · log n) sorting algorithm. The following procedure 'heapsort' sorts n elements stored in the array h into decreasing order.

```
procedure heapsort(n: addr);   { sort elements stored in h[1 .. n] }
var  i: addr;
begin  { heap creation phase: the heap is built up }
   for  i := n div 2  downto  1  do  restore(i, n);
   { shift-up phase: elements are extracted from heap in increasing
order }
   for  i := n  downto  2  do  { h[i] :=: h[1];  restore(1, i - 1) }
end;
```

Each of the for loops is executed less than n times, and the time complexity of restore is O(log n). Thus heapsort always Works in time $O(n \cdot \log n)$.

Exercises and programming projects

1. Block-diagonal matrices are composed of smaller matrices that line up along the diagonal and have 0 elements everywhere else, as shown in Exhibit 20.12. Show how to store an arbitrary block-diagonal matrix in a minimal storage area, and write down the corresponding address computation formulas.

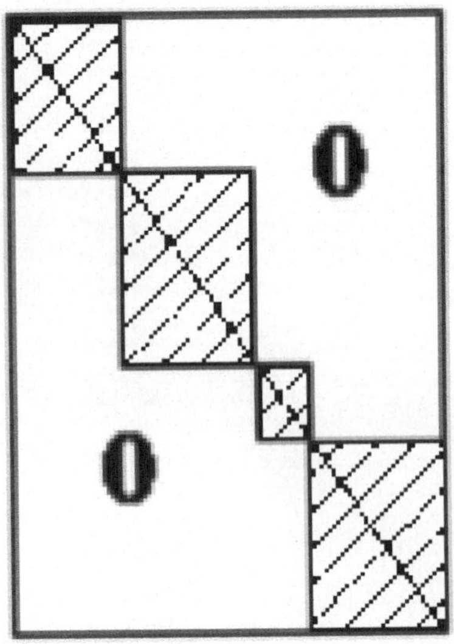

Exhibit 20.12: Structure of a block-diagonal matrix.

- Let A be an antisymmetric n × n-matrix (i. e., all elements of the matrix satisfy Aij = –Aji).
- What values do the diagonal elements Aii of the matrix have?
- How can A be stored in a linear array c of minimal size? What is the size of c?
- Write a

```
function A(i, j: 1 .. n): real;
```

which returns the value of the corresponding matrix element.

- Show that the product of two n × n matrices of width 2 · b + 1 (b = 0, 1, ...) is again a band matrix. What is the width of the product matrix? Write a procedure that computes the product of two band matrices both having the same width and stores the result as a band matrix of minimal width.
- Implement a double-ended queue (deque) by a circular buffer.
- What are the minimum and maximum numbers of elements in a heap of height h?
- Determine the time complexities of the following operations performed on a heap storing n elements. (a) Searching any element. (b) Searching the largest element (i.e. the element with lowest priority).
- Implement heapsort and animate the sorting process, for example as shown in the snapshots in "Algorithm animation". Compare the number of comparisons and exchange operations needed by heapsort and other sorting algorithms (e.g. quicksort) for different input configurations.
- What is the running time of heapsort on an array h[1 .. n] that is already sorted in increasing order? What about decreasing order?
- In a k-ary heap, nodes have k children instead of 2 children.
- How would you represent a k-ary heap in an array?
- What is the height of a k-ary heap in terms of the number of elements n and k?
- Implement a priority queue by a k-ary heap. What are the time complexities of the operations 'insert' and 'delete' in terms of n and k?

21

List Structures

Lists, memory management, pointer variables

The spectrum of data structures ranges from static objects, such as a table of constants, to dynamic structures, such as lists. A list is designed so that not only the data values stored in it, but its *size* and *shape* can change at run time, due to insertions, deletions, or rearrangement of data elements. Most of the data structures discussed so far can change their size and shape to a limited extent. A circular buffer, for example, supports insertion at one end and deletion at the other, and can grow to a predeclared maximal size. A heap supports deletion at one end and insertion anywhere into an array. In a list, any local change can be done with an effort that is independent of the size of the list - provided that we know the memory locations of the data elements involved. The key to meeting this requirement is the idea of abandoning memory allocation in large contiguous chunks, and instead allocating it dynamically in the smallest chunk that will hold a given object. Because data elements are stored randomly in memory, not contiguously, an insertion or deletion into a list does not propagate a ripple effect that shifts other elements around. An element inserted is allocated anywhere in memory where there is space and tied to other elements by **pointers** (i.e. addresses of the memory locations where these elements happen to be stored at the moment). An element deleted does not leave a gap that needs to be filled as it would in an array. Instead, it leaves some free space that can be reclaimed later by a memory management process. The element deleted is likely to break some chains that tie other elements together; if so, the broken chains are relinked according to rules specific to the type of list used.

Pointers are the language feature used in modern programming languages to capture the equivalent of a memory address. A pointer value is essentially an address, and a pointer variable ranges over addresses. A pointer, however, may contain more information than merely an address. In Pascal and other strongly typed languages, for example, a pointer also references the type definition of the objects it can point to - a feature that enhances the compiler's ability to check for consistent use of pointer variables.

Let us illustrate these concepts with a simple example: a **one-way linear list** is a sequence of cells each of which (except the last) points to its successor. The first cell is the head of the list, the last cell is the tail. Since the tail has no successor, its pointer is assigned a predefined value 'nil', which differs from the address of any cell. Access to the list is provided by an external pointer 'head'. If the list is empty, 'head' has the value 'nil'. A cell stores an element xi and a pointer to the successor cell (Exhibit 21.1):

```
type cptr =  ^cell;
     cell = record  e: elt;  next: cptr  end;
```

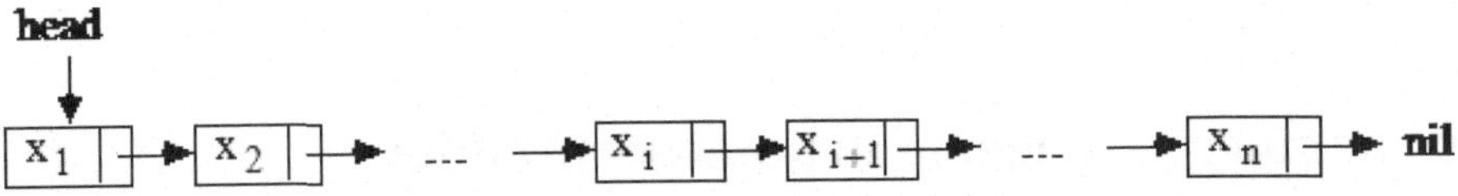

Exhibit 21.1: A one-way linear list.

Local operations, such as insertion or deletion at a position given by a pointer p, are efficient. For example, the following statements insert a new cell containing an element y as successor of a cell being pointed at by p (Exhibit 21.2):

```
new(q);  q^.e := y;  q^.next := p^.next;  p^.next := q;
```

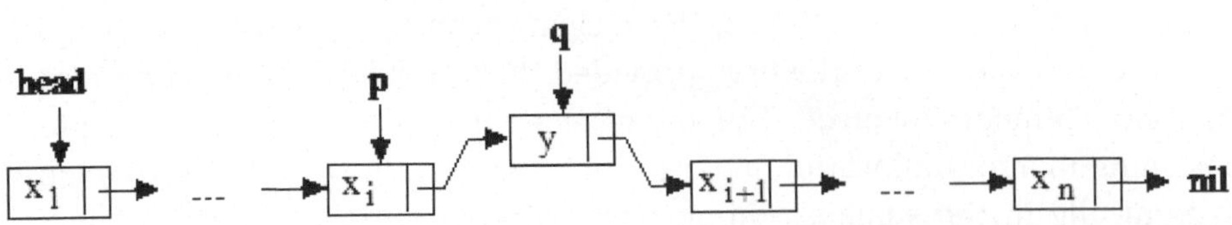

Exhibit 21.2: Insertion as a local operation.

The successor of the cell pointed at by p is deleted by a single assignment statement (Exhibit 21.3):

```
p^.next := p^.next^.next;
```

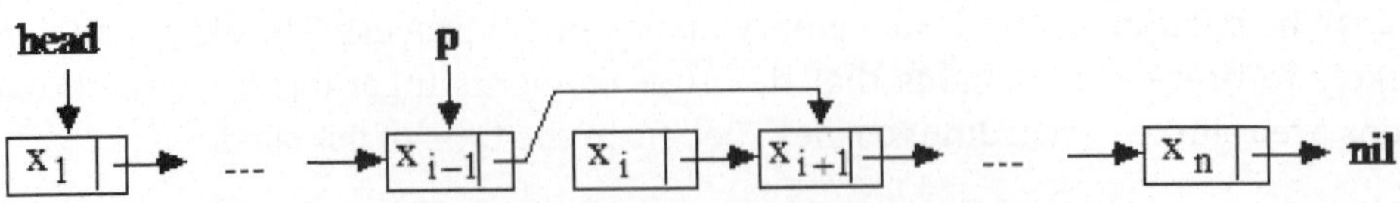

Exhibit 21.3: Deletion as a local operation.

An insertion or deletion at the head or tail of this list is a special case to be handled separately. To support insertion at the tail, an additional pointer variable 'tail' may be set to point to the tail element, if it exists.

A one-way linear list sometimes is handier if the tail points back to the head, making it a **circular list**. In a circular list, the head and tail cells are replaced by a single entry cell, and any cell can be reached from any other without having to start at the external pointer 'entry' (Exhibit 21.4).

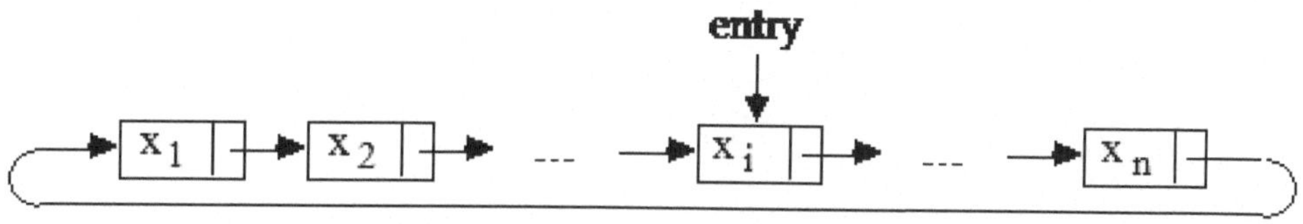

Exhibit 21.4: A circular list combines head and tail into a single entry point

In a *two-way* (or *doubly linked*) *list* each cell contains two pointers, one to its successor, the other to its predecessor. The list can be traversed in both directions. Exhibit 21.5 shows a circular two-way list.

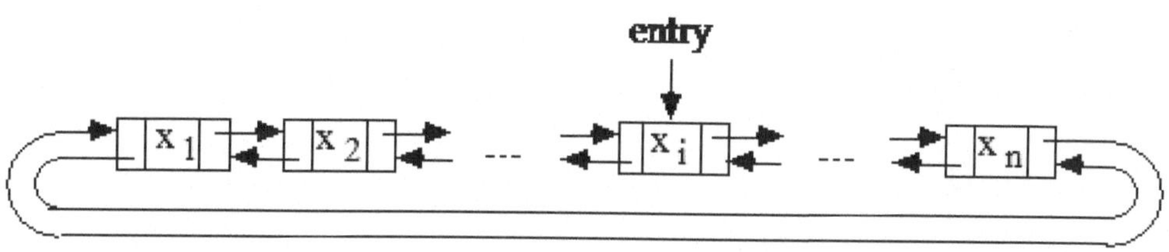

Exhibit 21.5: A circular two-way or doubly-linked list

Exercise: traversal of a singly linked list in both directions

Write a recursive

```
procedure traverse(p: cptr);
```

to traverse a singly linked list from the head to the tail and back again. At each visit of a node, call the

```
procedure visit(p: cptr);
```

Solve the same problem iteratively without using any additional storage beyond a few local pointers. Your traversal procedure may modify the structure of the list temporarily.

Solution

```
(a)     procedure traverse(p: cptr);
  begin  if  p ≠ nil  then  { visit(p);  traverse(p^.next);
visit(p) }  end;

  The initial call of this procedure is
  traverse(head);

(b)     procedure traverse(p: cptr);
  var  o, q: cptr;  i: integer;
  begin
    for  i := 1  to  2  do  { forward and back again }  begin
      o := nil;
      while  p ≠ nil  do  begin
        visit(p);  q := p^.next;  p^.next := o;
          o := p;  p := q  { the fork advances }
      end;
      p := o
    end
  end;
```

Traversal becomes simpler if we let the 'next' pointer of the tail cell point to this cell itself:

```
  procedure traverse(p: cptr);
var o, q: cptr;
begin
  o := nil;
  while  p ≠ nil  do  begin
    visit(p);  q := p^.next;  p^.next := o;
    o := p;  p := q  { the fork advances }
  end
end;
```

The fifo queue implemented as a one-way list

It is natural to implement a fifo queue as a one-way linear list, where each element points to the next one "in line". The operation 'dequeue' occurs at the pointer 'head', and 'enqueue' is made fast by having an external pointer 'tail' point to the last element in the queue. A crafty implementation of this data structure involves an empty cell, called a *sentinel*, at the tail of the list. Its purpose is to make the list-handling procedures simpler and faster by making the empty queue look more like all other states of the queue. More precisely, when the queue is empty, the external pointers 'head' and 'tail' both point to the sentinel rather than having the value 'nil'. The sentinel allows insertion into the empty queue, and deletion that results in an empty queue, to be handled by the same code that handles the general case of 'enqueue' and 'dequeue'. The reader should verify our claim that a sentinel simplifies the code by programming the plausible, but less efficient, procedures which assume that an empty queue is represented by head = tail = nil.

The queue is empty if and only if 'head' and 'tail' both point to the sentinel (i.e. if head = tail). An 'enqueue' operation is performed by inserting the new element into the sentinel cell and then creating a new sentinel.

```
type cptr = ^cell;
  cell = record  e: elt;  next: cptr  end;
  fifoqueue = record  head, tail: cptr  end;

procedure create(var f: fifoqueue);
begin  new(f.head);  f.tail := f.head  end;

function empty(f: fifoqueue): boolean;
begin  return(f.head = f.tail)  end;

procedure enqueue(var f: fifoqueue; x: elt);
begin  f.tail^.e := x;  new(f.tail^.next);  f.tail := f.tail^.next
end;

function front(f: fifoqueue): elt;
{ not to be called if the queue is empty }
begin  return(f.head^.e)  end;

procedure dequeue(var f: fifoqueue);
{ not to be called if the queue is empty }
begin  f.head := f.head^.next  end;
```

Tree traversal

When we speak of trees in computer science, we usually mean **rooted, ordered trees**: they have a distinguished node called the root, and the subtrees of any node are ordered. Rooted, ordered trees are best defined recursively: a tree T is either empty, or it is a tuple (N, T1, ... , Tk), where N is the root of the tree, and T1, ... , Tk is a sequence of trees. Binary trees are the special case k = 2.

Trees are typically used to organize data or activities in a hierarchy: a top-level data set or activity is composed of a next level of data or activities, and so on. When one wishes to gather or survey all of the data or activities, it is necessary to traverse the tree, visiting (i.e. processing) the nodes in some systematic order. The visit at each node might be as simple as printing its contents or as complicated as computing a function that depends on all nodes in the tree. There are two major ways to traverse trees: breadth first and depth first.

Breadth-first traversal visits the nodes level by level. This is useful in heuristic search, where a node represents a partial solution to a problem, with deeper levels representing more complete solutions. Before pursuing any one solution to a great depth, it may be advantageous to assess all the partial solutions at the present level, in order to pursue the most promising one. We do not discuss breadth-first traversal further, we merely suggest the following:

Exercise: breadth-first traversal

Decide on a representation for trees where each node may have a variable number of children. Write a procedure for breadth-first traversal of such a tree. *Hint:* use a fifo queue to organize the traversal. The node to be visited is removed from the head of the queue, and its children are enqueued, in order, at the tail end.

Depth-first traversal always moves to the first unvisited node at the next deeper level, if there is one. It turns out that depth-first better fits the recursive definition of trees than breadth-first does and orders nodes in ways that are more often useful. We discuss depth-first for binary trees and leave the generalization to other trees to the reader. Depth-first can generate three basic orders for traversing a binary tree: **preorder**, **inorder**, and **postorder**, defined recursively as:

```
preorder  Visit root, traverse left subtree, traverse right subtree.
Inorder   Traverse left subtree, visit root, traverse right subtree.
postorder Traverse left subtree, traverse right subtree, visit root.
```

For the tree in Exhibit 21.6we obtain the orders shown.

Breadth-first:	4, 2, 6, 1, 3, 5, 7
Depth-first	
- preorder:	4, 2, 1, 3, 6, 5, 7
- inorder:	1, 2, 3, 4, 5, 6, 7
- postorder:	1, 3, 2, 5, 7, 6, 4

Exhibit 21.6: Standard orders defined on a binary tree

An arithmetic expression can be represented as a binary tree by assigning the operands to the leaves and the operators to the internal nodes. The basic traversal orders correspond to different notations for representing arithmetic expressions. By traversing the expression tree (Exhibit 21.7) in preorder, inorder, or postorder, we obtain the **prefix**, **infix**, or **suffix** notation, respectively.

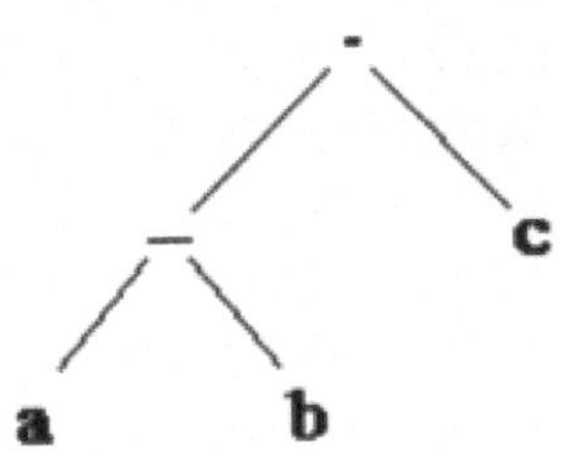

preorder (prefix):	· – a b c
inorder (infix):	(a – b) · c
postorder (postfix):	a b – c ·

Exhibit 21.7: Standard traversal orders correspond to different notations for arithmetic expressions

A binary tree can be implemented as a list structure in many ways. The most common way uses an external pointer 'root' to access the root of the tree and represents each node by a cell that contains a field for an element to be stored, a pointer to the root of the left subtree, and a pointer to the root of the right subtree (Exhibit 21.8). An empty left or right subtree may be represented by the pointer value 'nil', or by pointing at a sentinel, or, as we shall see, by a pointer that points to the node itself.

```
type nptr = ^node;
  node = record  e: elt;  L, R: nptr  end;
var  root: nptr;
```

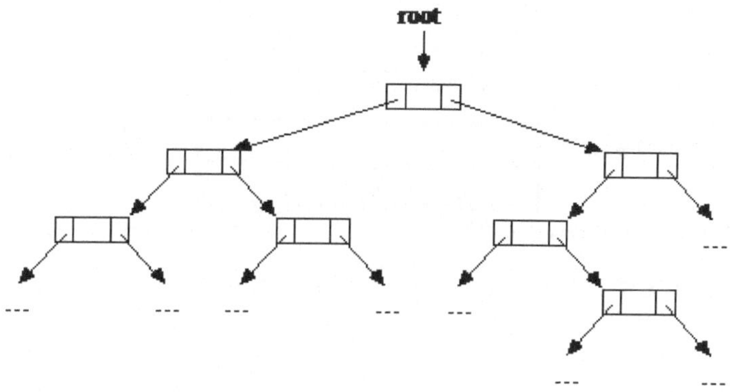

Exhibit 21.8: Straightforward implementation of a binary tree

The following procedure 'traverse' implements any or all of the three orders preorder, inorder, and postorder, depending on how the procedures 'visit1', 'visit2', and 'visit3' process the data in the node referenced by the pointer p. The root of the subtree to be traversed is passed through the formal parameter p. In the simplest case, a visit does nothing or simply prints the contents of the node.

```
procedure traverse(p: nptr);
begin
   if  p ≠ nil  then  begin
     visit1(p);   { preorder }
     traverse(p^.L);
     visit2(p);   { inorder }
     traverse(p^.R);
     visit3(p)   { postorder }
   end
end;
```

Traversing a tree involves both advancing from the root toward the leaves, and backing up from the leaves toward the root. Recursive invocations of the procedure 'traverse' build up a stack whose entries contain references to the nodes for which 'traverse' has been called. These entries provide a means of returning to a node after the traversal of one of its subtrees has been finished. The bookkeeping done by a stack or equivalent auxiliary structure can be avoided if the tree to be traversed may be modified temporarily.

The following **triple-trec traversal** algorithm provides an elegant and efficient way of traversing a binary tree without using any auxiliary memory (i.e. no stack is used and it is not assumed that a node contains a pointer to its parent node). The data structure is modified temporarily to retain the information needed to find the way back up the tree and to restore each subtree to its initial condition after traversal. The triple-tree traversal algorithm assumes that an empty subtree is encoded not by a 'nil' pointer, but rather by an L (left) or R (right) pointer that points to the node itself, as shown in Exhibit 21.9.

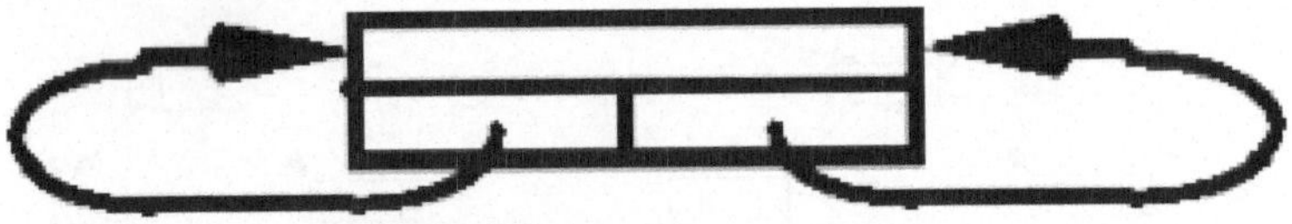

Exhibit 21.9: Coding of a leaf used in procedure TTT

```
procedure TTT;
var  o, p, q: nptr;
begin
   o := nil;   p:= root;
   while  p ≠ nil   do  begin
      visit(p);
      q := p^.L;
      p^.L := p^.R;    { rotate left pointer }
      p^.R := o;   { rotate right pointer }
      o := p;
      p := q
   end
end;
```

In this procedure the pointers p ("present") and o ("old") serve as a two-pronged fork. The tree is being traversed by the pointer p and the companion pointer o, which always lags one step behind p. The two pointers form a two-pronged fork that runs around the tree, starting in the initial condition with p pointing to the root of the tree, and o = nil. An auxiliary pointer q is needed temporarily to advance the fork. The while loop in 'TTT' is executed as long as p points to a node in the tree and is terminated when p assumes the value 'nil'. The initial value of the o pointer gets saved as a temporary value. First it is assigned to the R pointer of the root, later to the L pointer. Finally, it gets assigned to p, the fork exits from the root of the tree, and the traversal of the tree is complete. The correctness of this algorithm is proved by induction on the number of nodes in the tree.

Induction hypothesis H: if at the beginning of an iteration of the while loop, the fork pointer p points to the root of a subtree with $n > 0$ nodes, and o has a value x that is different from any pointer value inside this subtree, then after $3 \cdot n$ iterations the subtree will have been traversed in triple order (visiting each node exactly three times), all tree pointers in the subtree will have been restored to their original value, and the fork pointers will have been reversed (i.e. p has the value x and o points to the root of the subtree).

Base of induction: H is true for $n = 1$.

Proof: The smallest tree we consider has exactly one node, the root alone. Before the while loop is executed for this subtree, the fork and the tree are in the initial state shown inExhibit 21.10. Exhibit 21.11 shows the state of the fork and the tree after each iteration of the while loop. The node is visited in each iteration.

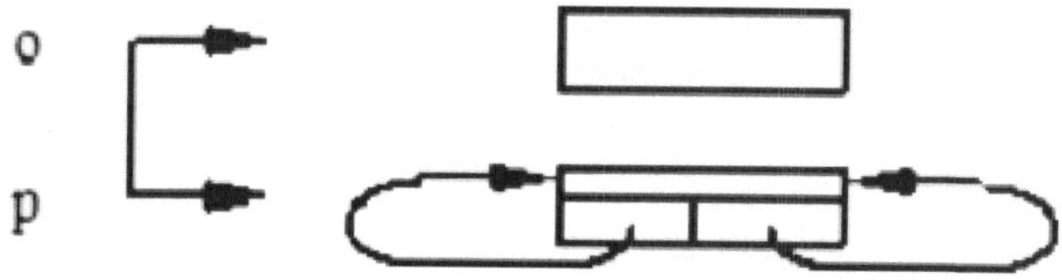

Exhibit 21.10 : Initial configuration for traversing a tree consisting of a single node

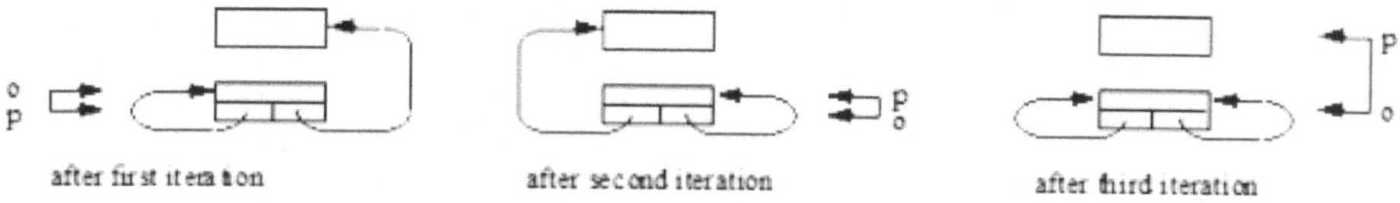

Exhibit 21.11: Tracing procedure TTT while traversing the smallest tree

Induction step: If H is true for all n, 0 < n ≤ k, H is also true for k + 1.

Proof: Consider a tree T with k + 1 nodes. T consists of a root and k nodes shared among the left and right subtrees of the root. Each of these subtrees has ≤ k nodes, so we apply the induction hypothesis to each of them. The following is a highly compressed account of the proof of the induction step, illustrated by Exhibit 21.12. Consider the tree with k + 1 nodes shown in state 1. The root is a node with three fields; the left and right subtrees are shown as triangles. The figure shows the typical case when both subtrees are nonempty. If one of the two subtrees is empty, the corresponding pointer points back to the root; these two cases can be handled similarly to the case n = 1. The fork starts out with p pointing at the root and o pointing at anything outside the subtree being traversed. We want to show that the initial state 1 is transformed in 3 · (k + 1) iterations into the final state 6. In the final state the subtrees are shaded to indicate that they have been correctly traversed; the fork has exited from the root, with p and o having exchanged values. To show that the algorithm correctly transforms state 1 into state 6, we consider the intermediate states 2 to 5, and what happens in each transition.

- → 2 One iteration through the while loop advances the fork into the left subtree and rotates the pointers of the root.
- → 3 H applied to the left subtree of the root says that this subtree will be correctly traversed, and the fork will exit from the subtree with pointers reversed.
- → 4 This is the second iteration through the while loop that visits the root. The fork advances into the right subtree, and the pointers of the root rotate a second time.
- → 5 H applied to the right subtree of the root says that this subtree will be correctly traversed, and the fork will exit from the subtree with pointers reversed.

❖ → 6 This is the third iteration through the while loop that visits the root. The fork moves out of the tree being traversed; the pointers of the root rotate a third time and thereby assume their original values.

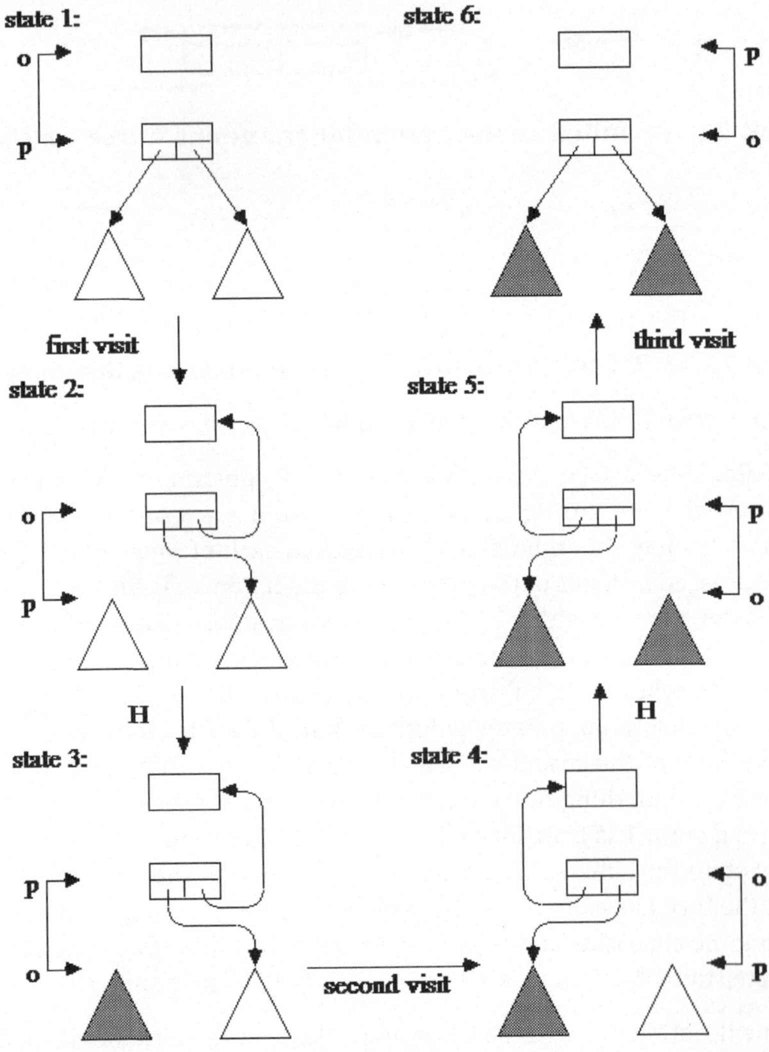

Exhibit 21.12: Trace of procedure TTT, invoking the induction hypothesis

Exercise: binary trees

Consider a binary tree declared as follows:

```
type nptr = ^node;
  node = record  L, R: nptr  end;
var  root: nptr;
```

❖ If a node has no left or right subtree, the corresponding pointer has the value 'nil'. Prove that a binary tree with n nodes, n > 0, has n + 1 'nil' pointers.

❖ Write a function nodes(...): integer; that returns the number of nodes, and a function depth(...): integer; that returns the depth of a binary tree. The depth of the root is defined to be 0; the depth of any other node is the depth of its parent increased by 1. The depth of the tree is the maximum depth of its nodes.

Solution

❖ Each node contains two pointers, for a total of 2 · n pointers in the tree. There is exactly one pointer that points to each of n – 1 nodes, none points to the root. Thus 2 · n – (n – 1) = n + 1 pointers are 'nil'. This can also be proved by induction on the number of nodes in the tree.

❖ function nodes(p: nptr): integer;

```
begin
   if  p = nil  then
      return(0)
   else
      return(nodes(p^.L) + nodes(p^.R) + 1)
end;

function depth(p: nptr): integer;
begin
   if  p = nil then return (-1)
        else return(1 + max(depth(p^.L), depth(p^.R)))
end;
```

where 'max' is

```
function max(a, b: integer): integer;
begin  if  a > b  then  return(a)  else  return(b)  end;
```

Exercise: list copying

Effective memory management sometimes makes it desirable or necessary to copy a list. For example, performance may improve drastically if a list spread over several pages can be compressed into a single page. List copying involves a traversal of the original concurrently with a traversal of the copy, as the latter is being built up.

❖ Consider binary trees built from nodes of type 'node' and pointers of type 'nptr'. A tree is accessed through a pointer to the root, which is 'nil' for an empty tree

```
type nptr =  ^ node;
       node = record  e: elt;  L, R: nptr  end;
```

Write a recursive to copy a tree given by a pointer p to its root, and return a pointer to the root of the copy.

```
function cptree(p: nptr): nptr;
```

❖ Consider arbitrary graphs built from nodes of a type similar to the nodes in (a), but they have an additional pointer field cn, intended to point to the copy of a node:

```
type  node = record  e: elt;  L, R: nptr;  cn: nptr  end;
```

A graph is accessed through a pointer to a node called the origin, and we are only concerned with nodes that can be reached from the origin; this access pointer is 'nil' for an empty graph. Write a recursive to copy a graph given by a pointer p to its origin, and return a pointer to the origin of the copy. Use the field cn, assuming that its initial value is 'nil' in every node of the original graph; set it to 'nil' in every node of the copy.

```
function cpgraph(p: nptr): nptr;
```

Solution

```
(a)     function cptree(p: nptr): nptr;
   var  cp: nptr;
   begin
     if  p = nil  then
       return(nil)
     else  begin
       new(cp);
       cp^.e := p^.e;  cp^.L := cptree(p^.L);  cp^.R := cptree(p^.R);
       return(cp)
     end
   end;
(b)     function cpgraph(p: nptr): nptr;
   var  cp: nptr;
   begin
     if  p = nil  then
       return(nil)
     elsif  p^.cn ≠ nil  then  { node has already been copied }
       return(p^.cn)
     else  begin
       new(cp);  p^.cn := cp;  cp^.cn := nil;
       cp^.e := p^.e;  cp^.L := cpgraph(p^.L);  cp^.R := cpgraph(p^.R);
       return(cp)
     end
   end;
```

Exercise: list copying with constant auxiliary memory

Consider binary trees as in part (a) of the preceding exercise. Memory for the stack implied by the recursion can be saved by writing an iterative tree copying procedure that uses only a constant amount of auxiliary memory. This requires a trick, as any depth-first traversal must be able to back up from the leaves toward the root. In the triple- tree traversal procedure, the return path is temporarily encoded in the tree being traversed. This idea can again be used here, but there is a simpler solution: The return path is temporarily encoded in the R-fields of the copy; the L- fields of certain nodes of the copy point back to the corresponding node in the original. Work out the details of a tree-copying procedure that works with O(1) auxiliary memory.

Exercise: traversing a directed acyclic graph

A directed graph consists of nodes and directed arcs, where each arc leads from one node to another. A directed graph is *acyclic* if the arcs form no cycles. One way to ensure that a graph is acyclic is to label nodes with distinct integers and to draw each arc from a lower number to a higher number. Consider a binary directed acyclic graph, where each node has two pointer fields, L and R, to represent at most two arcs that lead out of that node. An example is shown in Exhibit 21.13.

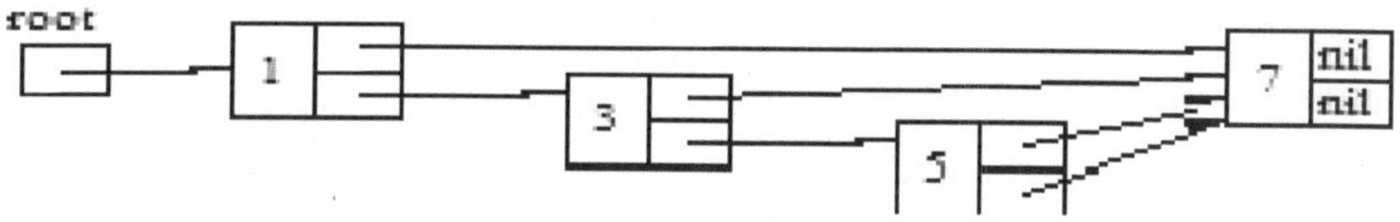

Exhibit 21.13: A rooted acyclic graph.

Write a program to visit every node in a directed acyclic graph reachable from a pointer called 'root'. You are free to execute procedure 'visit' for each node as often as you like.

Write a program similar to (a) where you are required to execute procedure 'visit' exactly once per node.

Hint: Nodes may need to have additional fields.

Exercise: counting nodes on a square grid

Consider a network superimposed on a square grid: each node is connected to at most four neighbors in the directions east, north, west, south (Exhibit 21.14):

```
type nptr =  ^node;
   node =  record  E, N, W, S: nptr;  status: boolean  end;
var  origin: nptr;
```

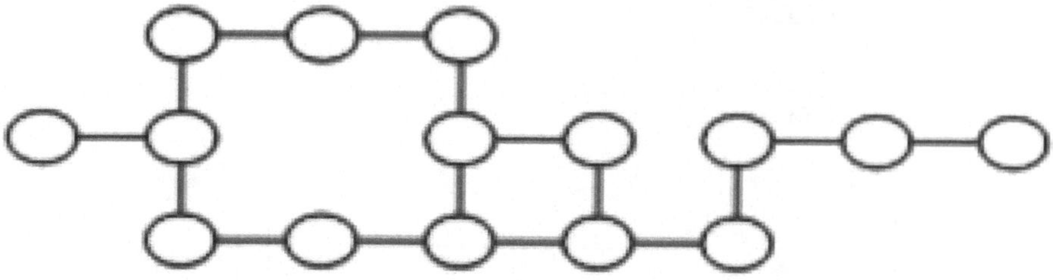

Exhibit 21.14: A graph embedded in a square grid.

A 'nil' pointer indicates the absence of a neighbor. Neighboring nodes are doubly linked: if a pointer in node p points to node q, the reverse pointer of q points to p; (e.g., p^.W = q and q^.E = p). The pointer 'origin' is 'nil' or points to a node. Consider the problem of counting the number of nodes that can be reached from 'origin'. Assume that the status field of all nodes is initially set to false. How do you use this field? Write a function nn(p: nptr): integer; to count the number of nodes.

Solution

```
function nn(p: nptr): integer;
begin
  if  p = nil cor p^.status  then
    return(0)
  else  begin
    p^.status:= true;
    return(1 + nn(p^.E) + nn(p^.N) + nn(p^.W) + nn(p^.S))
  end
end;
```

Exercise: counting nodes in an arbitrary network

We generalize the problem above to arbitrary directed graphs, such as that of Exhibit 21.15, where each node may have any number of neighbors. This graph is represented by a data structure defined by Exhibit 21.16 and the type definitions below. Each node is linked to an arbitrary number of other nodes.

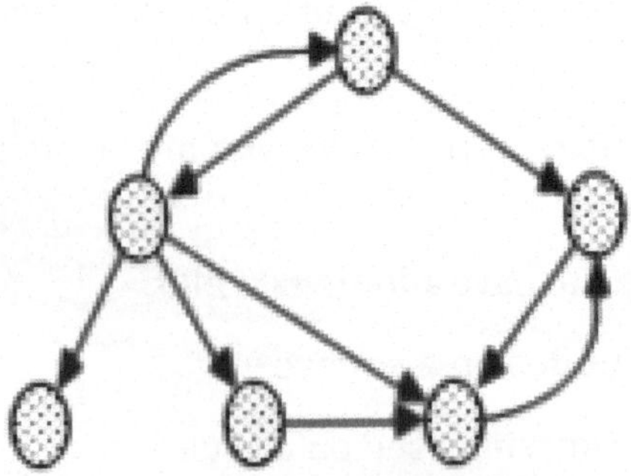

Exhibit 21.15: An arbitrary (cyclic) directed graph.

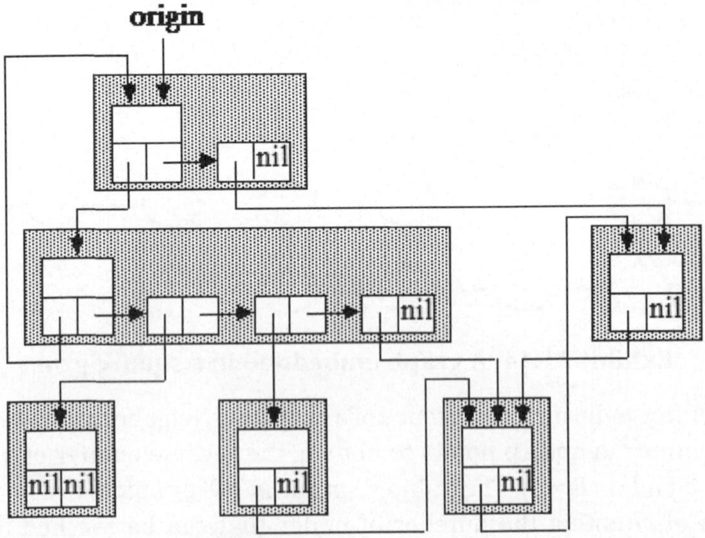

Exhibit 21.16: A possible implementation as a list structure.

```
type nptr = ^node;  cptr = ^cell;
  node = record  status: boolean;  np: nptr;  cp: cptr  end;
  cell = record  np: nptr;  cp: cptr  end;
var  origin: nptr;
```

The pointer 'origin' has the value 'nil' or points to a node. Consider the problem of counting the number n of nodes that can be reached from 'origin'. The status field of all nodes is initially set to false. How do you use it? Write a function nn(p: nptr): integer; that returns n.

Binary search trees

A **binary search tree** is a binary tree T where each node N stores a data element e(N) from a domain X on which a total order ≤ is defined, subject to the following order condition: For every node N in T, all elements in the left subtree L(N) of N are < e(N), and all elements in the right subtree R(N) of N are > e(N). Let x1, 2, ... , xn be n elements drawn from the domain X.

Definition: A binary search tree for x1, x2, ... , xn is a binary tree T with n nodes and a one-to-one mapping between the n given elements and the n nodes, such that

∀ N in T ∀ N' ∈ L(N) ∀ N" ∈ R(N): e(N') < e(N) < e(N")

Exercise

Show that the following statement is equivalent to this order condition: The inorder traversal of the nodes of T coincides with the natural order < of the elements assigned to the nodes.

Remark: The order condition can be relaxed to e(N') ≤ e(N) < e(N") to accommodate multiple occurrences of the same value, with only minor modifications to the statements and algorithms presented in this section. For simplicity's sake we assume that all values in a tree are distinct.

The order condition permits binary search and thus guarantees a worst-case search time O(h) for a tree of height

h. Trees that are well balanced (in an intuitive sense; see the next section for a definition), that have not degenerated into linear lists, have a height h = O(log n) and thus support search in logarithmic time.

Basic operations on binary search trees are most easily implemented as recursive procedures. Consider a tree represented as in the preceding section, with empty subtrees denoted by 'nil'. The following function 'find' searches for an element x in a subtree pointed to by p. It returns a pointer to a node containing x if the search is successful, and 'nil' if it is not.

```
function find(x: elt; p: nptr): nptr;
begin
  if  p = nil  then  return(nil)
  elsif  x < p^.e  then  return(find(x, p^.L))
  elsif  x > p^.e  then  return(find(x, p^.R))
  else  { x = p^.e }  return(p)
end;
```

The following procedure 'insert' leaves the tree alone if the element x to be inserted is already stored in the tree.

The parameter p initially points to the root of the subtree into which x is to be inserted.

```
  procedure insert(x: elt; var p: nptr);
  begin
    if  p = nil  then  { new(p);  p^.e := x;  p^.L := nil;  p^.R :=
nil }
    elsif  x < p^.e  then  insert(x, p^.L)
    elsif  x > p^.e  then  insert(x, p^.R)
  end;

Initial call:
  insert(x, root);
```

To delete an element x, we first have to find the node N that stores x. If this node is a leaf or semileaf (a node with only one subtree), it is easily deleted; but if it has two subtrees, it is more efficient to leave this node in place and to replace its element x by an element found in a leaf or semileaf node, and delete the latter (Exhibit 21.17). Thus we distinguish three cases:

- If N has no child, remove N.
- If N has exactly one child, replace N by this child node.
- If N has two children, replace x by the largest element y in the left subtree, or by the smallest element z in the right subtree of N. Either of these elements is stored in a node with at most one child, which is removed as in case (1) or (2).

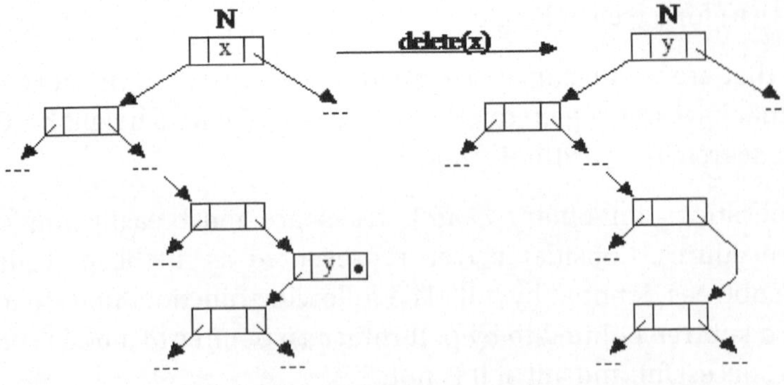

Exhibit 21.17: Element x is deleted while preserving its node N. Node N is filled with a new value y, whose old node is easier to delete.

A sentinel is again the key to an elegant iterative implementation of binary search trees. In a node with no left or right child, the corresponding pointer points to the sentinel. This sentinel is a node that contains no element; its left pointer points to the root and its right pointer points to itself. The root, if it exists, can only be accessed through the left pointer of the sentinel. The empty tree is represented by the sentinel alone (Exhibit 21.18). A typical tree is shown in Exhibit 21.19.

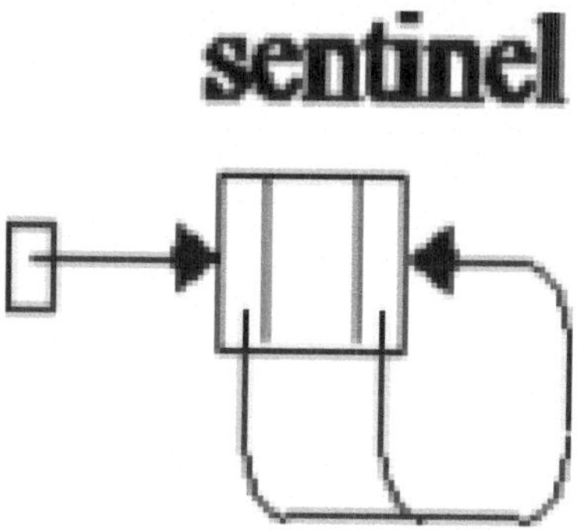

Exhibit 21.18: The empty binary tree is represented by the sentinel which points to itself.

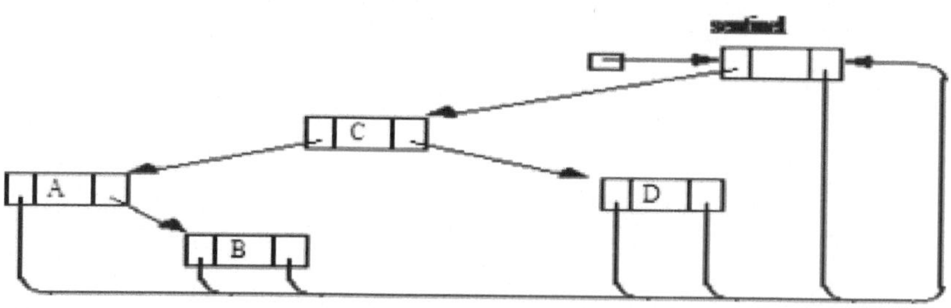

Exhibit 21.19: A binary tree implemented as a list structure with sentinel.

The following implementation of a dictionary as a binary search tree uses a sentinel accessed via the variable d:

```
type nptr = ^node;
  node = record  e: elt;  L, R: nptr  end;
  dictionary = nptr;

procedure create(var d: dictionary);
begin  {create sentinel }  new(d);  d^.L := d;  d^.R := d  end;

function member(d: dictionary; x: elt): boolean;
var  p: nptr;
begin
  d^.e := x;   { initialize element in sentinel }
  p := d^.L;   { point to root, if it exists }
  while  x ≠ p^.e  do
    if  x < p^.e  then  p := p^.L  else  { x > p^.e }  p := p^.R;
  return(p ≠ d)
end;
```

Procedure 'find' searches for x. If found, p points to the node containing x, and q to its parent. If not found, p points to the sentinel and q to the parent-to-be of a new node into which x will be inserted.

```
procedure find(d: dictionary; x: elt; var p, q: nptr);
begin
  d^.e := x;  p := d^.L;  q := d;
  while  x ≠ p^.e  do  begin
    q := p;
    if  x < p^.e  then  p := p^.L  else  { x > p^.e }  p := p^.R
  end
end;
procedure insert(var d: dictionary; x: elt);
var  p, q: nptr;
begin
  find(d, x, p, q);
  if  p = d  then  begin  { x is not yet in the tree }
    new(p);  p^.e := x;  p^.L := d;  p^.R := d;
    if  x ≤ q^.e  then  q^.L := p  else  { x > q^.e }  q^.R := p
  end
end;
  procedure delete(var d: dictionary; x: elt);
  var  p, q, t: nptr;
  begin
    find(d, x, p, q);
    if  p ≠ d  then  { x has been found }
      if  (p^.L ≠ d) and (p^.R ≠ d)  then  begin
        { p has left and right children; find largest element in left
subtree }
        t := p^.L;  q:= p;
        while  t^.R ≠ d  do  { q := t;  t := t^.R };
        if  t^.e < q^.e  then  q^.L := t^.L  else  { t^.e > q^.e }
q^.R := t^.L
        p^.e := t^.e;
      end
      else  begin  { p has at most one child }
        ifp^.L ≠ d  then{ left child only }  p := p^.L
        elsif  p^.R ≠ d  then{ right child only } p := p^.R
        else   { p has no children }p := d;
        if  x ≤ q^.e  then  q^.L := p  else  { x > q^.e }  q^.R := p
      end
  end;
```

In the best case of a completely balanced binary search tree for n elements, all leaves are on levels [$\log_2$ n] or [$\log_2$ n]– 1, and the search tree has the height [log2 n]. The cost for performing the 'member', 'insert', or 'delete' operation is bounded by the longest path from the root to a leaf (i.e. the height of the tree) and is therefore O(log n).

Without any further provisions, a binary search tree can degenerate into a linear list in the worst case. Then the cost for each of the operations would be O(n).

What is the expected average cost for the search operation in a *randomly generated* binary search tree? "Randomly generated" means that each permutation of the n elements to be stored in the binary search tree has the same probability of being chosen as the input sequence. Furthermore, we assume that the tree is generated by insertions only. Therefore, each of the n elements is equally likely to be chosen as the root of the tree. Let pn be the expected path length of a randomly generated binary search tree storing n elements. Then

$$p_n = \frac{1}{n}\sum_{k=1}^{n}(p_{k-1} + p_{n-k}) + (n-1) = \frac{2}{n}\sum_{k=0}^{n-1} p_k + (n-1).$$

As shown in chapter 16 in the section "Recurrence relations", this recurrence relation has the solution

$$p_n = (\ln 4) \cdot n \cdot \log_2 n + g(n) \quad \text{with} \quad g(n) \in O(n).$$

Since the average search time in randomly generated binary search trees, measured in terms of the number of nodes visited, is pn / n and ln 4 ≈ 1.386, it follows that the cost is O(log n) and therefore only about 40 per cent higher than in the case of completely balanced binary search trees.

Balanced trees: general definition

If insertions and deletions occurred at random, and the assumption of the preceding section was realistic, we could let search trees grow and shrink as they please, incurring a modest increase of 40 per cent in search time over completely balanced trees. But real data are not random: they are typically clustered, and long runs of monotonically increasing or decreasing elements occur, often as the result of a previous processing step. Unfortunately, such deviation from randomness degrades the performance of search trees.

To prevent search trees from degenerating into linear lists, we can monitor their shape and restructure them into a more balanced shape whenever they have become too skewed. Several classes of balanced search trees guarantee that each operation 'member', 'insert', and 'delete' can be performed in time O(log n) in the worst case. Since the work to be done depends directly on the height of the tree, such a class B of search trees must satisfy the following two conditions (hT is the height of a tree T, nT is the number of nodes in T):

Balance condition: $\exists c > 0\ \forall T \forall B$: $h_T \leq c \cdot \log_2 nT$

Rebalancing condition: If an 'insert' or 'delete' operation, performed on a tree T ∈ B, yields a tree T' ∉ B, it must be possible to rebalance T' in time O(log n) to yield a tree T'' ∈ B.

Example: almost complete trees

The class of almost complete binary search trees satisfies the balance condition but not the restructuring condition. In the worst case it takes time O(n) to restructure such a binary search tree (Exhibit 21.20), and if 'insert' and 'delete' are defined to include any rebalancing that may be necessary, these operations cannot be guaranteed to run in time O(log n).

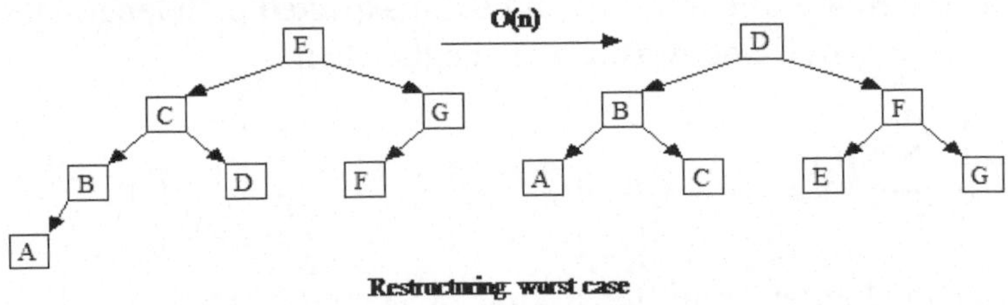

Exhibit 21.20: Restructuring: worst case

In the next two sections we present several classes of balanced trees that meet both conditions: the height- balanced or AVL-trees (G. Adel'son-Vel'skii and E. Landis, 1962) [AL 62] and various multiway trees, such as B- trees [BM 72, Com 79] and their generalization, (a,b)-trees [Meh 84a].

AVL-trees, with their small nodes that hold a single data element, are used primarily for storing data in main memory. Multiway trees, with potentially large nodes that hold many elements, are also useful for organizing data on secondary storage devices, such as disks, that allow direct access to sizable physical data blocks. In this case, a node is typically chosen to fill a physical data block, which is read or written in one access operation.

Height-balanced trees

Definition: A binary tree is *height-balanced* if, for each node, the heights of its two subtrees differ by at most one. Height-balanced search trees are also called *AVL-trees*. Exhibit 21.21 to Exhibit 21.23 show various AVL-trees, and one that is not.

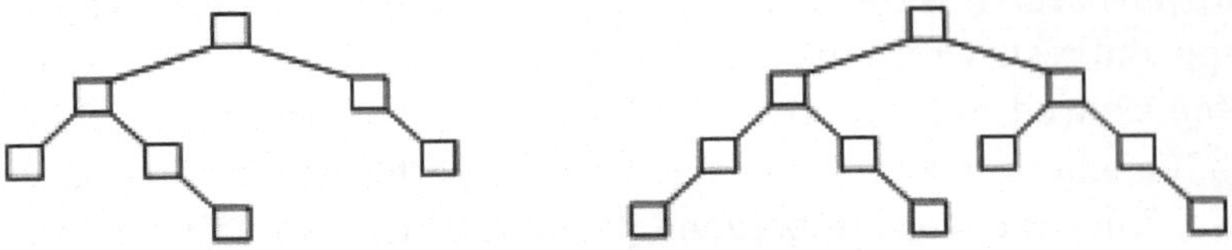

Exhibit 21.21: Examples of height-balanced trees

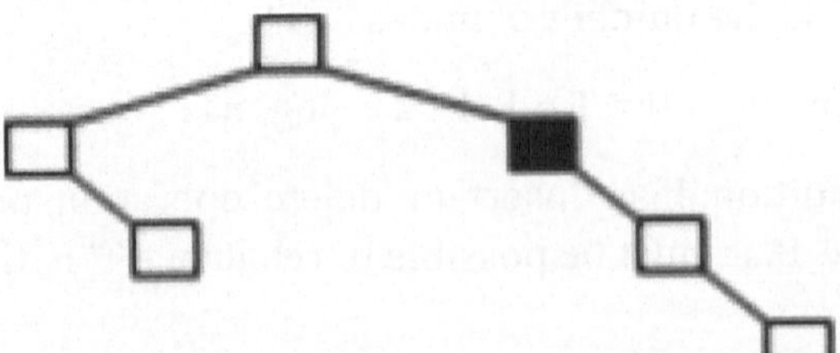

Exhibit 21.22: Example of a tree not height-balanced;the marked node violates the balance condition.

A "most-skewed" AVL-tree T_h is an AVL-tree of height h with a minimal number of nodes. Starting with T_0 and T_1 shown in Exhibit 21.23, T_h is obtained by attaching T_{h-1} and T_{h-2} as subtrees to a new root.

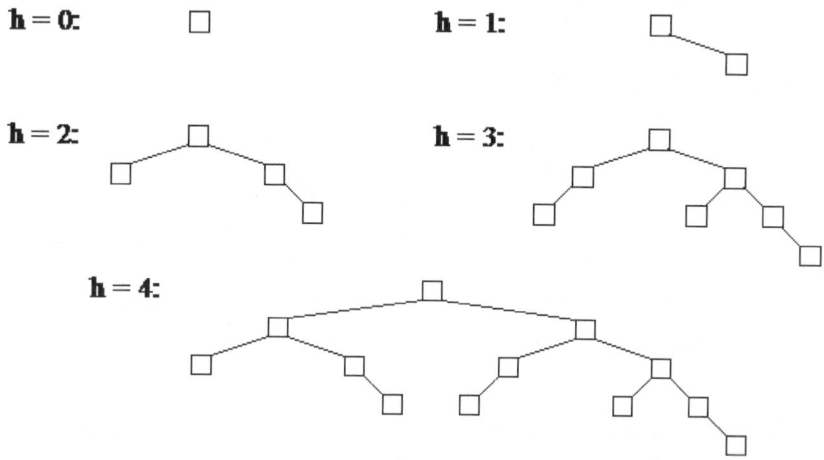

Exhibit 21.23: Most skewed AVL trees of heights h = 0 through h = 4

The number of nodes in a most-skewed AVL-tree of height h is given by the recurrence relation

$n_h = n_{h-1} + n_{h-2} + 1, n_0 = 1, n_1 = 2.$

In the section on recurrence relations in the chapter entitled "The mathematics of algorithm analysis", it has been shown that the recurrence relation

$m_h = m_{h-1} + m_{h-2}, m_0 = 0, m_1 = 1$ has the solution

$$m_h = \frac{1}{\sqrt{5}} \cdot \left(\frac{1+\sqrt{5}}{2}\right)^h - \frac{1}{\sqrt{5}} \cdot \left(\frac{1-\sqrt{5}}{2}\right)^h.$$

Since nh = $m_{h+3} - 1$ we obtain

$$n_h = \left(1+\frac{2}{\sqrt{5}}\right) \cdot \left(\frac{1+\sqrt{5}}{2}\right)^h + \left(1-\frac{2}{\sqrt{5}}\right) \cdot \left(\frac{1-\sqrt{5}}{2}\right)^h - 1.$$

Since

$$\left|\frac{1-\sqrt{5}}{2}\right| < 1,$$

it follows that

$$\left(1-\frac{2}{\sqrt{5}}\right)\cdot\left(\frac{1-\sqrt{5}}{2}\right)^{h}-1 \in O(1),$$

and therefore nh behaves asymptotically as

$$n_h \approx \left(1+\frac{2}{\sqrt{5}}\right)\cdot\left(\frac{1+\sqrt{5}}{2}\right)^{h}.$$

Applying the logarithm results in

$$\log_2 n_h \approx \log_2\left(1+\frac{2}{\sqrt{5}}\right)+h\cdot\log_2\left(\frac{1+\sqrt{5}}{2}\right).$$

Therefore, the height of a worst-case AVL-tree with n nodes is about 1.44 · log2 n. Thus the class of AVL-trees satisfies the balance condition, and the 'member' operation can always be performed in time O(log n).

We now show that the class of AVL-trees also satisfies the rebalancing condition. Thus AVL-trees support insertion and deletion in time O(log n). Each node N of an AVL-tree has one of the balance properties / (left- leaning), \ (right-leaning), or – (horizontal), depending on the relative height of its two subtrees.

Two local tree operations, **rotation** and **double rotation**, allow the restructuring of height-balanced trees that have been disturbed by an insertion or deletion. They split a tree into subtrees and rebuild it in a different way. Exhibit 21.24 shows a node, marked black, that got out of balance, and how a local transformation builds an equivalent tree (for the same elements, arranged in order) that is balanced. Each of these transformations has a mirror image that is not shown. The algorithms for insertion and deletion use these rebalancing operations as described below.

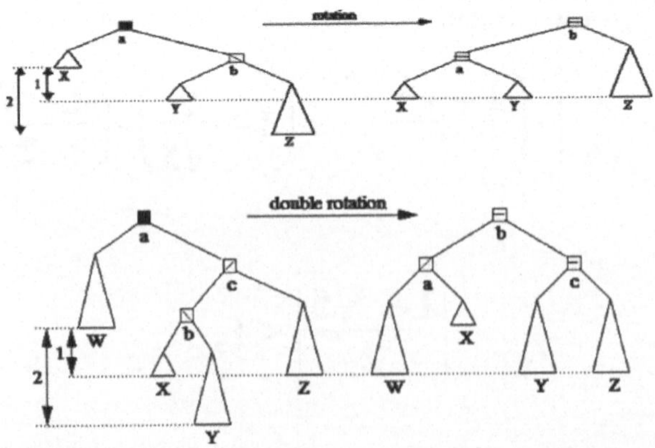

Exhibit 21.24: Two local rebalancing operations

Insertion

A new element is inserted as in the case of a binary search tree. The balance condition of the new node becomes – (horizontal). Starting at the new node, we walk toward the root of the tree, passing along the message that the height of the subtree rooted at the current node has increased by one. At each node encountered along this path, an operation determined by the following rules is performed. These rules depend on the balance condition of the node before the new element was inserted, and on the direction from which the node was entered (i.e. from its left or right child).

Rule I_1: If the current node has balance condition –, change it to / or \ depending on whether we entered from the node's left or from its right child. If the current node is the root, terminate; if not, continue to follow the path upward.

Rule I_2: If the current node has balance condition / or \ and is entered from the subtree that was previously shorter, change the balance condition to—and terminate (the height of the subtree rooted at the current node has not changed).

Rule I_3: If the current node has balance condition / or \ and is entered from the subtree that was previously taller, the balance condition of the current node is violated and gets restored as follows:

- If the last two steps were in the same direction (both from left children, or both from right children), an appropriate rotation restores all balances and the procedure terminates.
- If the last two steps were in opposite directions (one from a left child, the other from a right child), an appropriate double rotation restores all balances and the procedure terminates.

The initial insertion travels along a path from the root to a leaf, and the rebalancing process travels back up along the same path. Thus the cost of an insertion in an AVL-tree is O(h), or O(log n) in the worst case. Not ice that an insertion calls for at most one rotation or double rotation, as shown in the example in Exhibit 21.25.

Example

Insert 1, 2, 5, 3, 4, 6, 7 into an initially empty AVL-tree (Exhibit 21.25). The balance condition of a node is shown below it. Boldfaced nodes violate the balance condition.

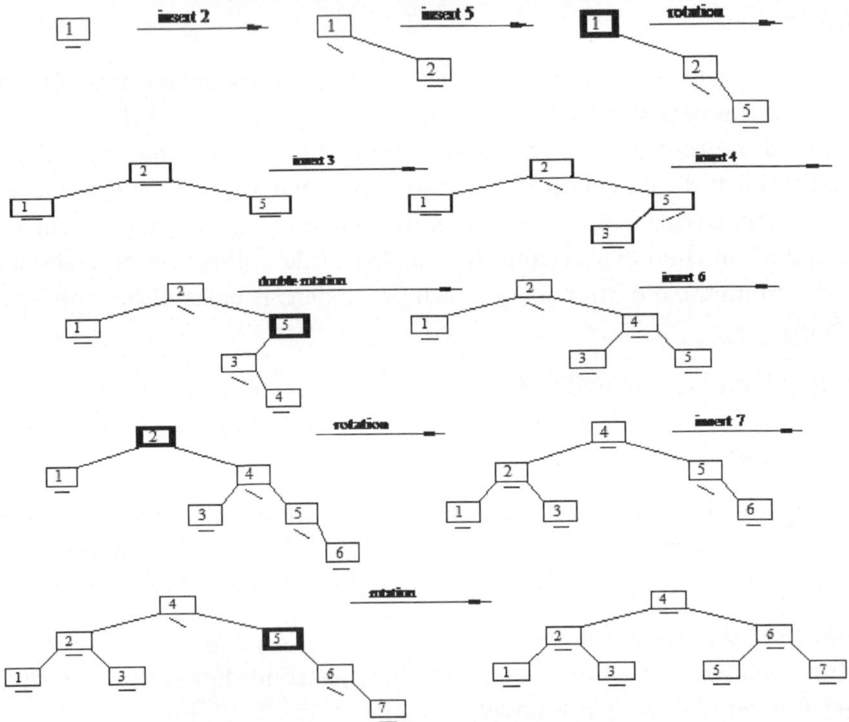

Exhibit 21.25: Trace of consecutive insertions and the rebalancings they trigger

Deletion

An element is deleted as in the case of a binary search tree. Starting at the parent of the deleted node, walk towards the root, passing along the message that the height of the subtree rooted at the current node has decreased by one. At each node encountered, perform an operation according to the following rules. These rules depend on the balance condition of the node before the deletion and on the direction from which the current node and its child were entered.

Rule D_1: If the current node has balance condition –, change it to \ or / depending on whether we entered from the node's left or from its right child, and terminate (the height of the subtree rooted at the current node has not changed).

Rule D_2: If the current node has balance condition / or \ and is entered from the subtree that was previously taller, change the balance condition to – and continue upward, passing along the message that the subtree rooted at the current node has been shortened.

Rule D_3: If the current node has balance condition / or \ and is entered from the subtree that was previously shorter, the balance condition is violated at the current node. We distinguish three subcases according to the balance condition of the other child of the current node (consider also the mirror images of the following illustrations):

(a)

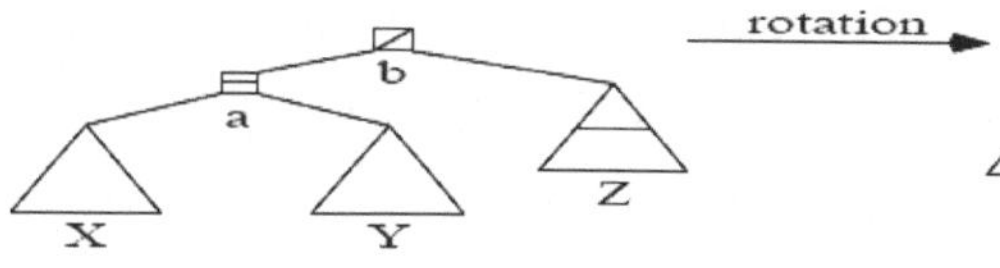

An appropriate rotation restores the balance of the current node without changing the height of the subtree rooted at this node. Terminate.

(b)

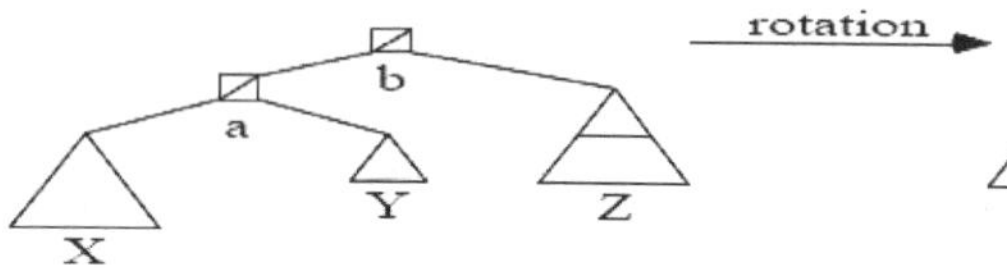

A rotation restores the balance of the current node. Continue upward, passing along the message that the subtree rooted at the current node has been shortened.

(c)

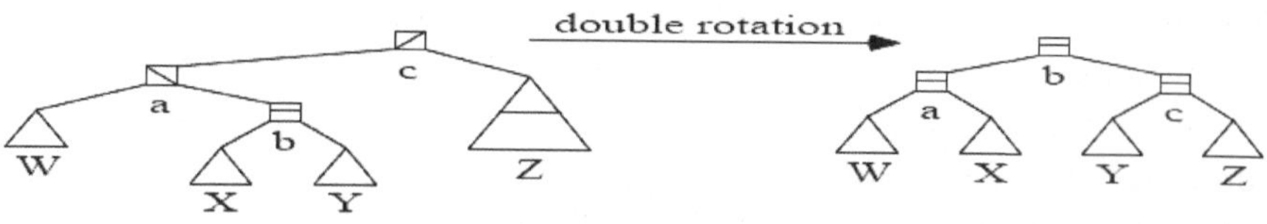

A double rotation restores the balance of the current node. Continue upward, passing along the message that the subtree rooted at the current node has been shortened. Similar transformations apply if either X or Y, but not both, are one level shorter than shown in this figure. If so, the balance conditions of some nodes differ from those shown, but this has no influence on the total height of the subtree. In contrast to insertion, deletion may require more than one rotation or double rotation to restore all balances. Since the cost of a rotation or double rotation is constant, the worst-case cost for rebalancing the tree depends only on the height of the tree, and thus the cost of a deletion in an AVL-tree is O(log n) in the worst case.

Multiway trees

Nodes in a multiway tree may have a variable number of children. As we are interested in balanced trees, we add two restrictions. First, we insist that all leaves (the nodes without children) occur at the same depth. Second, we constrain the number of children of all internal nodes by a lower bound a and an upper bound b. Many varieties of multiway trees are known; they differ in details, but all are based on similar ideas. For example, (2,3)-trees are defined by the requirement that all internal nodes have either two or three children. We generalize this concept and discuss (a,b)-trees.

Definition: Consider a domain X on which a total order $\leq$ is defined. Let a and b be integers with $2 \leq a$ and $2 \cdot a - 1 \leq b$. Let c(N) denote the number of children of node N. An *(a,b)-tree* is an ordered tree with the following properties:

- All leaves are at the same level
- $2 \leq c(\text{root}) \leq b$
- For all internal nodes N except the root, $a \leq c(N) \leq b$

A node with k children contains k – 1 elements $x_1 < x_2 < \ldots < x_{k-1}$ drawn from X; the subtrees corresponding to the k children are denoted by $T_1, T_2, \ldots, T_k$. An (a,b)-tree supports "c(N) search" in the same way that a binary tree supports binary search, thanks to the following order condition:

- $y \leq xi$ for all elements y stored in subtrees $T_1, \ldots, T_i$
- $x_i < z$ for all elements z stored in subtrees $T_{i+1}, \ldots, T_k$

Definition: (a,b)-trees with $b = 2 \cdot a - 1$ are known as *B-trees* [BM 72, Com 79].

The algorithms we discuss operate on internal nodes, shown in white in Exhibit 21.26, and ignore the leaves, shown in black. For the purpose of understanding search and update algorithms, leaves can be considered fictitious entities used only for counting. In practice, however, things are different. The internal nodes merely constitute a directory to a file that is stored in the leaves. A leaf is typically a physical storage unit, such as a disk block, that holds all the records whose key values lie between two (adjacent) elements stored in internal nodes.

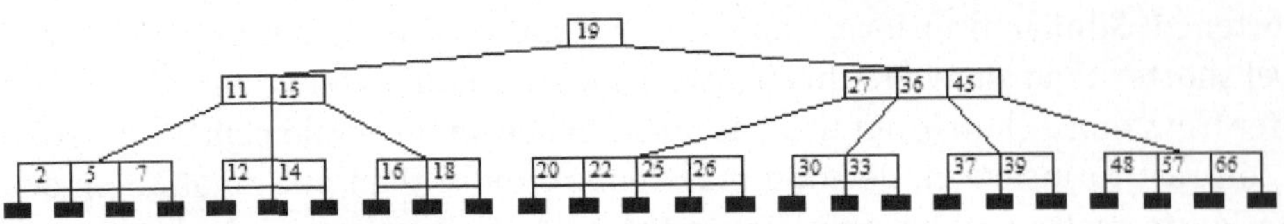

Exhibit 21.26: Example of a (3,5)-tree

The number n of elements stored in the internal nodes of an (a,b)-tree of height h is bounded by

$$2 \cdot a^{h-1} - 1 \leq n \leq b^h - 1,$$

and thus this shows that the class of (a,b)-trees satisfies the balance condition h = O(log n). We show that this class also meets the rebalancing condition, namely, that (a,b)-trees support insertion and deletion in time O(log n).

$$\log_b(n+1) \le h \le 1 + \log_a \frac{n+1}{2}.$$

Insertion

Insertion of a new element x begins with a search for x that terminates unsuccessfully at a leaf. Let N be the parent node of this leaf. If N contained fewer than b – 1 elements before the insertion, insert x into N and terminate. If N was full, we imagine b elements temporarily squeezed into the overflowing node N. Let m be the median of these b elements, and use m to split N into two: a left node NL populated by the (b – 1) / 2 elements smaller than m, and a right node NR populated by the (b – 1) / 2 elements larger than m. The condition $2 \cdot a - 1 \le b$ ensures that $\lfloor (b - 1) / 2 \rfloor \ge a - 1$, in other words, that each of the two new nodes contains at least a – 1 elements.

The median element m is pushed upward into the parent node, where it serves as a separator between the two new nodes NL and NR that now take the place formerly inhabited by N. Thus the problem of insertion into a node at a given level is replaced by the same problem one level higher in the tree. The new separator element may be absorbed in a nonfull parent, but if the parent overflows, the splitting process described is repeated recursively. At worst, the splitting process propagates to the root of the tree, where a new root that contains only the median element is created. (a,b)-trees grow at the root, and this is the reason for allowing the root to have as few as two children.

Deletion

Deletion of an element x begins by searching for it. As in the case of binary search trees, deletion is easiest at the bottom of the tree, at a node of maximal depth whose children are leaves. If x is found at a higher level of the tree, in a node that has internal nodes as children, x is the separator between two subtrees TL and TR. We replace x by another element z, either the largest element in TL or the smallest element in TR, both of which are stored in a node at the bottom of the tree. After this exchange, the problem is reduced to deleting an element z from a node N at the deepest level.

If deletion (of x or z) leaves N with at least a – 1 elements, we are done. If not, we try to restore N's occupancy condition by stealing an element from an adjacent sibling node M. If there is no sibling M that can spare an element, that is, if M is minimally occupied, M and N are merged into a single node L. L contains the a – 2 elements of N, the a – 1 elements of M, and the separator between M and N which was stored in their parent node, for a total of $2 \cdot (a - 1) \le b - 1$ elements. Since the parent (of the old nodes M and N, and of the new node L) lost an element in this merger, the parent may underflow. As in the case of insertion, this underflow can propagate to the root and may cause its deletion. Thus (a,b)-trees grow and shrink at the root.

Both insertion and deletion work along a single path from the root down to a leaf and (possibly) back up. Thus their time is bounded by O(h), or equivalently, by O(log n): (a,b)-trees can be rebalanced in logarithmic time.

Amortized cost. The performance of (a,b)-trees is better than the worst-case analysis above suggests. It can be shown that the total cost of *any sequence of s insertions and deletions* into an initially empty (a,b)-tree is linear in the length s of the sequence: whereas the worst-case cost of a single operation is O(log n), the *amortized cost* per operation is O(1) [Meh 84a]. Amortized cost is a complexity measure that involves both an average and a worst-case consideration. The average is taken over all operations in a sequence; the worst case is taken over all sequences. Although any one operation may take time O(log n), we are guaranteed that the total of all s operations in any sequence of length s can be done in time O(s), as if each single operation were done in time O(1).

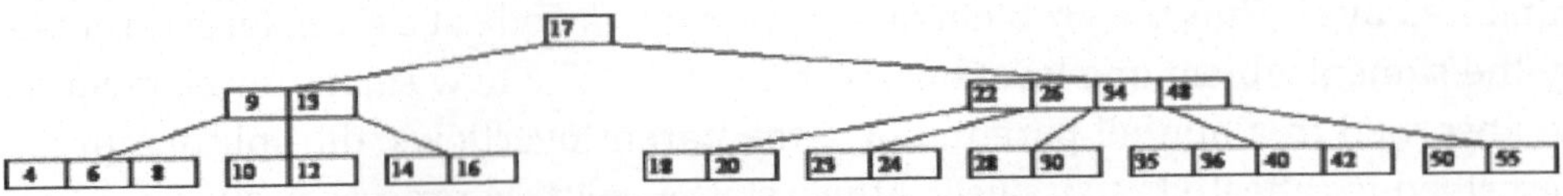

Exhibit 21.27: A slightly skewed (3,5)-tree.

Exercise: insertion and deletion in a (3,5)-tree

Starting with the (3,5)-tree shown in Exhibit 21.27, perform the sequence of operations: insert 38, delete 10, delete 12, delete 50. Draw the tree after each operation.

Solution

Inserting 38 causes a leaf and its parent to split (Exhibit 21.28). Deleting 10 causes underflow, remedied by borrowing an element from the left sibling (Exhibit 21.29). Deleting 12 causes underflow in both a leaf and its parent, remedied by merging (Exhibit 21.30). Deleting 50 causes merging at the leaf level and borrowing at the parent level (Exhibit 21.31).

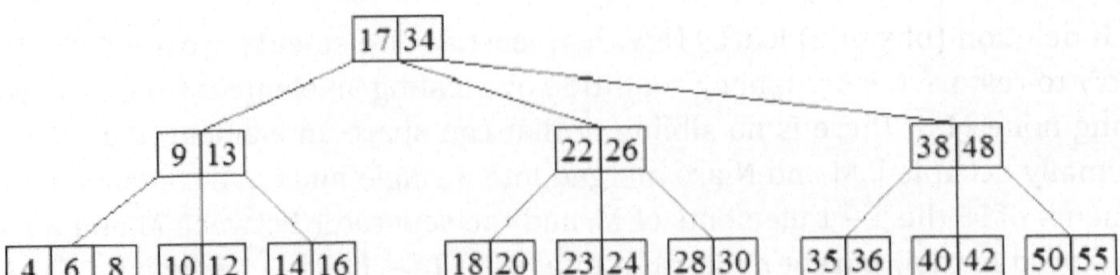

Exhibit 21.28: Node splits propagate towards the root

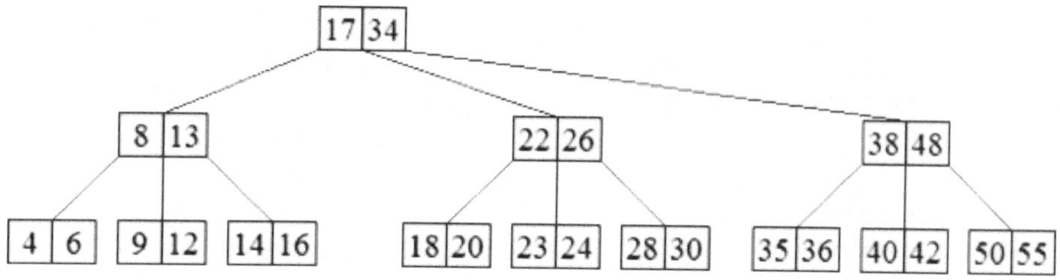

Exhibit 21.29: A deletion is absorbed by borrowing

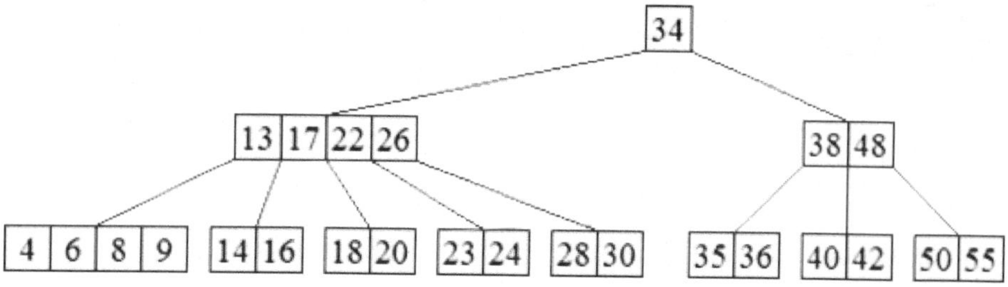

Exhibit 21.30: Another deletion propagates node merges towards the root

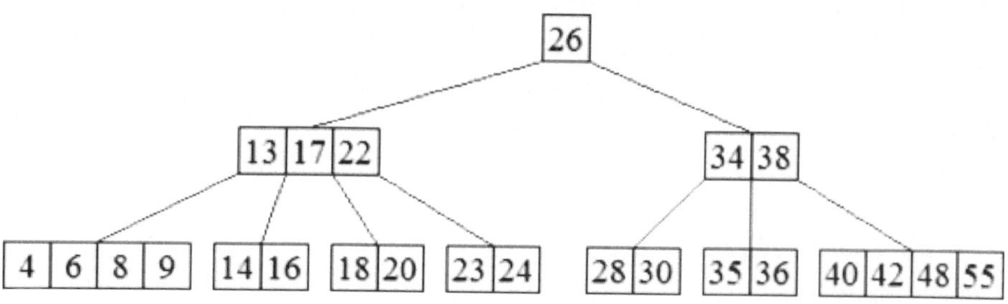

Exhibit 21.31: Node merges and borrowing combined

(2,3)-trees are the special case a = 2, b = 3: each node has two or three children. Exhibit 21.32 omits the leaves. Starting with the tree in state 1 we insert the value 9: the rightmost node at the bottom level overflows and splits, the median 8 moves up into the parent. The parent also overflows, and the median 6 generates a new root (state 2). The deletion of 1 is absorbed without any rebalancing (state 3). The deletion of 2 causes a node to underflow, remedied by stealing an element from a sibling: 2 is replaced by 3 and 3 is replaced by 4 (state 4). The deletion of 3 triggers the merger of the nodes assigned to 3 and 5; this causes an underflow in their parent, which in turn propagates to the root and results in a tree of reduced height (state 5).

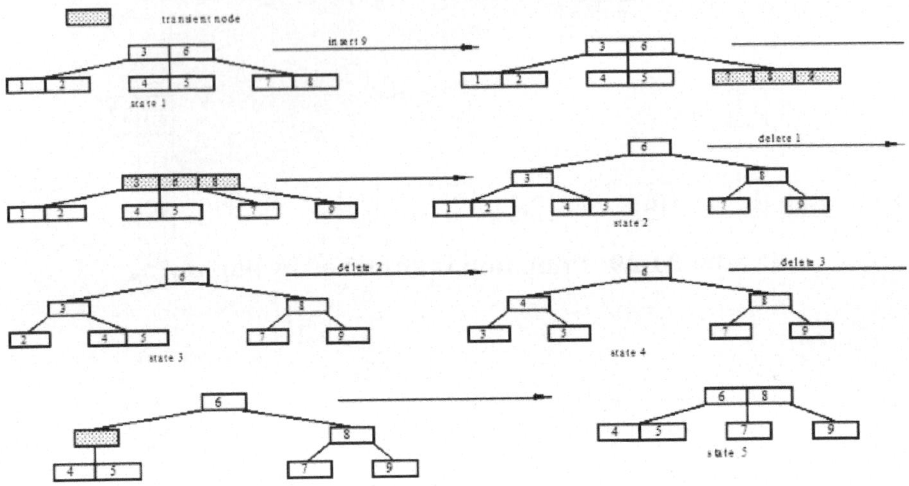

Exhibit 21.32: Tracing insertions and deletions in a (2,3)-tree

As mentioned earlier, multiway trees are particularly useful for managing data on a disk. If each node is allocated to its own disk block, searching for a record triggers as many disk accesses as there are levels in the tree. The depth of the tree is minimized if the maximal fan-out b is maximized. We can pack more elements into a node by shrinking their size. As the records to be stored are normally much larger than their identifying keys, we store keys only in the internal nodes and store entire records in the leaves (which we had considered to be empty until now). Thus the internal nodes serve as an index that assigns to a key value the path to the corresponding leaf.

Exercises and programming projects

- Design and implement a list structure for storing a sparse matrix. Your implementation should provide procedures for inserting, deleting, changing, and reading matrix elements.
- Implement a fifo queue by a circular list using only one external pointer f and a sentinel. f always points to the sentinel and provides access to the head and tail of the queue.
- Implement a double-ended queue (deque) by a doubly linked list.
- *Binary search trees and sorting* A binary search tree given by the following declarations is used to manage a set of integers:

```
type nptr = ^node
  node = record  L, R: nptr;  x: integer  end;
var  root: nptr;
```

The empty tree is represented as root = nil.

- Draw the result of inserting the sequence 6, 15, 4, 2, 7, 12, 5, 18 into the empty tree.
- Write a procedure smallest(var x: integer); which returns the smallest number stored in the tree, and a procedure remove smallest; which deletes it. If the tree is empty both procedures should call a procedure message('tree is empty');
- Write a procedure sort; that sorts the numbers stored in var a: array[1 .. n] of integer; by inserting the numbers into a binary search tree, then writing them back to the array in sorted order as it traverses the tree.
- Analyze the asymptotic time complexity of 'sort' in a typical and in the worst case.
- Does this approach lead to a sorting algorithm of time complexity Θ (ν •λογ ν)
- Extend the implementation of a dictionary as a binary search tree in the "Binary search trees" section to support the operations 'succ' and 'pred' as defined in chapter 19 in the section "Dictionary".
- *Insertion and deletion in AVL-trees:* Starting with an empty AVL-tree, insert 1, 2, 5, 6, 7, 8, 9, 3, 4, in this order. Draw the AVL-tree after each insertion. Now delete all elements in the opposite order of insertion (i.e. in last-in-first-out order). Does the AVL-tree go through the same states as during insertion but in reverse order?
- Implement an AVL-tree supporting the dictionary operations 'insert', 'delete', 'member', 'pred', and 'succ'.
- Explain how to find the smallest element in an (a,b)-tree and how to find the predecessor of a given element in an (a,b)-tree.
- Implement a dictionary as a B-tree.

22

Address Computation

Concepts and terminology

The term *address computation* (also *hashing, hash coding, scatter storage*, or *key-to-address transformations*) refers to many search techniques that aim to assign an *address* of a storage cell to any *key value* x by means of a formula that depends on x only. Assigning an address to x independently of the presence or absence of other key values leads to faster access than is possible with the comparative search techniques discussed in earlier chapters. Although this goal cannot always be achieved, address computation does provide the fastest access possible in many practical situations.

We use the following concepts and terminology (Exhibit 22.1). The *home address* a of x is obtained by means of a *hash function* h that maps the *key domain* X into the *address space* A [i.e. a = h(x)]. The address range is A = {0, 1, ... , m − 1}, where m is the number of storage cells available. The storage cells are represented by an array T[0 .. m − 1], the *hash table*; T[a] is the cell addressed by a ∈ A. T[h(x)] is the cell where an element with key value x is *preferentially* stored, but alas, not necessarily.

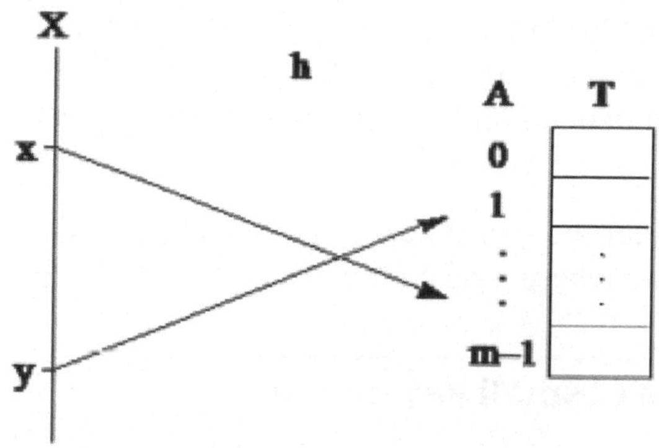

Exhibit 22.1: The hash function h maps a (typically large) key domain X into a (much smaller) address space A.

Each cell has a capacity of b > 0 elements; b stands for *bucket capacity*. The number n of elements to be stored is therefore bounded by m · b. Two cases are usefully distinguished, depending on whether the hash table resides on disk or in central memory:

- Disk or other secondary storage device: Considerations of efficiency suggest that a bucket be identified with a physical unit of transfer, typically a disk block. Such a unit is usually large compared to the size of an element, and thus b > 1.
- Main memory: Cell size is less important, but the code is simplest if a cell can hold exactly one element (i.e. b = 1).

For simplicity of exposition we assume that b = 1 unless otherwise stated; the generalization to arbitrary b is straightforward.

The key domain X is normally much larger than the number n of elements to be stored and the number m of available cells T[a]. For example, a table used for storing a few thousand identifiers might have as its key domain the set of strings of length at most 10 over the alphabet {'a', 'b', ... , 'z', '0', ... , '9'}; its cardinality is close to 3610. Thus in general the function h is many-to-one: Different key values map to the same address.

The content to be stored is a sample from the key domain: It is not under the programmer's control and is usually not even known when the hash function and table size are chosen. Thus we must expect *collisions*, that is, events where more than b elements to be stored are assigned the same address. *Collision resolution* methods are designed to handle this case by storing some of the colliding elements elsewhere. The more collisions that occur, the longer the search time. Since the number of collisions is a random event, the search time is a random variable. Hash tables are known for excellent average performance and for terrible worst-case performance, which, one hopes, will never occur.

Address computation techniques support the operations 'find' and 'insert' (and to a lesser extent also 'delete') in expected time O(1). This is a remarkable difference from all other data structures that we have discussed so far, in that the average time complexity does not depend on the number n of elements stored, but on the *load factor* $\lambda = n / (m \cdot b)$, or, for the special case b = 1: $\lambda = n / m$. Note that $0 \le \lambda \le 1$.

Before we consider the typical case of a hash table, we illustrate these concepts in two special cases where everything is simple; these represent ideals rarely attainable.

The special case of small key domains

If the number of possible key values is less than or equal to the number of available storage cells, h can map X one-to-one into or onto A. Everything is simple and efficient because collisions never occur. Consider the following example:

X = {'a', 'b', ... , 'z'}, A = {0, ... , 25}

h(x) = ord(x) – ord('a'); that is,

h('a') = 0, h('b') = 1, h('c') = 2, ... , h('z') = 25.

Since h is one-to-one, each key value x is implied by its address h(x). Thus we need not store the key values explicitly, as a single bit (present / absent) suffices:

```
var  T: array[0 .. 25] of boolean;

function member(x): boolean;
begin  return(T[h(x)])  end;
  procedure insert(x);
  begin  T[h(x)] := true  end;

  procedure delete(x);
  begin  T[h(x)] := false  end;
```

The idea of collision-free address computation can be extended to large key domains through a combination of address computation and list processing techniques, as we will see in the chapter "Metric data structures".

The special case of perfect hashing: table contents known a priori

Certain common applications require storing a set of elements that never changes. The set of reserved words of a programming language is an example; when the lexical analyzer of a compiler extracts an identifier, the first issue to be determined is whether this is a reserved word such as 'begin' or 'while', or whether it is programmer defined. The special case where the table contents are known a priori, and no insertions or deletions occur, is handled more efficiently by special-purpose data structures than by a general dictionary.

If the elements $x_1, x_2, \ldots, x_n$ to be stored are known before the hash table is designed, the underlying key domain is not as important as the set of actually occurring key values. We can usually find a table size m, not much larger than the number n of elements to be stored, and an easily evaluated hash function h that assigns to each xi a unique address from the address space {0, ... , m – 1}. It takes some trial and error to find such a *perfect hash function* h for a given set of elements, but the benefit of avoiding collisions is well worth the effort —the code that implements a collision-free hash table is simple and fast. A perfect hash function works for a static table only —a single insertion, after h has been chosen, is likely to cause a collision and destroy the simplicity of the concept and efficiency of the implementation. Perfect hash functions should be generated automatically by a program.

The following unrealistically small example illustrates typical approaches to designing a perfect hash table. The task gets harder as the number m of available storage cells is reduced toward the minimum possible, that is, the number n of elements to be stored.

Example

In designing a perfect hash table for the elements 17, 20, 24, 38, and 51, we look for arithmetic patterns. These are most easily detected by considering the binary representations of the numbers to be stored:

	5	4	3	2	1	0	bit position
17	0	1	0	0	0	1	
20	0	1	0	1	0	0	
24	0	1	1	0	0	0	
38	1	0	0	1	1	0	
51	1	1	0	0	1	1	

We observe that the least significant three bits identify each element uniquely. Therefore, the hash function h(x) = x mod 8 maps these five elements collision-free into the address space A = {0, ... , 6}, with m = 7 and two empty cells. An attempt to further economize space leads us to observe that the bits in positions 1, 2, and 3, with weights 2, 4, and 8 in the binary number representation, also identify each element uniquely, while ranging over the address space of minimal size A = {0, ... , 4}. The function h(x) = (x div 2) mod 8 extracts these three bits and assigns the following addresses:

```
X: 17 20 24 38 51

A: 0  2  4  3  1
```

A perfect hash table has to store each element explicitly, not just a bit (present/absent). In the example above, the elements 0, 1, 16, 17, 32, 33, ... all map into address 0, but only 17 is present in the table. The access function 'member(x)' is implemented as a single statement:

```
return ((h(x) ≤ 4) cand (T[h(x)] = x));
```

The boolean operator 'cand' used here is understood to be the *conditional and*: Evaluation of the expression proceeds from left to right and stops as soon as its value is determined. In our example, h(x) > 4 suffices to assign 'false' to the expression (h(x) ≤ 4) and (T[h(x)] = x). Thus the 'cand' operator guarantees that the table declared as:

```
var   T: array[0 .. 4] of element;
```

is accessed within its index bounds.

For table contents of realistic size it is impractical to construct a perfect hash function manually—we need a program to search exhaustively through the large space of functions. The more slack m – n we allow, the denser is the population of perfect functions and the quicker we will find one. [Meh 84a] presents analytical results on the complexity of finding perfect hash functions.

Exercise: perfect hash tables

Design several perfect hash tables for the content {3, 13, 57, 71, 82, 93}.

Solution

Designing a perfect hash table is like answering a question of the type: What is the next element in the sequence 1, 4, 9, ... ? There are infinitely many answers, but some are more elegant than others. Consider:

h	3	13	57	71	82	93	Address range
(x div 3) mod 7	1	4	5	2	6	3	[1 .. 6]
x mod 13	3	0	5	6	4	2	[0 .. 6]
(x div 4) mod 8	0	3	6	1	4	7	[0 .. 7]
if x = 71 then 4 else x mod 7	3	6	1	**4**	5	2	[1 .. 6]

Conventional hash tables: collision resolution

In contrast to the special cases discussed, most applications of address computation present the data structure designer with greater uncertainties and less favorable conditions. Typically, the underlying key domain is much larger than the available address range, and not much is known about the elements to be stored. We may have an upper bound on n, and we may know the probability distribution that governs the random sample of elements to be stored. In setting up a customer list for a local business, for example, the number of customers may be bounded by the population of the town, and the distribution of last names can be obtained from the telephone directory —many names will start with H and S, hardly any with Q and Y. On the basis of such information, but in ignorance of the actual table contents to be stored, we must choose the size m of the hash table and design the hash function h that maps the key domain X into the address space A= {0, ... , m – 1}. We will then have to live with the consequences of these decisions, at least until we decide to *rehash*: that is, resize the table, redesign the hash function, and reinsert all the elements that we have stored so far.

Later sections present some pragmatic advice on the choice of h; for now, let us assume that an appropriate hash function is available. Regardless of how smart a hash function we have designed, collisions (more than b elements share the same home address of a bucket of capacity b) are inevitable in practice. Thus hashing requires techniques for handling collisions. We present the three major *collision resolution* techniques in use: separate chaining, coalesced chaining, and open addressing. The two techniques called *chaining* call upon list processing techniques to organize overflowing elements. *Separate chaining* is used when these lists live in an overflow area distinct from the hash table proper; *coalesced chaining* when the lists live in unused parts of the table. *Open addressing* uses address computation to organize overflowing elements. Each of these three techniques comes in different variations; we illustrate one typical choice.

Separate chaining

The memory allocated to the table is split into a *primary* and an *overflow* area. Any overflowing cell or bucket in the primary area is the head of a list, called the *overflow chain*, that holds all elements that overflow from that bucket. Exhibit 22.2 shows six elements inserted in the order x1, x2, ... The first arrival resides at its home address; later ones get appended to the overflow chain.

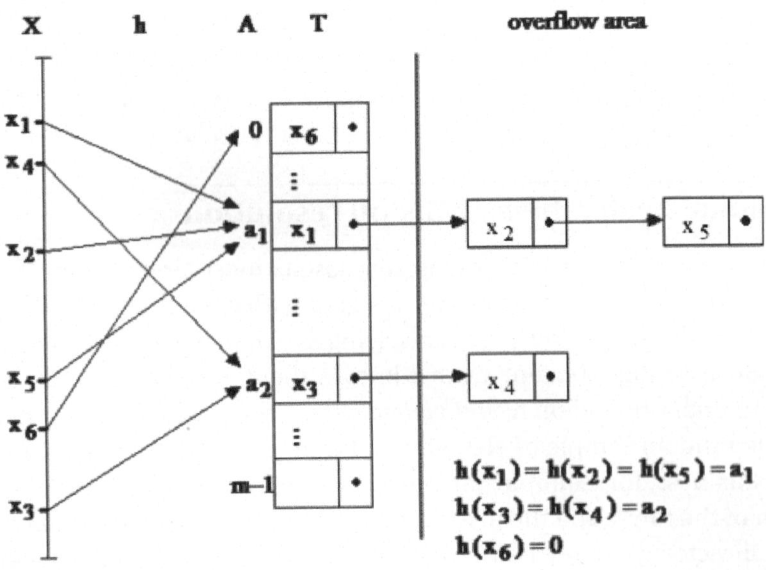

Exhibit 22.2: Separate chaining handles collisions in a separate overflow area.

Separate chaining is easy to understand: insert, delete, and search operations are simple. In contrast to other collision handling techniques, this hybrid between address computation and list processing has two major advantages: (1) deletions do not degrade the performance of the hash table, and (2) regardless of the number m of home addresses, the hash table will not overflow until the entire memory is exhausted. The size m of the table has a critical influence on the performance. If m « n, overflow chains are long and we have essentially a list processing technique that does not support direct access. If m » n, overflow chains are short but we waste space in the table. Even for the practical choice m ≈ n, separate chaining has some disadvantages:

- Two different accessing techniques are required.
- Pointers take up space; this may be a significant overhead for small elements.
- Memory is partitioned into two separate areas that do not share space: If the overflow area is full, the entire table is full, even if there is still space in the array of home cells. This consideration leads to the next technique.

Coalesced chaining

The chains that emanate from overflowing buckets are stored in the empty space in the hash table rather than in a separate overflow area (Exhibit 22.3). This has the advantage that all available space is utilized fully (except for the overhead of the pointers). However, managing the space shared between the two accessing techniques gets complicated.

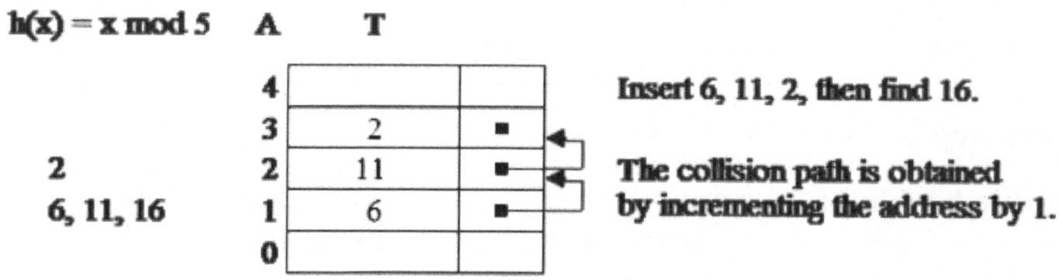

Exhibit 22.3: Coalesced chaining handles collisions by building lists that share memory with the hash table.

The next technique has similar advantages (in addition, it incurs no overhead for pointers) and disadvantages; all things considered, it is probably the best collision resolution technique.

Open addressing

Assign to each element $x \in X$ a *probe sequence* a0 = h(x), a_1, a_2, ... of addresses that fills the entire address range A. The intention is to store x preferentially at a0, but if T[a0] is occupied then at a1, and so on, until the first empty cell is encountered along the probe sequence. The occupied cells along the probe sequence are called the *collision path* of x—note that the collision path is a prefix of the probe sequence. If we enforce the *invariant*:

If x is in the table at T[a] and if i precedes a in the probe sequence for x, then T[i] is occupied. The following fast and simple loop that travels along the collision path can be used to search for x:

```
a := h(x);
while  T[a] ≠ x  and  T[a] ≠ empty  do
   a := (next address in probe sequence);
```

Let us work out the details so that this loop terminates correctly and the code is as concise and fast as we can make it.

The probe sequence is defined by formulas in the program (an example of an implicit data structure) rather than by pointers in the data as is the case in coalesced chaining.

Example: linear probing

$a_{i+1} = (a_i + 1) \bmod m$ is the simplest possible formula. Its only disadvantage is a phenomenon called *clustering*.

Clustering arises when the collision paths of many elements in the table overlap to a large extent, as is likely to happen in linear probing. Once elements have collided, linear probing will store them in consecutive cells. All elements that hash into this block of contiguous occupied cells travel along the same collision path, thus lengthening this block; this in turn increases the probability that future elements will hash into this block. Once this positive feedback loop gets started, the cluster keeps growing.

Double hashing is a special type of open addressing designed to alleviate the clustering problem by letting different elements travel with steps of different size. The probe sequence is defined by the formulas

$a_0 = h(x),\ \delta = g(x) > 0,\ a_{i+1} = (ai + \delta) \bmod m$, m prime

g is a second hash function that maps the key space X into [1 .. m – 1].

Two important important details must be solved:

- The probe sequence of each element must span the entire address range A. This is achieved if m is relatively prime to every step size δ, and the easiest way to guarantee this condition is to choose m prime.
- The termination condition of the search loop above is: T[a] = x or T[a] = empty. An unsuccessful search (x not in the table) can terminate only if an address a is generated with T[a] = empty. We have already insisted that each probe sequence generates all addresses in A. In addition, we must guarantee that the table contains at least one empty cell at all times—this serves as a sentinel to terminate the search loop.

The following declarations and procedures implement double hashing. We assume that the comparison operators = and ≠ are defined on X, and that X contains a special value 'empty', which differs from all values to be stored in the table. For example, a string of blanks might denote 'empty' in a table of identifiers. We choose to identify an unsuccessful search by simply returning the address of an empty cell.

```
const   m = ... ;    { size of hash table - must be prime! }
   empty = ... ;
type key = ... ;   addr = 0 .. m - 1;   step = 1 .. m - 1;
var  T: array[addr] of key;
   n: integer;   { number of elements currently stored in T }

function h(x: key): addr;   { hash function for home address }

function g(x: key): step;   { hash function for step }

procedure init;
var  a: addr;
begin
   n := 0;
   for a := 0 to m - 1 do  T[a] := empty
end;

function find(x: key): addr;
var  a: addr;  d: step;
begin
   a := h(x);  d := g(x);
   while  (T[a] ≠ x) and (T[a] ≠ empty)  do  a := (a + d) mod m;
   return(a)
end;

function insert(x: key): addr;
var  a: addr;  d: step;
begin
   a := h(x);  d := g(x);
   while  T[a] ≠ empty  do  begin
      if  T[a] = x  then  return(a);
      a := (a + d) mod m
   end;
   if  n < m - 1  then  { n := n + 1;  T[a] := x }  else  err-
msg('table is full');
   return(a)
end;
```

Deletion of elements creates problems, as is the case in many types of hash tables. An element to be deleted cannot simply be replaced by 'empty', or else it might break the collision paths of other elements still in the table— recall the basic invariant on which the correctness of open addressing is based. The idea of rearranging elements in the table so as to refill a cell that was emptied but needs to remain full is quickly abandoned as too complicated —if deletions are numerous, the programmer ought to choose a data structure that fully supports deletions, such as balanced trees implemented as list structures. A limited number of deletions can be accommodated in an open address hash table by using the following technique.

At any time, a cell is in one of three states:

- empty (was never occupied, the initial state of all cells)
- occupied (currently)
- deleted (used to be occupied but is currently free)

A cell in state 'empty' terminates the find loop; a cell in state 'empty' or in state 'deleted' terminates the insert loop. The state diagram shown in Exhibit

22.4 describes the transitions possible in the lifetime of a cell. Deletions degrade the performance of a hash table, because a cell, once occupied, never returns to the virgin state 'empty' which alone terminates an unsuccessful find. Even if an equal number of insertions and deletions keeps a hash table at a low load factor λ, unsuccessful finds will ultimately scan the entire table, as all cells drift into one of the states 'occupied' or 'deleted'. Before this occurs, the table ought to be rehashed; that is, the contents are inserted into a new, initially empty table.

Exhibit 22.4: This state diagram describes possible life cycles of a cell: Once occupied, a cell will never again be as useful as an empty cell.

Exercise: hash table with deletions

Modify the program above to implement double hashing with deletions.

Choice of hash function: randomization

In conventional terminology, hashing is based on the concept of **randomization**. The purpose of randomizing is to transform an **unknown distribution** over the key domain X into a uniform distribution, and to turn consecutive samples that may be dependent into independent samples. This task appears to call for magic, and indeed, there is little or no mathematics that applies to the construction of hash functions; but there are commonsense observations worth remembering. These observations are primarily "don'ts". They stem from properties that sets of elements we wish to store frequently possess, and thus are based on *some* knowledge about the populations to be stored. If we assumed strictly nothing about these populations, there would be little to say about hash functions: an order-preserving proportional mapping of X into A would be as good as any other function. But in practice it is not, as the following examples show.

- A Fortran compiler might use a hash table to store the set of identifiers it encounters in a program being compiled. The rules of the language and human habits conspire to make this set a highly biased sample from the set of legal Fortran identifiers. *Example:* Integer variables begin with I, J, K, L, M, N; this convention is likely to generate a cluster of identifiers that begin with one of these letters. *Example:* Successive identifiers encountered cannot be considered independent samples: If X and Y have occurred, there is a higher chance for Z to follow than for WRKHG. *Example:* Frequently, we see sequences of identifiers or statement numbers whose character codes form arithmetic progressions, such as A_1, A_2, A_3, ... or 10, 20, 30, ...

- All file systems require or encourage the use of naming conventions, so that most file names begin or end with one of just a few prefixes or suffixes, such as ⋯.SYS, ⋯.BAK, ⋯.OBJ. An individual user, or a user community, is likely to generate additional conventions, so that most file names might begin, for example, with the initials of the names of the people involved. The files that store this text, for example, are structured according to 'part' and 'chapter', so we are currently in file P5 C22. In some directories, file names might be sorted alphabetically, so if they are inserted into a table in order, we process a monotonic sequence.

The purpose of a hash function is to break up all regularities that might be present in the set of elements to be stored. This is most reliably achieved by "hashing" the elements, a word for which the dictionary offers the following explanations: (1) from the French *hache*, "battle-ax"; (2) to chop into small pieces; (3) to confuse, to muddle. Thus, to approximate the elusive goal of randomization, a hash function destroys patterns, including, unfortunately, the order $<$ defined on X. Hashing typically proceeds in two steps.

- Convert the element x into a number #(x). In most cases #(x) is an integer, occasionally, it is a real number $0 \le \#(x) < 1$. Whenever possible, this conversion of x into #(x) involves no action at all: The representation of x, whatever type x may be, is reinterpreted as the representation of the number #(x). When x is a variable-length item, for example a string, the representation of x is partitioned into pieces of suitable length that are "folded" on top of each other. For example, the four-letter word x = 'hash' is encoded one letter per byte using the 7-bit ASCII code and a leading 0 as 01101000 01100001 01110011 01101000. It may be folded to form a 16-bit integer by exclusive-or of the leading pair of bytes with the trailing pair of bytes:

  0110100001100001

  xor 0111001101101000

  0001101100001001 which represents $\#(x) = 27 \cdot 2^8 + 9 = 6921$.

Such folding, by itself, is not hashing. Patterns in the representation of elements easily survive folding. For example, the leading 0 we have used to pad the 7-bit ASCII code to an 8-bit byte remains a zero regardless of x. If we had padded with a trailing zero, all #(x) would be even. Because #(x) often has the same representation as x, or a closely related one, we drop #() and use x slightly ambiguously to denote both the original element and its interpretation as a number.

- Scramble x [more precisely, #(x)] to obtain h(x). Any scrambling technique is a sensible try, as long as it avoids fairly obvious pitfalls. Rules of thumb:

- Each bit of an address h(x) should depend on all bits of the key value x. In particular, don't ignore any part of x in computing h(x). Thus h(x) = x mod 213 is suspect, as only the least significant 13 bits of x affect h(x).

- Make sure that arithmetic progressions such as Ch1, Ch2, Ch3, ... get broken up rather than being mapped into arithmetic progressions. Thus h(x) = x mod k, where k is significantly smaller than the table size m, is suspect.

- Avoid any function that cannot produce a uniform distribution of addresses. Thus h(x) = x2 is suspect; if x is uniformly distributed in [0, 1], the distribution of x2 is highly skewed.

A hash function must be fast and simple. All of the desiderata above are obtained by a hash function of the type: h(x) = x mod m

where m is the table size and a prime number, and x is the key value interpreted as an integer.

No hash function is guaranteed to avoid the worst case of hashing, namely, that all elements to be stored collide on one address (this happens here if we store only multiples of the prime m). Thus a hash function must be judged in relation to the data it is being asked to store, and usually this is possible only after one has begun using it. Hashing provides a perfect example for the injunction that the programmer must think about the data, analyze its statistical properties, and adapt the program to the data if necessary.

Performance analysis

We analyze open addressing without deletions assuming that each address αi is chosen independently of all other addresses from a uniform distribution over A. This assumption is reasonable for double hashing and leads to the conclusion that the average cost for a search operation in a hash table is O(1) if we consider the load factor λ to be constant. We analyze the average number of probes executed as a function of λ in two cases: U(λ) for an unsuccessful search, and S(λ) for a successful search.

Let p_i denote the probability of using *exactly* i probes in an unsuccessful search. This event occurs if the first I – 1 probes hit occupied cells, and the i-th probe hits an empty cell: pi = λi–1 · (1 – λ). Let qi denote the probability that *at least* i probes are used in an unsuccessful search; this occurs if the first i – 1 inspected cells are occupied: qi = λi–1. qi can also be expressed as the sum of the probabilities that we probe exactly j cells, for j running from i to m. Thus we obtain

$$U(\lambda) = \sum_{i=1}^{m} i \cdot p_i = \sum_{i=1}^{m} \sum_{j=i}^{m} p_j = \sum_{i=1}^{m} q_i = \sum_{i=1}^{m} \lambda^{i-1} \approx \frac{1}{1-\lambda}.$$

The number of probes executed in a successful search for an element x equals the number of probes in an unsuccessful search for the same element x before it is inserted into the hash table. [Note: This holds only when elements are never relocated or deleted]. Thus the average number of probes needed to search for the i-th element inserted into the hash table is U((i – 1) / m), and S(λ) can be computed as the average of U(μ), for μ increasing in discrete steps from 0 to λ. It is a reasonable approximation to let μ vary continuously in the range from 0 to λ:

$$S(\lambda) \approx \frac{1}{\lambda}\int_0^{\lambda} U(\mu)\,d\mu = \frac{1}{\lambda}\int_0^{\lambda} \frac{d\mu}{1-\mu} = \frac{1}{\lambda}\ln\frac{1}{1-\lambda}.$$

Exhibit 22.5 suggests that a reasonable operating range for a hash table keeps the load factor λ between 0.25 and 0.75. If λ is much smaller, we waste space, if it is larger than 75 per cent, we get into a domain where the performance degrades rapidly. Note: If all searches are successful, a hash table performs well even if loaded up to 95 per cent—unsuccessful searching is the killer!

Table 22.1: The average number of probes per search grows rapidly as the load factor approaches 1.

λ	0.25	0.5	0.75	0.9	0.95	0.99
U(λ)	1.3	2.0	4.0	10.0	20.0	100.0
S(λ)	1.2	1.4	1.8	2.6	3.2	4.7

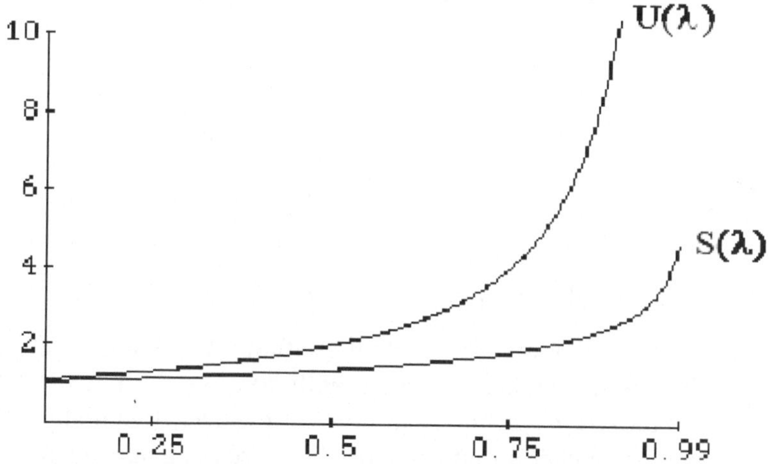

Exhibit 22.5: The average number of probes per search grows rapidly as the load factor approaches 1.

Thus the hash table designer should be able to estimate n within a factor of 2—not an easy task. An incorrect guess may waste memory or cause poor performance, even table overflow followed by a crash. If the programmer becomes aware that the load factor lies outside this range, she may rehash—change the size of the table, change the hash function, and reinsert all elements previously stored.

Extendible hashing

In contrast to standard hashing methods, extendible forms of hashing allow for the dynamic extension or shrinkage of the address range into which the hash function maps the keys. This has two major advantages: (1) Memory is allocated only as needed (it is unnecessary to determine the size of the address range a priori), and (2) deletion of elements does not degrade performance. As the address range changes, the hash function is changed in such a way that only a few elements are assigned a new address and need to be stored in a new bucket. The idea that makes this possible is to map the keys into a very large address space, of which only a portion is active at any given time.

Various extendible hashing methods differ in the way they represent and manage a smaller *active address range* of variable size that is a subrange of a larger *virtual address range*. In the following we describe the method of extendible hashing that is especially well suited for storing data on secondary storage devices; in this case an address points to a physical block of secondary storage that can contain more than one element. An address is a bit string of maximum length k; however, at any time only a prefix of d bits is used. If all bit strings of length k are represented by a so-called *radix tree* of height k, the active part of all bit strings is obtained by using only the upper d levels of the tree (i.e. by cutting the tree at level d). Exhibit 22.6 shows an example for d = 3.

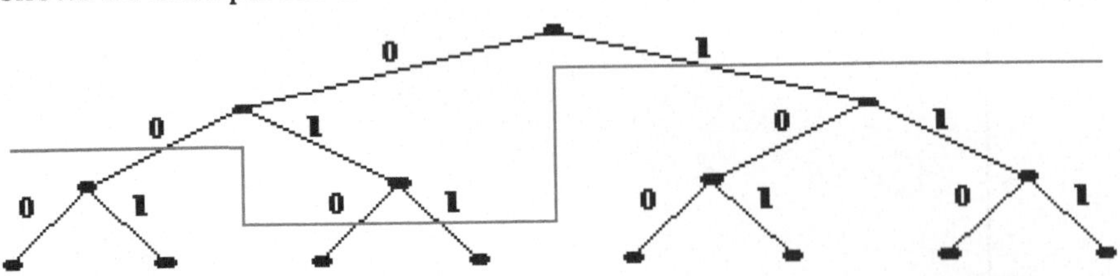

Exhibit 22.6: Address space organized as a binary radix tree.

The radix tree shown in Exhibit 22.6 (without the nodes that have been clipped) describes an active address range with addresses {00, 010, 011, 1} that are considered as bit strings or binary numbers. To each active node with address s there corresponds a bucket B that can store b records. If a new element has to be inserted into a full bucket B, then B is split: Instead of B we find two twin buckets B_0 and B_1 which have a one bit longer address than B, and the elements stored in B are distributed among B_0 and B_1 according to this bit. The new radix tree now has to point to the two data buckets B_0 and B_1 instead of B; that is, the active address range must be extended locally (by moving the broken line in Exhibit 22.6). If the block with address 00 overflows, two new twin blocks with addresses 000 and 001 will be created which are represented by the corresponding nodes in the tree. If the overflowing bucket B has depth d, then d is incremented by 1 and the radix tree grows by one level.

In extendible hashing the clipped radix tree is represented by a directory that is implemented by an array. Let d be the maximum number of bits that are used in one of the bit strings for forming an address; in the example above, d = 3. Then the directory consists of 2d entries. Each entry in this directory corresponds to an address and points to a physical data bucket which contains all elements that have been assigned this address by the hash function h. The directory for the radix tree in Exhibit 22.6 looks as shown in Exhibit 22.7.

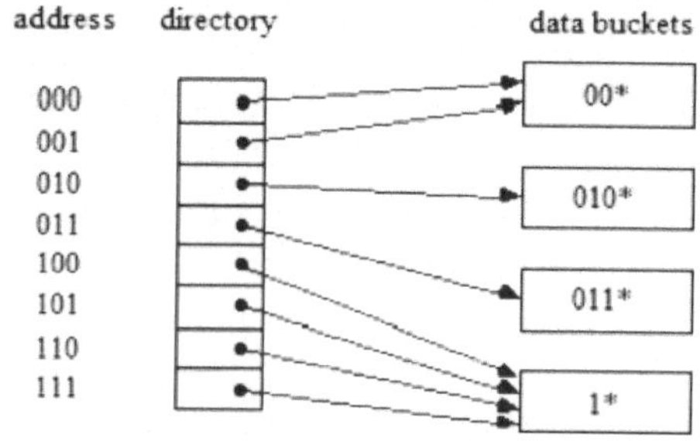

Exhibit 22.7: The active address range of the tree in Exhibit 22.6 implemented as an array.

The bucket with address 010 corresponds to a node on level 3 of the radix tree, and there is only one entry in the directory corresponding to this bucket. If this bucket overflows, the directory and data buckets are reorganized as shown in Exhibit 22.8. Two twin buckets that jointly contain fewer than b elements are merged into a single bucket. This keeps the average bucket occupancy at a high 70 per cent even in the presence of deletions, as probabilistic analysis predicts and simulation results confirm. Bucket merging may lead to halving the directory. A formerly large file that shrinks to a much smaller size will have its directory shrink in proportion. Thus extendible hashing, unlike conventional hashing, suffers no permanent performance degradation under deletions.

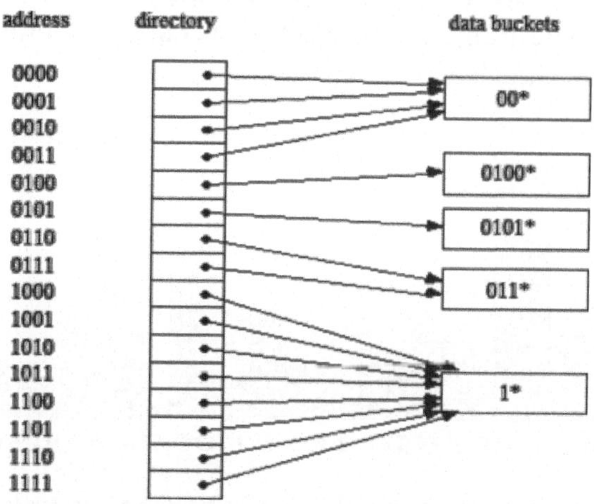

Exhibit 22.8: An overflowing bucket may trigger doubling of the directory.

A virtual radix tree: order-preserving extendible hashing

Hashing, in the usual sense of the word, destroys structure and thus buys uniformity at the cost of order. Extendible hashing, on the other hand, is practical without randomization and thus needs not accept its inevitable consequence, the destruction of order. A uniform distribution of elements is not nearly as important: Nonuniformity causes the directory to be deeper and thus larger than it would be for a uniform distribution, but it affects neither access time nor bucket occupancy. And the directory is only a small space overhead on top of the space required to store the data: It typically contains only one or a few pointers, say a dozen bytes, per data bucket of, say 1k bytes; it adds perhaps a few percent to the total space requirement of the table, so its growth is not critical. Thus extendible hashing remains feasible when the identity is used as the address computation function h, in which case data is accessible and can be processed sequentially in the order ≤ defined on the domain X.

When h preserves order, the word *hashing* seems out of place. If the directory resides in central memory and the data buckets on disk, what we are implementing is a virtual memory organized in the form of a radix tree of unbounded size. In contrast to conventional virtual memory, whose address space grows only at one end, this address space can grow anywhere: It is a virtual radix tree.

As an example, consider the domain X of character strings up to length 32, say, and assume that elements to be stored are sampled according to the distribution of the first letter in English words. We obtain an approximate distribution by counting pages in a dictionary (Exhibit 22.9). Encode the blank as 00000, 'a' as 00001, up to 'z' as 11011, so that 'aah', for example, has the code 00001 00001 01000 00000 ... (29 quintuples of zeros pad 'aah' to32letters). This address computation function h is almost an identity: It maps {' ', 'a', ... , 'z'} 32 one-to-one into {0, 1}160. Such an order-preserving address computation function supports many useful types of operations: for example, range queries such as "list in alphabetic order all the words stored from 'unix' to 'xinu' ".

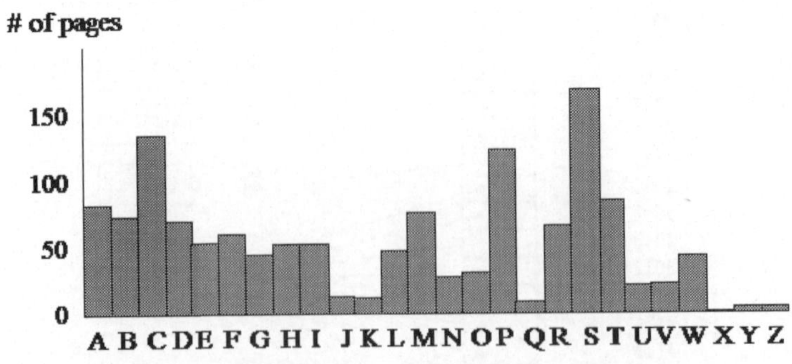

Exhibit 22.9: Relative frequency of words beginning with a given letter in Webster's dictionary.

If there is one page of words starting with X for 160 pages of words starting with S, this suggests that if our active address space is partitioned into equally sized intervals, some intervals may be populated 160 times more densely than others. This translates into a directory that may be 160 times larger than necessary for a uniform distribution, or, since directories grow as powers of 2, may be 128 or 256 times larger. This sounds like a lot but may well be bearable, as the following estimates show.

Assume that we store 105 records on disk, with an average occupancy of 100 records per bucket, requiring about 1000 buckets. A uniform distribution generates a directory with one entry per bucket, for a total of 1k entries, say 2k or 4k bytes. The nonuniform distribution above requires the same number of buckets, about 1,000, but generates a directory of 256k entries. If a pointer requires 2 to 4 bytes, this amounts to 0.5 to 1 Mbyte. This is less of a memory requirement than many applications require on today's personal computers. If the application warrants it (e.g. for an on-line reservation system) 1 Mbyte of memory is a small price to pay.

Thus we see that for large data sets, extendible hashing approximates the ideal characteristics of the special case we discussed in this chapter's section on "the special case of small key domains". All it takes is a disk and a central memory of a size that is standard today but was practically infeasible a decade ago, impossible two decades ago, and unthought of three decades ago.

Exercises and programming projects

- Design a perfect hash table for the elements 1, 10, 14, 20, 25, and 26.
- The six names AL, FL, GA, NC, SC and VA must be distinguished from all other ordered pairs of uppercase letters. To solve this problem, these names are stored in the array T such that they can easily be found by means of a hash function h.

```
type addr = 0 .. 7;
pair = record c1, c2: 'A' .. 'Z' end;
var T: array [addr] of pair;
```

- Write a which maps the six names onto different addresses in the range 'adr'.

```
function h (name: pair): adr;
```

- Write a which initializes the entries of the hash table T.

```
procedure initTable;
```

- Write a which returns for any pair of uppercase letters whether it is stored in T.

```
function member (name: pair): boolean;
```

- Consider the hash function h(x) = x mod 9 for a table having nine entries. Collisions in this hash table are resolved by coalesced chaining. Demonstrate the insertion of the elements 14, 19, 10, 6, 11, 42, 21, 8, and 1.
- Consider inserting the keys 14, 1, 19, 10, 6, 11, 42, 21, 8, and 17 into a hash table of length m = 13 using open addressing with the hash function h(x) = x mod m. Show the result of inserting these elements using
- Linear probing.
- Double hashing with the second hash function g(x) = 1 + x mod (m+1).
- Implement a dictionary supporting the operations 'insert', 'delete', and 'member' as a hash table with double hashing.
- Implement a dictionary supporting the operations 'insert', 'delete', 'member', 'succ', and 'pred' by order- preserving extendible hashing.

23

Metric Data Structures

Organizing the embedding space versus organizing its contents

Most of the data structures discussed so far organize the set of elements to be stored depending primarily, or even exclusively, on the relative values of these elements to each other and perhaps on their order of insertion into the data structure. Often, the only assumption made about these elements is that they are drawn from an ordered domain, and thus these structures support only *comparative search* techniques: the search argument is compared against stored elements. The shape of data structures based on comparative search varies dynamically with the set of elements currently stored; it does not depend on the static domain from which these elements are samples. These techniques organize the particular contents to be stored rather than the embedding space.

The data structures discussed in this chapter mirror and organize the domain from which the elements are drawn—much of their structure is determined before the first element is ever inserted. This is typically done on the basis of fixed points of reference which are independent of the current contents, as inch marks on a measuring scale are independent of what is being measured. For this reason we call data structures that organize the embedding space *metric data structures*. They are of increasing importance, in particular for *spatial data*, such as needed in computer-aided design or geographic data processing. Typically, these domains exhibit a much richer structure than a mere order: In two- or three-dimensional Euclidean space, for example, not only is *order* defined along *any line* (not just the coordinate axes), but also *distance* between any two points. Most queries about spatial data involve the absolute position of elements in space, not just their relative position among each other. A typical query in graphics, for example, asks for the first object intercepted by a given ray of light. Computing the answer involves absolute position (the location of the ray) and relative order (nearest along the ray). A data structure that supports direct access to objects according to their position in space can clearly be more efficient than one based merely on the relative position of elements.

The terms "organizing the embedding space" and "organizing its contents" suggest two extremes along a spectrum of possibilities. As we have seen in previous chapters, however, many data structures are hybrids that combine features from distinct types. This is particularly true of metric data structures: They always have aspects of address computation needed to locate elements in space, and they often use list processing techniques for efficient memory utilization.

Radix trees, tries

We have encountered binary radix trees, and a possible implementation, in chapter 22 in the section "Extendible hashing". Radix trees with a branching factor, or *fan-out*, greater than 2 are ubiquitous. The Dewey decimal classification used in libraries is a radix tree with a fan-out of 10. The hierarchical structure of many textbooks, including this one, can be seen as a radix tree with a fan-out determined by how many subsections at depth d + 1 are packed into a section at depth d.

As another example, consider *tries*, a type of radix tree that permits the retrieval of variable-length data. As we traverse the tree, we check whether or not the node we are visiting has any successors. Thus the trie can be very long along certain paths. As an example, consider a trie containing words in the English language. In Exhibit 23.1 below, the four words 'a', 'at', 'ate', and 'be' are shown explicitly. The letter 'a' is a word and is the first letter of other words. The field corresponding to 'a' contains the value 1, signaling that we have spelled a valid word, and there is a pointer to longer words beginning with 'a'. The letter 'b' is not a word, thus is marked by a 0, but it is the beginning of many words, all found by following its pointer. The string 'aa' is neither a word nor the beginning of a word, so its field contains 0 and its pointer is 'nil'.

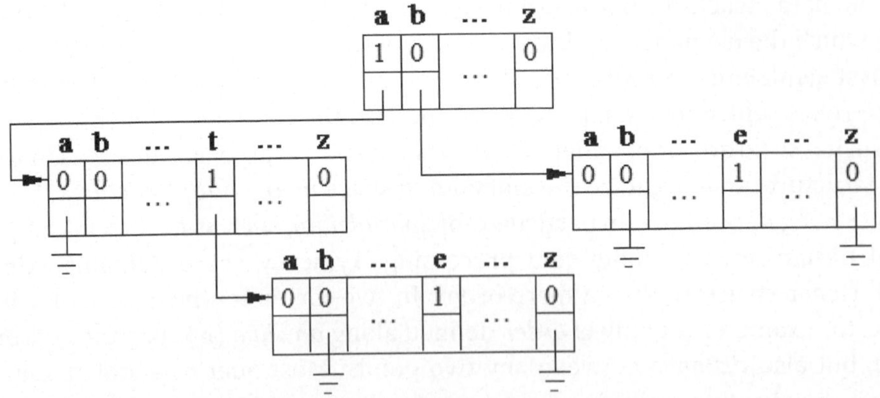

Exhibit 23.1: A radix tree over the alphabet of letters stores (prefixes of) words.

Only a few words begin with 'ate', but among these there are some long ones, such as 'atelectasis'. It would be wasteful to introduce eight additional nodes, one for each of the characters in 'lectasis', just to record this word, without making significant use of the fan-out of 26 provided at each node. Thus tries typically use an "overflow technique" to handle long entries: The pointer field of the prefix 'ate' might point to a text field that contains '(ate-)lectasis' and '(ate-)lier'.

Quadtrees and octtrees

Consider a square recursively partitioned into quadrants. Exhibit 23.2 23.2 shows such a square partitioned to the depth of 4. There are 4 quadrants at depth 1, separated by the thickest lines; 4 · 4 (sub-)quadrants separated by slightly thinner lines; 43 (sub-sub-)quadrants separated by yet thinner lines; and finally, 44 = 256 leaf quadrants separated by the thinnest lines. The partitioning structure described is a *quadtree*, a particular type of radix tree of fan-out 4. The root corresponds to the entire square, its 4 children to the 4 quadrants at depth 1, and so on, as shown in the Exhibit 23.2.

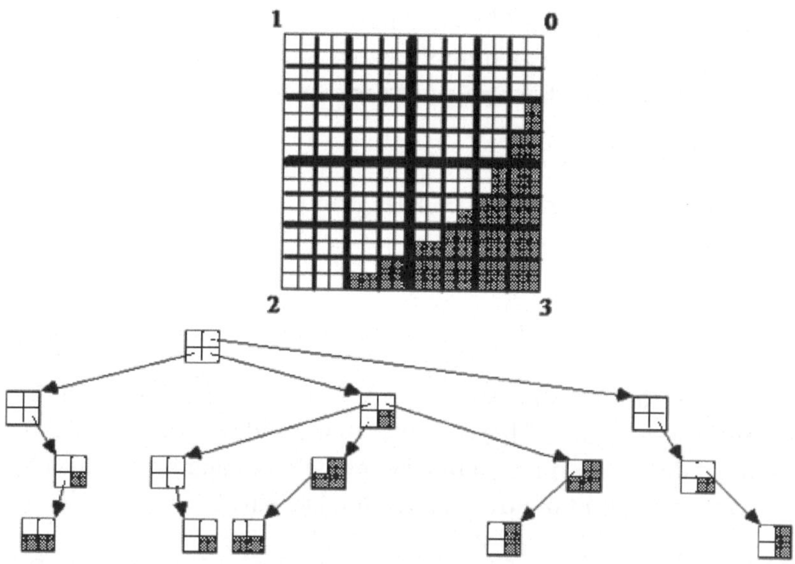

Exhibit 23.2: A quarter circle digitized on a 16 · 16 grid, and its representation as a 4-level quadtree.

A quadtree is the obvious two-dimensional analog of the one-dimensional binary radix tree we have seen. Accordingly, quadtrees are frequently used to represent, store, and process spatial data, such as images. The figure shows a quarter circle, digitized on a 16 · 16 grid of pixels. This image is most easily represented by a 16 · 16 array of bits. The quadtree provides an alternative representation that is advantageous for images digitized to a high level of resolution. Most graphic images in practice are digitized on rectangular grids of anywhere from hundreds to thousands of pixels on a side: for example, 512 · 512. In a quadtree, only the largest quadrants of constant color (black or white, in our example) are represented explicitly; their subquadrants are implicit.

The quadtree in Exhibit 23.2 is interpreted as follows. Of the four children of the root, the northwest quadrant, labeled 1, is simple: entirely white. This fact is recorded in the root. The other three children, labeled 0, 2, and 3, contain both black and white pixels. As their description is not simple, it is contained in three quadtrees, one for each quadrant. Pointers to these subquadtrees emanate from the corresponding fields of the root.

The southwestern quadrant labeled 2 in turn has four quadrants at depth 2. Three of these, labeled 2.0, 2.1, and 2.2, are entirely white; no pointers emanate from the corresponding fields in this node. Subquadrant 2.3 contains both black and white pixels; thus the corresponding field contains a pointer to a sub-subquadtree.

In this discussion we have introduced a notation to identify every quadrant at any depth of the quadtree. The root is identified by the null string; a quadrant at depth d is uniquely identified by a string of d radix-4 digits. This string can be interpreted in various ways as a number expressed in base 4. Thus accessing and processing a quadtree is readily reduced to arithmetic.

Breadth-first addressing

Label the root 0, its children 1, 2, 3, 4, its grand children 5 through 20, and so on, one generation after the other.

```
0
 1            2                3            4
    5 6 7 8    9 10 11 12      13 14 15 16    17 18 19 20
```

Notice that the children of any node i are $4 \cdot i + 1$, $4 \cdot i + 2$, $4 \cdot i + 3$, $4 \cdot i + 4$. The parent of node i is $(i - 1)$ div 4. This is similar to the address computation used in the heap of "Implicit data structures", a binary tree where each node i has children $2 \cdot i$ and $2 \cdot i + 1$; and the parent of node i is obtained as i div 2.

Exercise

The string of radix 4 digits along a path from the root to any node is called the *path address* of this node. Interpret the path address as an integer, most significant digit first. These integers label the nodes at depth $d > 0$ consecutively from 0 to 4d – 1. Devise a formula that transforms the path address into the breadth-first address. This formula can be used to store a quadtree as a one-dimensional array.

Data compression

The representation of an image as a quadtree is sometimes much more compact than its representation as a bit map. Two conditions must hold for this to be true:

1. The image must be fairly large, typically hundreds of pixels on a side.
2. The image must have large areas of constant value (color).

The quadtree for the quarter circle above, for example, has only 14 nodes. A bit map of the same image requires 256 bits. Which representation requires more storage? Certainly the quadtree. If we store it as a list, each node must be able to hold four pointers, say 4 or 8 bytes. If a pointer has value 'nil', indicating that its quadrant needs no refinement, we need a bit to indicate the color of this quadrant (white or black), or a total of 4 bits. If we store the quadtree breadth-first, no pointers are needed as the node relationships are expressed by address computation; thus a node is reduced to four three-valued fields ('white', 'black', or 'refine'), conveniently stored in 8 bits, or 1 byte. This implicit data structure will leave many unused holes in memory. Thus quadtrees do not achieve data compression for small images.

Octtrees

Exactly the same idea for three-dimensional space as quadtrees are for two-dimensional space: A cube is recursively partitioned into eight octants, using three orthogonal planes.

Spatial data structures: objectives and constraints

Metric data structures are used primarily for storing spatial data, such as points and simple geometric objects embedded in a multidimensional space. The most important objectives a spatial data structure must meet include:

- ❖ Efficient handling of large, dynamically varying data sets in interactive applications
- ❖ Fast access to objects identified in a fully specified query
- ❖ Efficient processing of proximity queries and region queries of arbitrary shape
- ❖ A uniformly high memory utilization

Achieving these objectives is subject to many constraints, and results in trade-offs.

Managing disks. By "large data set" we mean one that must be stored on disk; only a small fraction of the data can be kept in central memory at any one time. Many data structures can be used in central memory, but the choice is much more restricted when it comes to managing disks because of the well-known "memory speed gap" phenomenon. Central memory is organized in small physical units (a byte, a word) with access times of approximately 1 microsecond, 10–6 second. Disks are organizein large physical blocks (512 bytes to 5kilobytes) with access times ranging from 10 to 100 milliseconds (10–2 to 10–1 second). Compared to central memory, a disk delivers data blocks typically 103 times larger with a delay 104 times greater. In terms of the data rate delivered to the central processing unit:

$$\frac{\text{size of data block read}}{\text{access time}}$$

the disk is a storage device whose effectiveness is within an order of magnitude of that of central memory. The large size of a physical disk block is a potential source of inefficiency that can easily reduce the useful data rate of a disk a hundredfold or a thousandfold. Accessing a couple of bytes on disk, say a pointer needed to traverse a list, takes about as long as accessing the entire disk block. Thus the game of managing disks is about *minimizing the number of disk accesses*.

Dynamically varying data. The majority of computer applications today are interactive. That means that insertions, deletions, and modifications of data are at least as frequent as operations that merely process fixed data. Data structures that entail a systematic degradation of performance with continued use (such as ever-lengthening overflow chains, or an ever-increasing number of cells marked "deleted" in a conventional hash table) are unsuitable. Only structures that automatically adapt their shape to accommodate ever-changing contents can provide uniform response times.

Instantaneous response. Interactive use of computers sets another major challenge for data management: the goal of providing "instantaneous response" to a fully specified query. "Fully" specified means that every attribute relevant for the search has been provided, and that at most one element satisfies the query. Imagine the user clicking an icon on the screen, and the object represented by the icon appears instantaneously. In human terms, "instantaneous" is a well-defined physiological quantity, namely, about of a second, the limit of human time resolution. Ideally, an interactive system retrieves any single element fully specified in a query within 0.1 second.

Two-disk-access principle. We have already stated that in today's technology, a disk access typically takes from tens of milliseconds. Thus the goal of retrieving any single element in 0.1 second translates into "retrieve any element in at most a few disk accesses". Fortunately, it turns out that useful data structure can be designed that access data in a two-step process: (1) access the correct portion of a directory, and (2) access the correct data bucket. Under the assumption that both data and directory are so large that they are stored on disk, we call this the ***two-disk-access principle.***

Proximity queries and region queries of arbitrary shape. The simplest example of a proximity query is the operation 'next', which we have often encountered in one-dimensional data structure traversals: Given a pointer to an element, get the next element (the successor or the predecessor) accor ding to the order defined on the domain. Another simple example is an interval or range query such as "get all x between 13 and 17". This generalizes directly to k-dimensional *orthogonal* *range queries* such as the two-dimensional query "get all (x1, x2) with 13 ≤ x1 < 17 and 3 ≤ x2 < 4". In geometric computation, for example, many other instances of proximity queries are important, such as the "nearest neighbor" (in any direction), or intersection queriesamong objects. Region queries of arbitrary shape (not just rectangular) are able to express a variety of geometric conditions.

Uniformly high memory utilization. Any data structure that adapts its shape to dynamically changing contents is likely to leave "unused holes" in storage space: space that is currently unused, and that cannot conveniently be used for other purposes because it is fragmented. We have encountered this phenomenon in multiway trees such as B-trees and in hash tables. It is practically unavoidable that dynamic data structures use their allocated space to less than 100%, and an average space utilization of 50% is often tolerable. The danger to avoid is a built-in bias that drives space utilization toward 0 when the file shrinks—elements get deleted but their space is not relinquished. The grid file, to be discussed next, achieves an average memory utilization of about 70% regardless of the mix of insertions or deletions.

The grid file

The grid file is a metric data structure designed to store points and simple geometric objects in multidimensional space so as to achieve the objectives stated above. This section describes its architecture, access and update algorithms, and properties. More details can be found in [NHS 84] and [Hin 85].

Scales, directory, buckets

Consider as an example a two-dimensional domain: the Cartesian product X1 × X2, where X1 = 0 .. 1999 is a subrange of the integers, and X2 = a .. z is the ordered set of the 26 characters of the English alphabet. Pairs of theform (x1, x2), such as (1988, w), are elements from this domain.

The *bit map* is a natural data structure for storing a set S of elements from X1 × X2. It may be declared as

```
var  T: array[X1, X2] of boolean;
```

with the convention that

$$T[x_1, x_2] = \text{true} \Leftrightarrow (x_1, x_2) \in S.$$

Basic set operations are performed by direct access to the array element corresponding to an element: find(x_1, x_2) is simply the boolean expression T[x_1, x_2]; insert(x_1, x_2) is equivalent to T[x_1, x_2]:= 'true', delete(x_1, x_2) is equivalent to T[x_1, x_2] := 'false'. The bit map for our small domain requires an affordable 52k bits. Bit maps for realistic examples are rarely affordable, as the following reasoning shows. First, consider that x and y are just keys of records that hold additional data. If space is reserved in the array for this additional data, an array element is not a bit but as many bytes as are needed, and all the absent records, for elements $(x_1, x_2) \notin S$, waste a lot of storage. Second, most domains are much larger than the example above: the three-dimensional Euclidean space, for example, with elements (x, y, z) taken as triples of 32-bit integers, or 64-bit floating-point numbers, requires bit maps of about 1030 and 1060 bits, respectively. For comparison's sake: a large disk has about 1010 bits.

Since large bit maps are extremely sparsely populated, they are amenable to data compression. The grid file is best understood as a practical data compression technique that stores huge, sparsely populated bit maps so as to support direct access. Returning to our example, imagine a historical database indexed by the year of birth and the first letter of the name of scientists: thus we find 'John von Neumann' under (1903, v). Our database is pictured as a cloud of points in the domain shown in Exhibit 23.3; because we have more scientists (or at least, more records) in recent years, the density increases toward the right. Storing this database implies packing the records into buckets of fixed capacity to hold c (e.g. c = 3) records. The figure shows the domain partitioned by orthogonal hyperplanes into box-shaped grid cells, none of which contains more than c points.

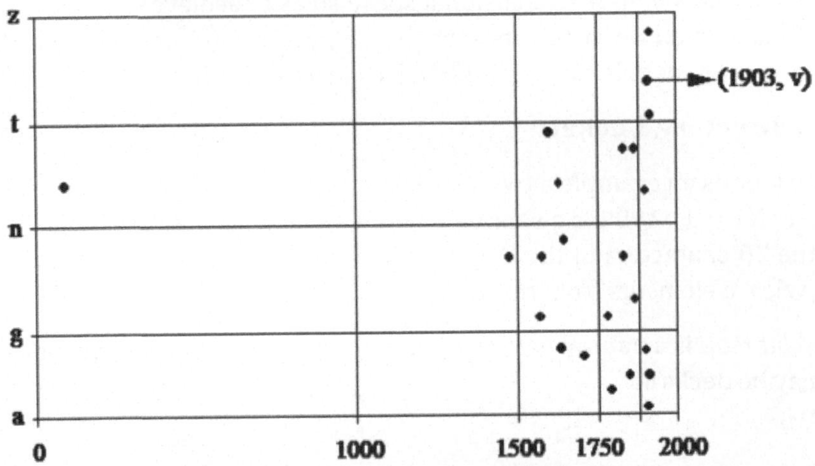

Exhibit 23.3: Cells of a grid partition adapt their size so that no cell is populated by more than c points.

A grid file for this database contains the following components:

- *Linear scales* show how the domain is currently partitioned.
- The *directory* is an array whose elements are in one-to-one correspondence with the grid cells; each entry points to a *data bucket* that holds all the records of the corresponding grid cell.

Access to the record (1903, v) proceeds through three steps:

- Scales transform key values to array indices: (1903, v) becomes (5, 4). Scales contain small amounts of data, which is kept in central memory; thus this step requires no disk access.
- The index tuple (5, 4) provides direct access to the correct element of the directory. The directory may be large and occupy many pages on disk, but we can compute the address of the correct directory page and in one disk access retrieve the correct directory element.
- The directory element contains a pointer (disk address) of the correct data bucket for (1903, v), and the second disk access retrieves the correct record: [(1903, v), John von Neumann ...].

Disk utilization

The grid file does not allocate a separate bucket to each grid cell—that would lead to an unacceptably low disk utilization. Exhibit 23.4 suggests, for example, that the two grid cells at the top right of the directory share the same bucket. How this bucket sharing comes about, and how it is maintained through splitting of overflowing buckets, and merging sparsely populated buckets, is shown in the following.

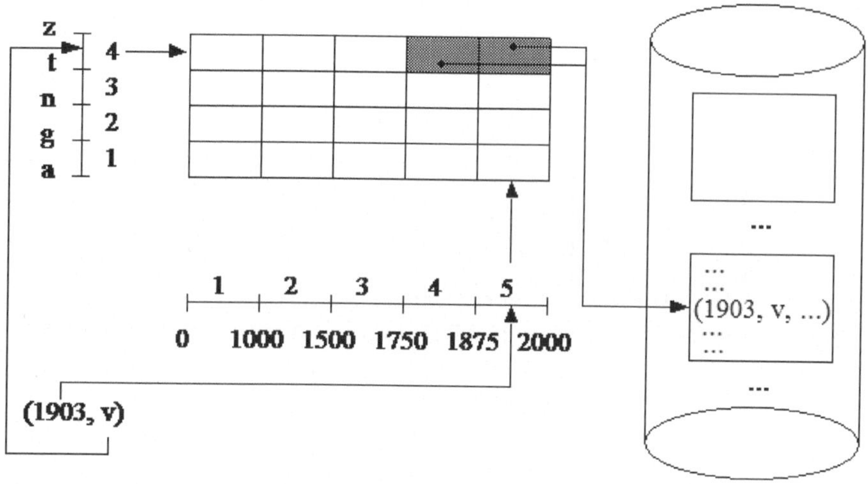

Exhibit 23.4: The search for a record with key values (1903, v) starts with the scales and proceeds via the directory to the correct data bucket on disk.

The dynamics of splitting and merging

The dynamic behavior of the grid file is best explained by tracing an example: we show the effect of repeated insertions in a two-dimensional file. Instead of showing the grid directory, whose elements are in one-to-one correspondence with the grid blocks, we draw the bucket pointers as originating directly from the grid blocks.

Initially, a single bucket A, of capacity c = 3 in our example, is assigned to the entire domain (Exhibit 23.5). When bucket A overflows, the domain is split, a new bucket B is made available, and those records that lie in one half of the space are moved from the old bucket to the new one (Exhibit 23.6). If bucket A overflows again, its grid block (i.e. the left half of the space) is split according to some splitting policy: We assume the simplest splitting policy of alternating directions. Those records of A that lie in the lower-left grid block of Exhibit 23.7 are moved to a new bucket C. Notice that as bucket B did not overflow, it is left alone: Its region now consists of two grid blocks. For effective memory utilization it is essential that in the process of refining the grid partition we need not necessarily split a bucket when its region is split.

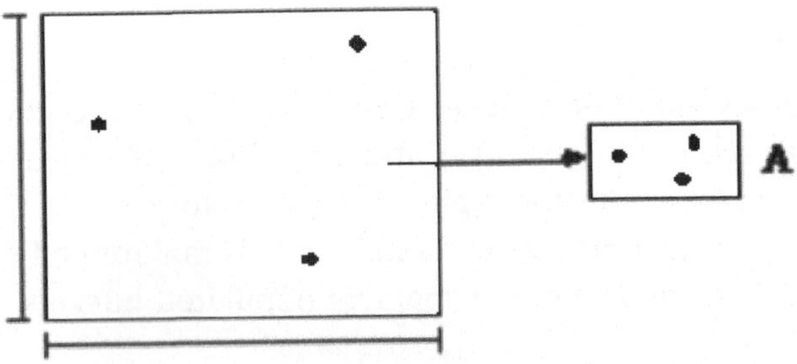

Exhibit 23.5: A growing grid file starts with a single bucket allocated to the entire key space.

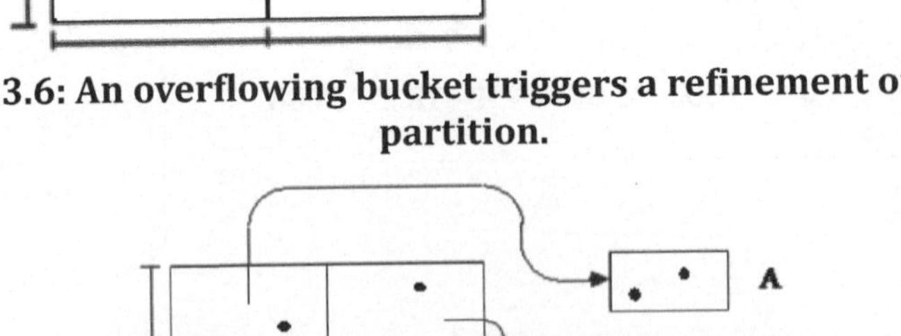

Exhibit 23.6: An overflowing bucket triggers a refinement of the space partition.

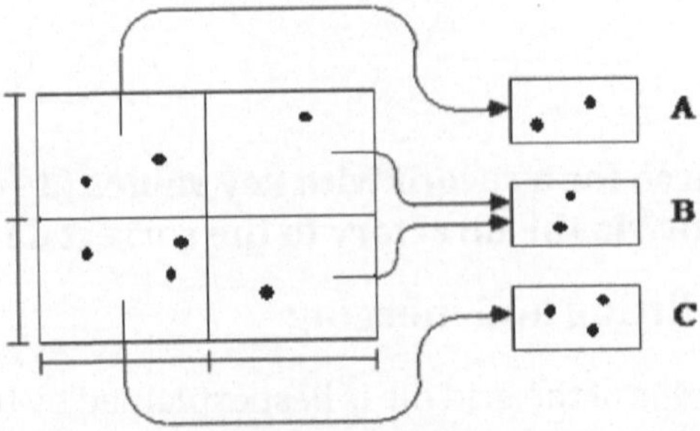

Exhibit 23.7: Bucket A has been split into A and C, but the contents of B remain unchanged.

Assuming that records keep arriving in the lower-left corner of the space, bucket C will overflow. This will trigger a further refinement of the grid partition as shown in Exhibit 23.8, and a splitting of bucket C into C and D. The history of repeated splitting can be represented in the form of a binary tree, which imposes on the set of buckets currently in use (and hence on the set of regions of these buckets) a *twin system* (also called a *buddy system*): Each bucket and its region have a unique twin from which it split off. In Exhibit 23.8, C and D are twins, the pair (C, D) is A's twin, and the pair (A, (C, D)) is B's twin.

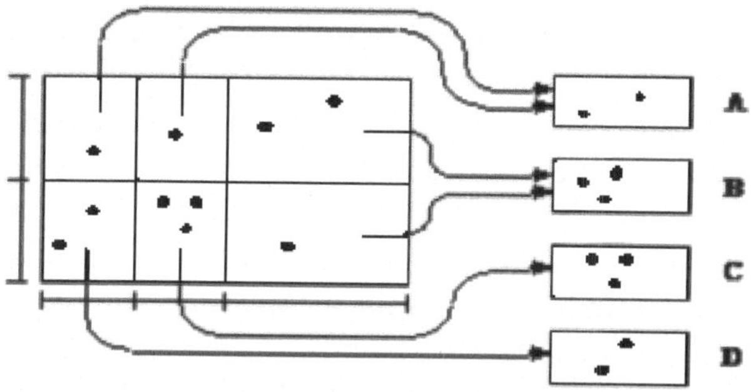

Exhibit 23.8: Bucket regions that span several cells ensure high disk utilization.

Deletions trigger merging operations. In contrast to one-dimensional storage, where it is sufficient to merge buckets that split earlier, merging policies for multidimensional grid files need to be more general in order to maintain a high occupancy.

Simple geometric objects and their parameter spaces

Consider a class of simple spatial objects, such as aligned rectangles in the plane (i.e. with sides parallel to the axes). Within its class, each object is defined by a small number of parameters. For example, an aligned rectangle is determined by its center (cx, cy) and the half-length of each side, dx and dy.

An object defined within its class by k parameters can be considered to be a point in a k-dimensional parameter space. For example, an aligned rectangle becomes a point in four-dimensional space. All of the geometric and topological properties of an object can be deduced from the class it belongs to and from the coordinates of its corresponding point in parameter space.

Different choices of the parameter space for the same class of objects are appropriate, depending on characteristics of the data to be processed. Some considerations that may determine the choice of parameters are:

- *Distinction between location parameters and extension parameters.* For some classes of simple objects it is reasonable to distinguish location parameters, such as the center (cx, cy) of an aligned rectangle, from extension parameters, such as the half-sides dx and dy. This distinction is always possible for objects that can be described as Cartesian products of spheres of various dimensions. For example, a rectangle is the product of two one-dimensional spheres, a cylinder the product of a one-dimensional and a two- dimensional sphere. Whenever this distinction can be made, cone-shaped search regions generated by proximity queries as described in the next section have a simple intuitive interpretation: The subspace of the location parameters acts as a "mirror" that reflects a query.

- *Independence of parameters, uniform distribution.* As an example, consider the class of all intervals on a straight line. If intervals are represented by their left and right endpoints, lx and rx, the constraint lx ≤ rx restricts all representations of these intervals by points (lx, rx) to the triangle above the diagonal. Any data structure that organizes the embedding space of the data points, as opposed to the particular set of points that must be stored, will pay some overhead for representing the unpopulated half of the embedding space. A coordinate transformation that distributes data all over the embedding space leads to more efficient storage. The phenomenon of nonuniform data distribution can be worse than this. In most applications, the building blocks from which complex objects are built are much smaller than the space in which they are embedded, as the size of a brick is small compared to the size of a house. If so, parameters such as lx and rx that locate boundaries of an object are highly dependent on each other. Exhibit 23.9 shows short intervals on a long line clustering along the diagonal, leaving large regions of a large embedding space unpopulated; whereas the same set of intervals represented by a location parameter cx and an extension parameter dx fills a smaller embedding space in a much more uniform way. With the assumption of bounded dx, this data distribution is easier to handle.

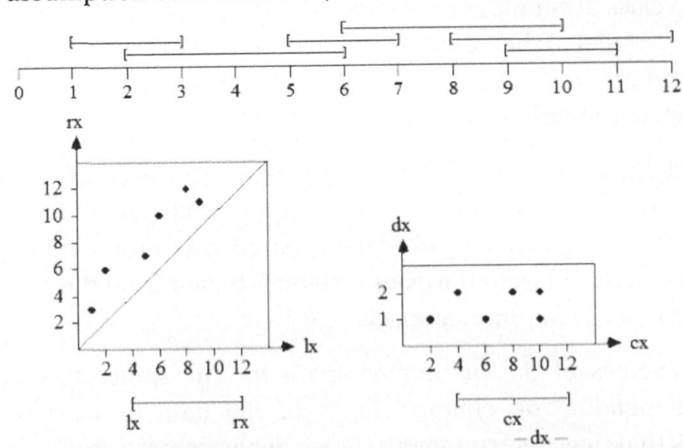

Exhibit 23.9: A set of intervals represented in two different parameter spaces.

Region queries of arbitrary shape

Intersection is a basic component of other proximity queries, and thus deserves special attention. CAD design rules, for example, often require different objects to be separated by some minimal distance. This is equivalent to requiring that objects surrounded by a rim do not intersect. Given a subset Γ of a class of simple spatial objects with parameter space H, we consider two types of queries:

- point query Given a query point q, find all objects A ∈ Γ for which q ∈ A.
- point set query Given a query set Q of points, find all objects A ∈ Γ that intersect Q.

Point query. For a query point q compute the region in H that contains all points representing objects in Γ that overlap q.

- Consider the class of intervals on a straight line. An interval given by its center cx and its half length dx overlaps a point q with coordinate qx if and only if cx – dx ≤ qx ≤ cx + dx.
- The class of aligned rectangles in the plane (with parameters cx, cy, dx, dy) can be treated as the Cartesian product of two classes of intervals, one along the x-axis, the other along the y-axis (Exhibit 23.10). All rectangles that contain a given point q are represented by points in four-dimensional space that lie in the Cartesian product of two point-in-interval query regions. The region is shown by its projections onto the cx- dx plane and the cy-dy plane.

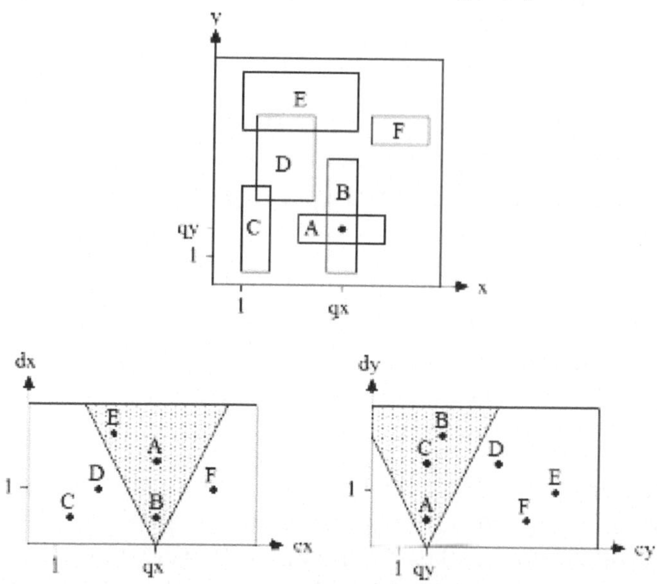

Exhibit 23.10: A set of aligned rectangles represented as a set of points in a four-dimensional parameter space. A point query is transformed into a cone-shaped region query.

Consider the class of circles in the plane. We represent a circle as a point in three-dimensional space by the coordinates of its center (cx, cy) and its radius r as parameters. All circles that overlap a point q are represented in the corresponding three-dimensional space by points that lie in the cone with vertex q shown in Exhibit 23.11. The axis of the cone is parallel to the r-axis (the extension parameter), and its vertex q is considered a point in the cx-cy plane (the subspace of the location parameters).

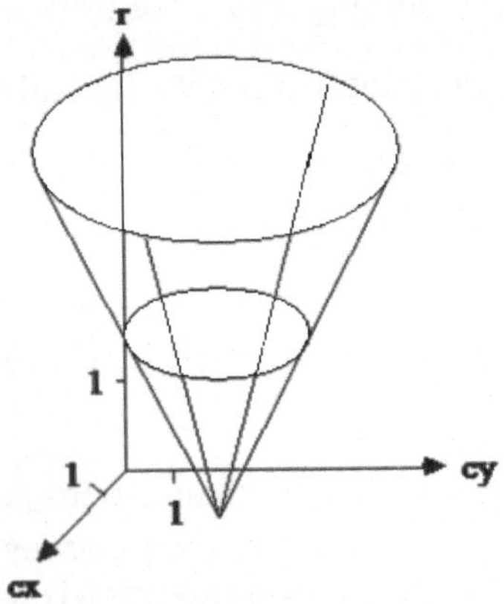

Exhibit 23.11: Search cone for a point query for circles in the plane.

Point set query. Given a query set Q of points, the region in H that contains all points representing objects A ∈ Γ that intersect Q is the union of the regions in H that results from the point queries for each point q ∈ Q. The union of cones is a particularly simple region in H if the query set Q is a simple spatial object.

- Consider the class of intervals on a straight line. An interval i = (cx, dx) intersects a query interval Q = (cq, dq) if and only if its representing point lies in the shaded region shown in Exhibit 23.12; this region is given by the inequalities cx – dx ≤ cq + dq and cx + dx ≥ cq – dq.
- The class of aligned rectangles in the plane is again treated as the Cartesian product of two classes of intervals, one along the x-axis, the other along the y-axis. If Q is also an aligned rectangle, all rectangles that intersect Q are represented by points in four-dimensional space lying in the Cartesian product of two interval intersection query regions.
- Consider the class of circles in the plane. All circles that intersect a line segment L are represented by points lying in the cone-shaped solid shown in Exhibit 23.13. This solid is obtained by embedding L in the cx-cy plane, the subspace of the location parameters, and moving the cone with vertex at q along L.

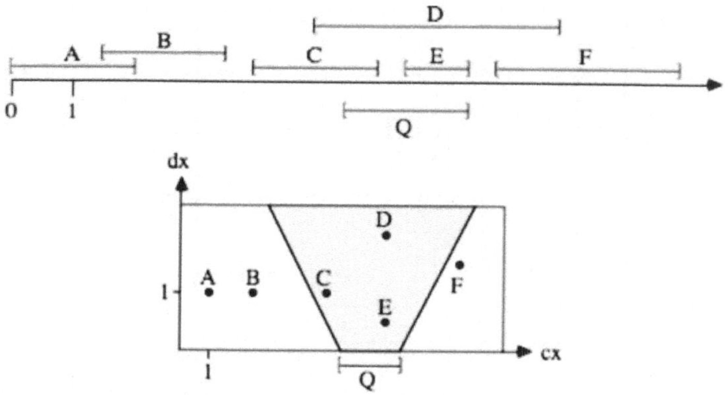

Exhibit 23.12: An interval query, as a union of point queries, again gets transformed into a search cone.

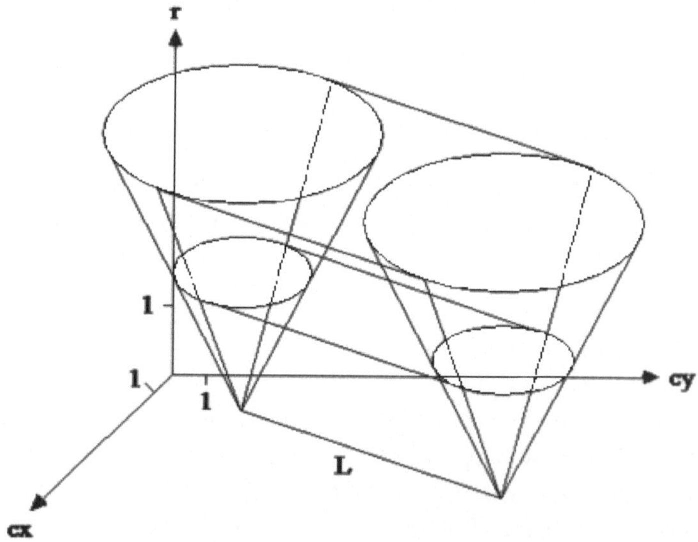

Exhibit 23.13: Search region as a union of cones.

Evaluating region queries with a grid file

We have seen that proximity queries on spatial objects lead to search regions significantly more complex than orthogonal range queries. The grid file allows the evaluation of irregularly shaped search regions in such a way that the complexity of the region affects CPU time but not disk accesses. The latter limits the performance of a data base implementation. A query region Q is matched against the scales and converted into a set I of index tuples that refer to entries in the directory. Only after this preprocessing do we access disk to retrieve the correct pages of the directory and the correct data buckets whose regions intersect Q (Exhibit 23.14).

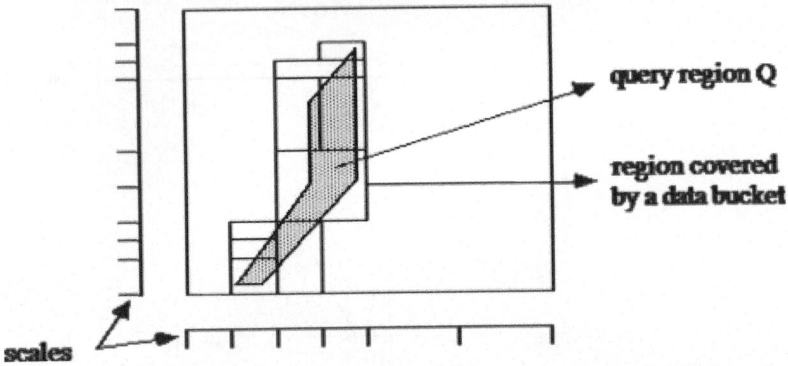

Exhibit 23.14: The cells of a grid partition that overlap an arbitrary query region Q are determined by merely looking up the scales.

Interaction between query processing and data access

The point of the two preceding sections was to show that in a metric data structure, intricate computations triggered by proximity queries can be preprocessed to a remarkable extent *before* the objects involved are retrieved.

Query preprocessing may involve a significant amount of computation based on small amounts of auxiliary data— the scales and the query—that are kept in central memory. The final access of data from disk is highly selective— data retrieved has a high chance of being part of the answer.

Contrast this to an approach where an object can be accessed only by its name (e.g. the part number) because the geometric information about its location in space is only included in the record for this object but is not part of the accessing mechanism. In such a database, all objects might have to be retrieved in order to determine which ones answer the query. Given that disk access is the bottleneck in most database applications, it pays to preprocess queries as much as possible in order to save disk accesses.

The integration of query processing and accessing mechanism developed in the preceding sections was made possible by the assumption of simple objects, where each instance is described by a small number of parameters. What can we do when faced with a large number of irregularly shaped objects?

Complex, irregularly shaped spatial objects can be represented or approximated by simpler ones in a variety of ways, for example: *decomposition*, as in a quad tree tessellation of a figure into disjoint raster squares; representation as a *cover* of overlapping simple shapes; and enclosing each object in a *container* chosen from a class of simple shapes. The container technique allows efficient processing of proximity queries because it preserves the most important properties for proximity-based access to spatial objects, in particular: It does not break up the object into components that must be processed separately, and it eliminates many potential tests as unnecessary (if two containers don't intersect, the objects within won't either). As an example, consider finding all polygons that intersect a given query polygon, given that each of them is enclosed in a simple container such as a circle or an aligned rectangle. Testing two polygons for intersection is an expensive operation compared to testing their containers for intersection. The cheap container test excludes most of the polygons from an expensive, detailed intersection check.

Any approximation technique limits the primitive shapes that must be stored to one or a few types: for example, aligned rectangles or boxes. An instance of such a type is determined by a few parameters, such as coordinates of its center and its extension, and can be considered to be a point in a (higher-dimensional) parameter space. This transformation reduces object storage to point storage, increasing the dimensionality of the problem without loss of information. Combined with an efficient multi-dimensional data structure for *point* storage it is the basis for an effective implementation of databases for spatial objects.

Exercises

❖ Draw three quadtrees, one for each of the 4 · 8 pixel rectangles A, B and C outlined in Exhibit 23.15.

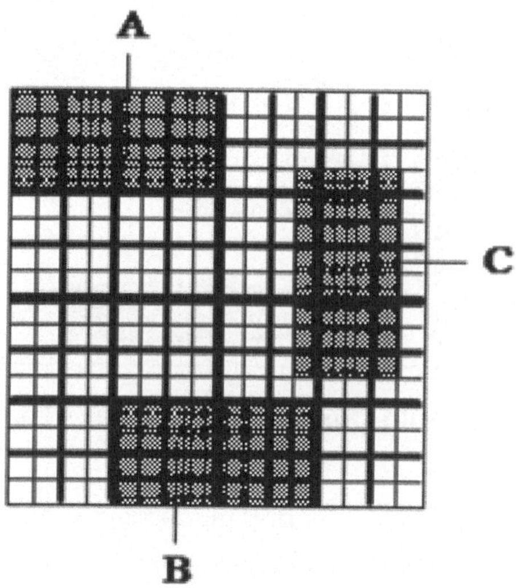

Exhibit 23.15: The location of congruent objects greatly affects the complexity of a quadtree representation.

❖ Consider a grid file that stores points lying in a two-dimensional domain: the Cartesian product X1 × X2, where X1 = 0 .. 15 and X2 = 0 .. 15 are subranges of the integers. Buckets have a capacity of two points.

❖ Insert the points (2, 3), (13, 14), (3, 5), (6, 9), (10, 13), (11, 5), (14, 9), (7, 3), (15, 11), (9, 9), and (11, 10) into the initially empty grid file and show the state of the scales, the directory, and the buckets after each insert operation. Buckets are split such that their shapes remain as quadratic as possible.

- Delete the points (10, 13), (9, 9), (11, 10), and (14, 9) from the grid file obtained in a) and show the state of the scales, the directory, and the buckets after each delete operation. Assume that after deleting a point in a bucket this bucket may be merged with a neighbor bucket if their joint occupancy does not exceed two points. Further, a boundary should be removed from its scale if there is no longer a bucket that is split with respect to this boundary.
- Without imposing further restrictions a deadlock situation may occur after a sequence of delete operations: No bucket can merge with any of its neighbors, since the resulting bucket region would no longer be rectangular. In the example shown in Exhibit 23.16 the shaded ovals represent bucket regions. Devise a merging policy that prevents such deadlocks from occurring in a two-dimensional grid file.

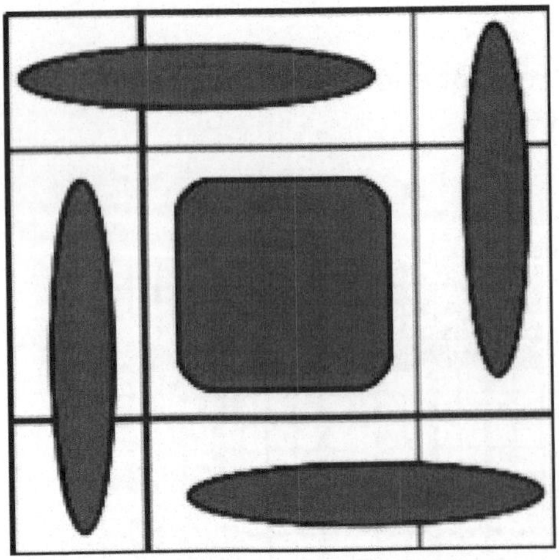

Exhibit 23.16: This example shows bucket regions that cannot be merged pairwise.

- Consider the class of circles in the plane represented as points in three-dimensional parameter space as proposed in chapter 23 in the section "Region queries of arbitrary shape". Describe the search regions in the parameter space (a) for all the circles intersecting a given circle C, (b) for all the circles contained in C, and (c) for all the circles enclosing .

Part-VI

Interaction between Algorithms and Data Structures: case studies in Geometric Computation

Organizing and processing Euclidean space

In Part III we presented a varied sample of algorithms that use simple, mostly static, data structures. Part V was dedicated to dynamic data structures, and we presented the corresponding access and update algorithms. In this final part we illustrate the use of these dynamic data structures by presenting algorithms whose efficiency depends crucially on them, in particular on priority queues and dictionaries. We choose these algorithms from computational geometry, a recently developed discipline of great practical importance with applications in computer graphics, computer-aided design, and geographic databases.

If data structures are tools for organizing sets of data and their relationships, geometric data processing poses one of the most challenging tests. The ability to organize data embedded in the Euclidean space in such a way as to reflect the rich relationships due to location (e.g. touching or intersecting, contained in, distance) is of utmost importance for the efficiency of algorithms for processing spatial data. Data structures developed for traditional commercial data processing were often based on the concept of one primary key and several subordinate secondary keys. This asymmetry fails to support the equal role played by the Cartesian coordinate axes x, y, z, ... of Euclidean space. If one spatial axis, say x, is identified as the *primary key*, there is a danger that queries involving the other axes, say y and z, become inordinately cumbersome to process, and therefore slow. For the sake of simplicity we concentrate on two-dimensional geometric problems, and in particular on the highly successful class of plane- sweep algorithms. Sweep algorithms do a remarkably good job at processing two-dimensional space efficiently using two distinct one-dimensional data structures, one for organizing the x-axis, the other for the y-axis.

24

Sample Problems and Algorithms

Geometry and geometric computation

Classical geometry, shaped by the ancient Greeks, is more axiomatic than constructive: It emphasizes axioms, theorems, and proofs, rather than algorithms. The typical statement of Euclidean geometry is an assertion about all geometric configurations with certain properties (e.g. the theorem of Pythagoras: "In a right-angled triangle, the square on the hypotenuse c is equal to the sum of the squares on the two catheti a and b: $c^2 = a^2 + b^2$") or an assertion of existence (e.g. the parallel axiom: "Given a line L and a point $P \notin L$, there is exactly one line parallel to L passing through P"). Constructive solutions to problems do occur, but the theorems about the *impossibility* of constructive solutions steal the glory: "You cannot trisect an arbitrary angle using ruler and compass only," and the proverbial "It is impossible to square the circle."

Computational geometry, on the other hand, starts out with problems of construction so simple that, until the 1970s, they were dismissed as trivial: "Given n line segments in the plane, are they free of intersections? If not, compute (construct) all intersections." This problem is only trivial with respect to the *existence* of a constructive solution. As we will soon see, the question is far from trivial if interpreted as: How *efficiently* can we obtain the answer?

Computational geometry has some appealing features that make it ideal for learning about algorithms and data structures: (a) The problem statements are easily understood, intuitively meaningful, and mathematically rigorous; right away the student can try his own hand at solving them, without having to worry about hidden subtleties or a lot of required background knowledge. (b) Problem statement, solution, and every step of the construction have natural visual representations that support abstract thinking and help in detecting errors of reasoning. (c) These algorithms are practical; it is easy to come up with examples where they can be applied.

Appealing as geometric computation is, writing geometric programs is a demanding task. Two traps lie hiding behind the obvious combinatorial intricacies that must be mastered, and they are particularly dangerous when they occur together: (a) degenerate configurations, and (b) the pitfalls of numerical computation due to discretization and rounding errors. Degenerate configurations, such as those we discussed in "Straight lines and circles" on intersecting line segments, are special cases that often require special code. It is not always easy to envision all the kinds of degeneracies that may occur in a given problem. A configuration may be degenerate for a specific algorithm, whereas it may be nondegenerate for a different algorithm solving the same problem. Rounding errors tend to cause more obviously disastrous consequences in geometric computation than, say, in linear algebra or differential equations. Whereas the traditional analysis of rounding errors focuses on bounding their cumulative value, geometry is concerned primarily with a stringent all-or-nothing question: Have errors impaired the topological consistency of the data? (Remember the pathology of the braided straight lines.)

In this Part VI we aim to introduce the reader to some of the central ideas and techniques of computational geometry. For simplicity's sake we limit coverage to two-dimensional Euclidean geometry - most problems become a lot more complicated when we go from two- to three-dimensional configurations. We focus on a type of algorithm that is remarkably well suited for solving two-dimensional problems efficiently: sweep algorithms. To illustrate their generality and effectiveness, we use plane-sweep to solve several rather distinct problems. We will see that sweep algorithms for different problems can be assembled from the same building blocks: a skeleton sweep program that sweeps a line across the plane based on a queue of events to be processed, and transition procedures that update the data structures (a dictionary or table, and perhaps other structures) at each event and maintain a geometric invariant. Sweeps show convincingly how the dynamic data structures of Part V are essential for the efficiency.

The problems and algorithms we discuss deal with very simple objects: points and line segments. Applications of geometric computation such as CAD, on the other hand, typically deal with very complex objects made up of thousands of polygons. The simplicity of these algorithms does not deter from their utility. Complex objects get processed by being broken into their primitive parts, such as points, line segments, and triangles. The algorithms we present are some of the most basic subroutines of geometric computation, which play a role analogous to that of a square root routine for numerical computation: As they are called untold times, they must be correct and efficient.

Convex hull: a multitude of algorithms

The problem of computing the convex hull H(S) of a set S consisting of n points in the plane serves as an example to demonstrate how the techniques of computational geometry yield the concise and elegant solution that we presented in "Algorithm animation". The convex hull of a set S of points in the plane is the smallest convex polygon that contains the points of S in its interior or on its boundary. Imagine a nail sticking out above each point and a tight rubber band surrounding the set of nails.

Many different algorithms solve this simple problem. Before we present in detail the algorithm that forms the basis of the program 'ConvexHull' of chapter 3, we briefly illustrate the main ideas behind three others. Most convex hull algorithms have an initialization step that uses the fact that we can easily identify two points of S that lie on the convex hull H(S): for example, two points Pmin and Pmax with minimal and maximal x-coordinate, respectively. Algorithms that grow convex hulls over increasing subsets can use the segment as a (degenerate) convex hull to start with. Other algorithms use the segment to partition S into an upper and a lower subset, and compute the upper and the lower part of the hull H(S) separately.

- *Jarvis's march* [Jar 73] starts at a point on H(S), say Pmin, and 'walks around' by computing, at each point P, the next tangent to S, characterized by the property that all points of S lie on the same side of $\overline{PQ}$

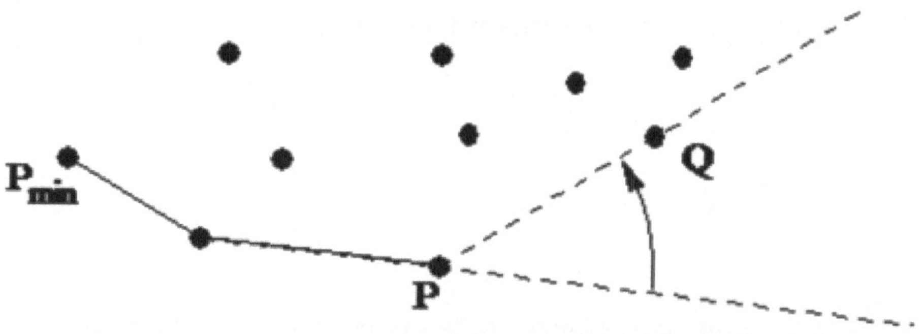

Exhibit 24.1: The "gift-wrapping" approach to building the convex hull.

- Divide-and-conquer comes to mind: Sort the points of S according to their x-coordinate, use the median xcoordinate to partition S into a left half SL and a right half SR, apply this convex hull algorithm recursively to each half, and merge the two solutions H(SL) and H(SR) by computing the two common exterior tangents to H(SL) and H(SR) (Exhibit 24.2). Terminate the recursion when a set has at most three points.

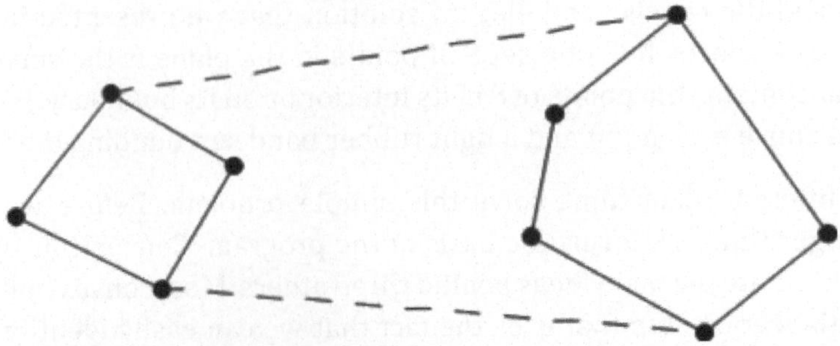

Exhibit 24.2: Divide-and-conquer applies to many problems on spatial data.

- *Quickhull* [Byk 78], [Edd 77], [GS 79] uses divide-and-conquer in a different way. We start with two points on the convex hull H(S), say Pmin and Pmax. In general, if we know ≥ 2 points on H(S), say P, Q, R in Exhibit 24.3, these define a convex polygon contained in H(S). (Draw the appropriate picture for just two points Pmin and Pmax on the convex hull.) There can be no points of S in the shaded sectors that extend outward from the vertices of the current polygon, PQR in the example. Any other points of S must lie either in the polygon PQR or in the regions extending outward from the sides.

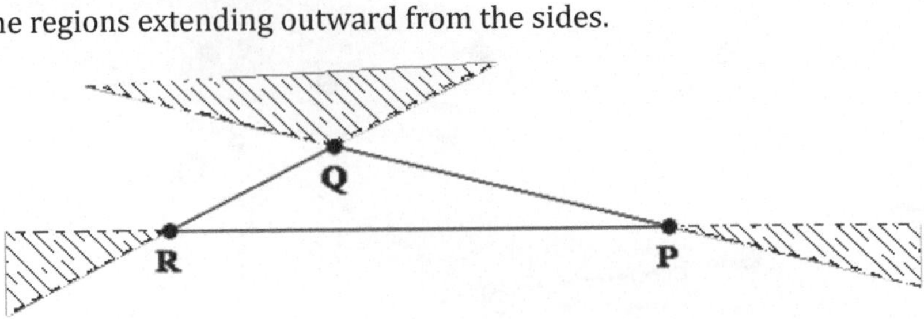

Exhibit 24.3: Three points known to lie on the convex hull identify regions devoid of points.

For each side, such as $\overline{PQ}$ in Exhibit 24.4, let T be a point *farthest* from $\overline{PQ}$ among all those in the region extending outward from $\overline{PQ}$, if there are any. T must lie on the convex hull, as is easily seen by considering the parallel to $\overline{PQ}$ that passes through T. Having processed the side $\overline{PQ}$, we extend the convex polygon to include T, and we now must process 2 additional sides, $\overline{PT}$ and $\overline{TQ}$. The reader will observe a formal analogy between quicksort ("Sorting and its complexity") and quickhull, which has given the latter its name.

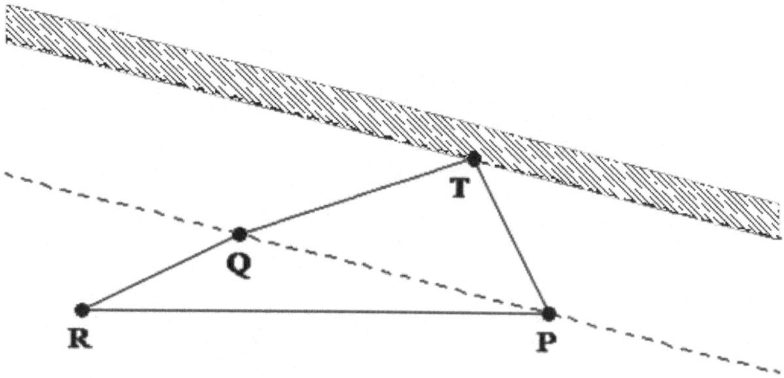

Exhibit 24.4: The point T farthest from identifies a new region of exclusion (shaded).

- In an *incremental scan* or *sweep* we sort the points of S according to their x-coordinates, and use the segment $\overline{P_{min}P_{max}}$ to partition S into an upper subset and a lower subset, as shown in Exhibit 24.5. For simplicity of presentation, we reduce the problem of computing H(S) to the two separate problems of computing the upper hull U(S) [i.e. the upper part of H(S)], shown in bold, and the lower hull L(S), drawn as a thin line. Our notation and pictures are chosen to describe U(S).

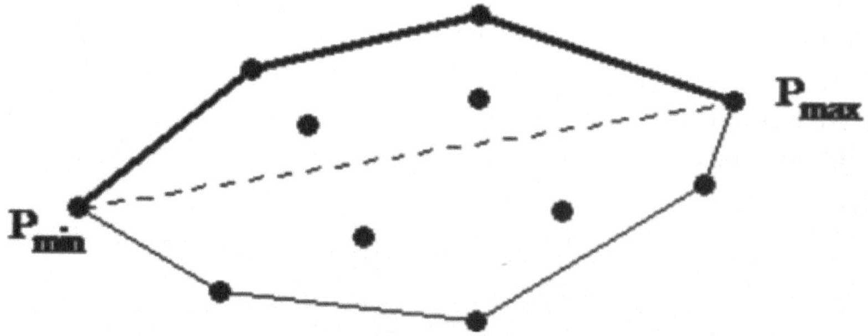

Exhibit 24.5: Separate computations for the upper hull and the lower hull.

Let $P_1, \ldots, P_n$ be the points of S sorted by x-coordinate, and let $U_i = U(P_1, \ldots, P_i)$ be the upper hull of the first i points. $U_1 = P_1$ may serve as an initialization. For i = 2 to n we compute U_i from U_{i-1}, as Exhibit 24.6 shows. Starting with the tentative tangent $\overline{P_iP_{i-1}}$ shown as a thin dashed line, we retrace the upper hull U_{i-1} until we reach the actual tangent: in our example, the bold dashed line $\overline{P_iP_2}$. The tangent is characterized by the fact that for j = 1, … , i−1, it minimizes the angle $A_{i,j}$ between P_iP_j and the vertical.

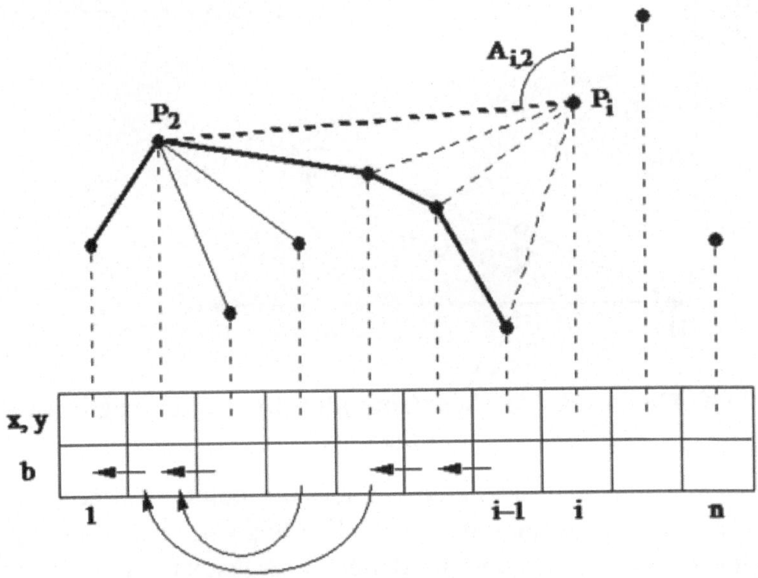

Exhibit 24.6: Extending the partial upper hull U(P1, ... , Pi–1) to the next point Pi

The program 'ConvexHull' presented in "Algorithm animation" as an example for algorithm animation is written as an on-line algorithm: Rather than reading all the data before starting the computation, it accepts one point at a time, which must lie to the right of all previous ones, and immediately extends the hull U i–1 to obtain Ui. Thanks to the input restriction that the points are entered in sorted order, 'ConvexHull' becomes simpler and runs in linear time. This explains the two-line main body:

```
  PointZero;  { sets first point and initializes all necessary
variables }
  while NextRight do  ComputeTangent;
```

There remain a few programming details that are best explained by relating Fig. 24.6 to the declarations:

```
var  x, y, dx, dy: array[0 .. nmax] of integer;
  b: array[0 .. nmax] of integer;  { backpointer }
  n: integer;  { number of points entered so far }
  px, py: integer;  { new point }
```

The coordinates of the points Pi are stored in the arrays x and y. Rather than storing angles such as Ai,j, we store quantities proportional to cos(Ai,j) and sin(Ai,j) in the arrays dx and dy. The array b holds back pointers for retracing the upper hull back toward the left: b[i] = j implies that Pj is the predecessor of Pi in Ui. This explains the key procedure of the program:

```
procedure ComputeTangent;    { from Pn = (px, py) to Un-1 }
var i: integer;
begin
  i := b[n];
  while  dy[n] · dx[i] > dy[i] · dx[n]  do  begin  { dy[n]/dx[n] >
dy[i]/dx[i] }
    i := b[i];
    dx[n] := x[n] - x[i];  dy[n] := y[n] - y[i];
    MoveTo(px, py);  Line(-dx[n], -dy[n]);
    b[n] := i
  end;
  MoveTo(px, py);  PenSize(2, 2);  Line(-dx[n], -dy[n]);  PenNormal
end;  { ComputeTangent }
```

The algorithm implemented by 'ConvexHull' is based on Graham's scan [Gra 72], where the points are ordered according to the angle as seen from a fixed internal point, and on [And 79].

The uses of convexity: basic operations on polygons

The convex hull of a set of points or objects (i.e. the smallest convex set that contains all objects) is a model problem in geometric computation, with many algorithms and applications. Why? As we stated in the introductory section, applications of geometric computation tend to deal with complex objects that often consist of thousands of primitive parts, such as points, line segments, and triangles. It is often effective to approximate a complex configuration by a simpler one, in particular, to package it in a container of simple shape. Many proximity queries can be answered by processing the container only. One of the most frequent queries in computer graphics, for example, asks what object, if any, is first struck by a given ray. If we find that the ray misses a container, we infer that it misses all objects in it without looking at them; only if the ray hits the container do we start the costly analysis of all the objects in it.

The convex hull is often a very effective container. Although not as simple as a rectangular box, say, convexity is such a strong geometric property that many algorithms that take time O(n) on an arbitrary polygon of n vertices require only time O(log n) on convex polygons. Let us list several such examples. We assume that a polygon G is given as a (cyclic) sequence of n vertices and/or n edges that trace a closed path in the plane. Polygons may be self- intersecting, whereas simple polygons may not. A simple polygon partitions the plane into two regions: the interior, which is simply connected, and the exterior, which has a hole.

Point-in-polygon test

Given a simple polygon G and a query point P (not on G), determine whether P lies inside or outside the polygon.

Two closely related algorithms that walk around the polygon solve this problem in time O(n). The first one computes the *winding number* of G around P. Imagine an observer at P looking at a vertex, say V, where the walk starts, and turning on her heels to keep watching the walker (Exhibit 24.7). The observer will make a first (positive) turn α, followed by a (negative) turn β, followed by ... , until the walker returns to the starting vertex V. The sum α + β + ... of all turning angles during one complete tour of G is: 2·π if P is inside G, and 0 if P is outside G.

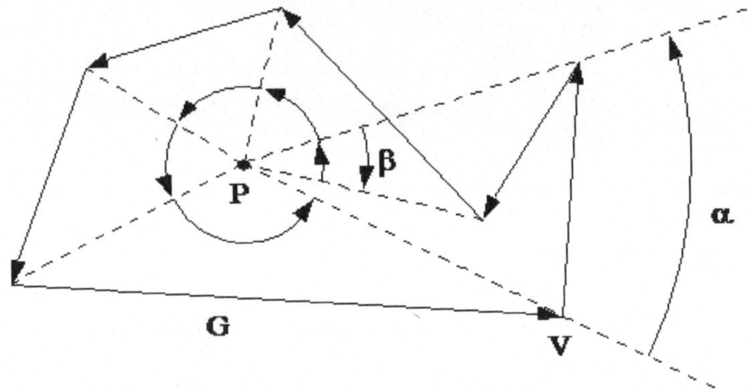

Exhibit 24.7: Point-in polygon test by adding up all turning angles.

The second algorithm computes the *crossing number* of G with respect to P. Draw a semi-infinite ray R from P in any direction (Exhibit 24.8). During the walk around the polygon G from an arbitrary starting vertex V back to V, keep track of whether the current oriented edge intersects R, and if so, whether the edge crosses R from below (+1) or from above (–1). The sum of all these numbers is +1 if P is inside G, and 0 if P is outside G.

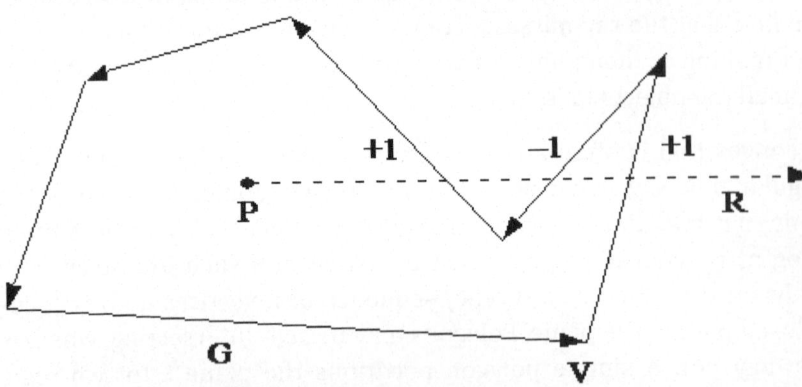

Exhibit 24.8: Point-in polygon test by adding up crossing numbers.

Point-in-convex-polygon test

For a convex polygon Q we use binary search to perform a point-in-polygon test in time O(log n). Consider the hierarchical decomposition of Q illustrated by the convex 12-gon shown in Exhibit 24.9. We choose three (approximately) equidistant vertices as the vertices of an innermost core triangle, painted black. "Equidistant" here refers not to any Euclidean distance, but rather to the number of vertices to be traversed by traveling along the perimeter of Q. For a query point P we first ask, in time O(1), which of the seven regions defined by the extended edges of this triangular core contains P. These seven regions shown in Exhibit 24.10 are all "triangles" (albeit six of them extend to infinity), in the sense that each one is defined as the intersection of three half-spaces. Four of these regions provide a definite answer to the query "Is P inside Q, or outside Q?" One region (shown hatched in Exhibit 24.10) provides the answer 'In', three the answer 'Out'. The remaining three regions, labeled 'Uncertain', lead recursively to a new point-in-convex-polygon test, for the same query point P, but a new convex polygon Q' which is the intersection of Q with one of the uncertain regions. As Q' has only about n / 3 vertices, the depth of recursion is O(log n). Actually, after the first comparison against the innermost triangular core of Q, we have no longer a general point-in-convex-polygon problem, but one with additional information that makes all but the first test steps of a binary search.

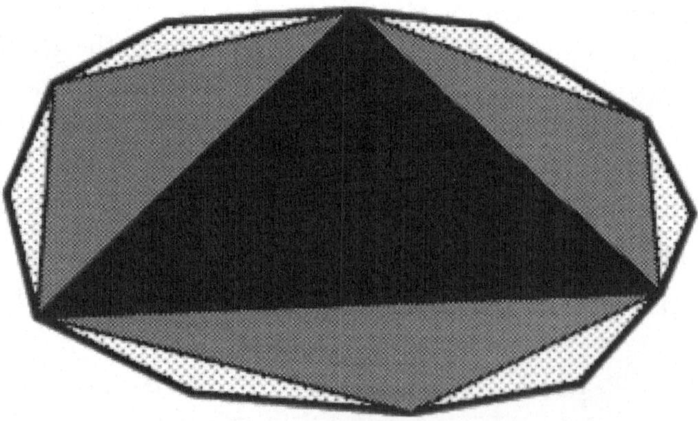

Exhibit 24.9: Hierarchical approximation of a convex 12-gon as a 3-level tree of triangles. The root is in black, its children are in dark grey, grandchildren in light grey.

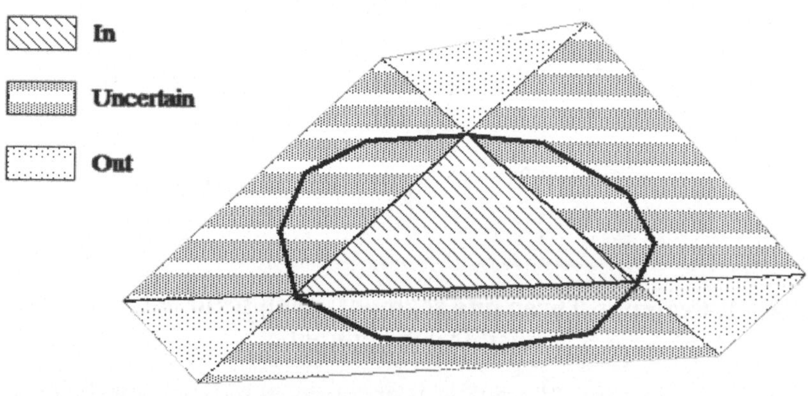

Exhibit 24.10: The plane partitioned into four regions of certainty and three of uncertainty.

The latter are processed recursively.

Visibility in the plane: a simple algorithm whose analysis is not

Many computer graphics programs are dominated by visibility problems: Given a configuration of objects in three-dimensional space, and given a point of view, what is visible? Dozens of algorithms for hidden-line or hidden- surface elimination have been developed to solve this everyday problem that our visual system performs "at a glance". In contrast to the problems discussed above, visibility is surprisingly complex. We give a hint of this complexity by describing some of the details buried below the smooth surface of a "simple" version: computing the visibility of line segments in the plane.

Problem: Given n line segments in the plane, compute the sequence of (sub) segments seen by an observer at infinity (say, at $y = -\infty$).

The complexity of this problem was unexpected until discovered in 1986 [WS 88]. Fortunately, this complexity is revealed not by requiring complicated algorithms, but in the analysis of the inherent complexity of the geometric problem. The example shown in Exhibit 24.11 illustrates the input data. The endpoints (P1, P10), (P2, P8), (P5, P12) of the three line segments labeled 1, 2, 3 are given; other points are computed by the algorithm. The required result is a list of visible segments, each segment described by its endpoints and by the identifier of the line of which it is a part:

(P1, P3, 1), (P3, P4, 2), (P5, P6, 3), (P6, P8, 2), (P7, P9, 3), (P9, P10, 1), (P11, P12, 3)

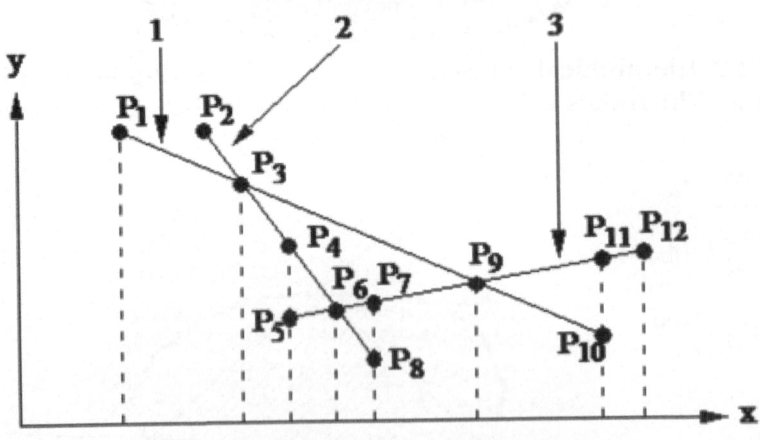

Exhibit 24.11: Example: Three line segments seen from below generate seven visible subsegments.

In search of algorithms, the reader is encouraged to work out the details of the first idea that might come to mind: For each of the n2 ordered pairs (Li, Lj) of line segments, remove from Li the subsegment occluded by Lj. Because Li can get cut into as many as n pieces, it must be managed as a sequence of subsegments. Finding the endpoints of Lj in this sequence will take time O(log n), leading to an overall algorithm of time complexity O(n2 · log n).

After the reader has mastered the sweep algorithm for line intersection presented in "Plane-sweep: a general- purpose algorithm for two-dimensional problems illustrated using line segment intersection", he will see that its straightforward application to the line visibility problem requires time O((n + k) · log n), where k ⍰ O(n2) is the number of intersections. Thus plane-sweep appears to do all the work the brute-force algorithm above does, organized in a systematic left-to-right fashion. It keeps track of all intersections, most of which may be invisible. It has the potential to work in time O(n · log n) for many realistic data configurations characterized by k ∈ O(n), but not in the worst case.

Divide-and-conquer yields a simple two-dimensional visibility algorithm with a better worst-case performance. If n = 0 or 1, the problem is trivial. If n > 1, partition the set of n line segments into two (approximate) halves, solve both subproblems, and merge the results. There is no constraint on how the set is halved, so the divide step is easy. The conquer step is taken care of by recursion. Merging amounts to computing the minimum of two piecewise (not necessarily continuous) linear functions, in time linear in the number of pieces. The example with n = 4 shown in Exhibit 24.12 illustrates the algorithm. f12 is the visible front of segments 1 and 2, f34 of segments 3 and 4, min(f12, f34) of all four segments (Exhibit 24.13).

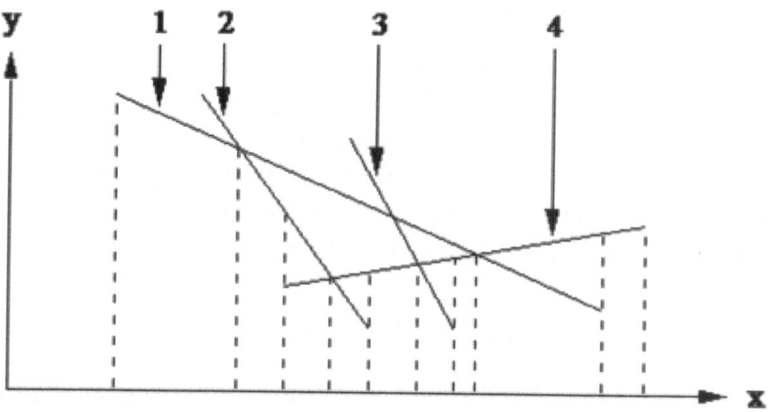

Exhibit 24.12: The four line segments will be partitioned into subsets {1, 2} and {3, 4}.

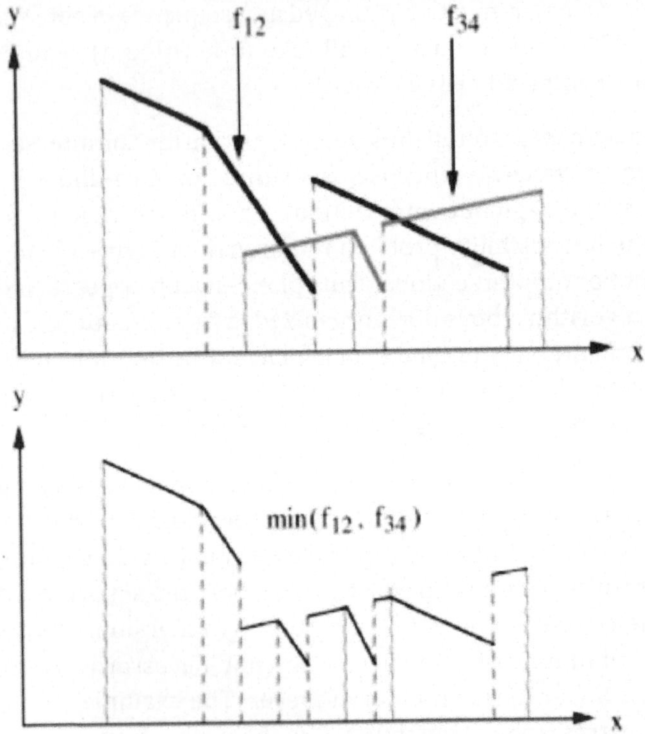

Exhibit 24.13: The min operation merges the solutions of this divide-and-conquer algorithm.

The time complexity of this divide-and-conquer algorithm is obtained as follows. Given that at each level of recursion the relevant sets of line segments can be partitioned into (approximate) halves, the depth of recursion is O(log n). A merge step that processes v visible subsegments takes linear time O(v). Together, all the merge steps at a given depth process at most V subsegments, where V is the total number of visible subsegments. Thus the total time is bounded by O(V · log n). How large can V be?

Surprising theoretical results

Let V(n) be the number of visible subsegments in a given configuration of n lines, i.e. the size of the output of the visibility computation. For tiny n, the worst cases [V(2) = 4, V(3) = 8] are shown in Exhibit 24.14. An attempt to find worst-case configurations for general n leads to examples such as that shown in Figure 24.15, with V(n) = 5·n – 8.

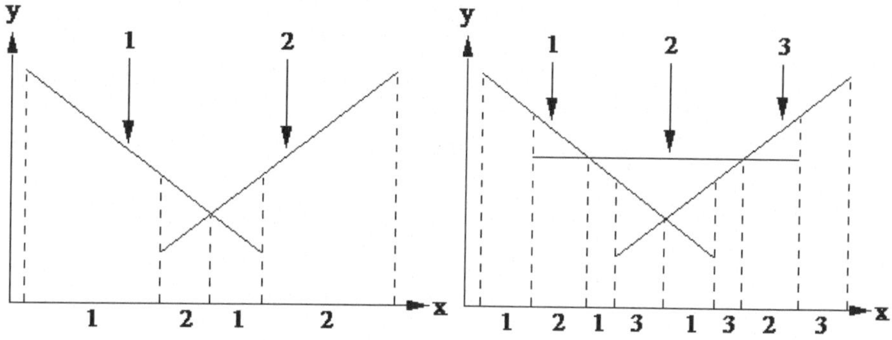

Exhibit 24.14: Configurations with the largest number of visible subsegments.

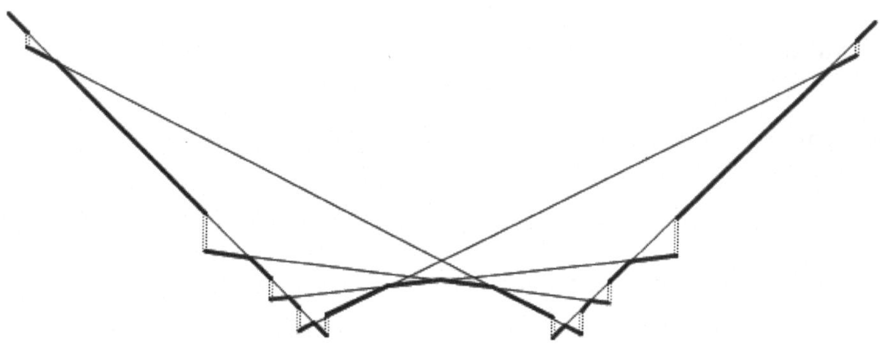

Figure 24.15: A family of configurations with 5·n – 8 visible subsegments.

You will find it difficult to come up with a class of configurations for which V(n) grows faster. It is tempting to conjecture that V(n) ∈ O(n), but this conjecture is very hard to prove - for the good reason that it is false, as was discovered in [WS 88]. It turns out that $V(n) \in \Theta(n \cdot \alpha(n))$, where $\alpha(n)$, the inverse of Ackermann's function (see "Computability and complexity", Exercise 2), is a monotonically increasing function that grows so slowly that for practical purposes it can be treated as a constant, call it α.

Let us present some of the steps of how this surprising result was arrived at. Occasionally, simple geometric problems can be tied to deep results in other branches of mathematics. We transform the two-dimensional visibility problem into a combinatorial string problem. By numbering the given line segments, walking along the x-axis from left to right, and writing down the number of the line segment that is currently visible, we obtain a sequence of numbers (Exhibit 24.16).

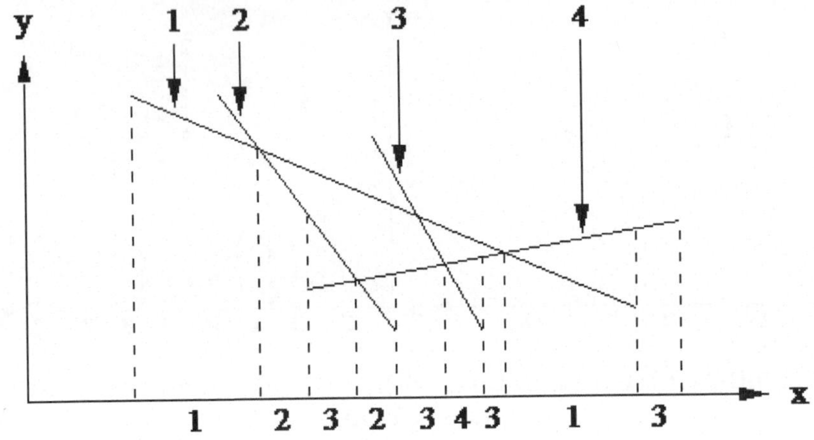

Exhibit 24.16: The Davenport-Schinzel sequence associated with a configuration of segments.

A geometric configuration gives rise to a sequence $u_1, u_2, \ldots, u_m$ with the following properties:

- $1 \le u_i \le n$ for $1 \le i \le m$ (numbers identify line segments).
- $u_i \ne u_{i+1}$ for $1 \le i \le m - 1$ (no two consecutive numbers are equal).
- There are no five indices $1 \le a < b < c < d < e \le m$ such that $u_a = u_c = u_e = r$ and $u_b = u_d = s$, $r \ne s$. This condition captures the geometric properties of two intersecting straight lines: If we ever see r, s, r, s (possibly separated), we will never see r again, as this would imply that r and s intersect more than once (Exhibit 24.17).

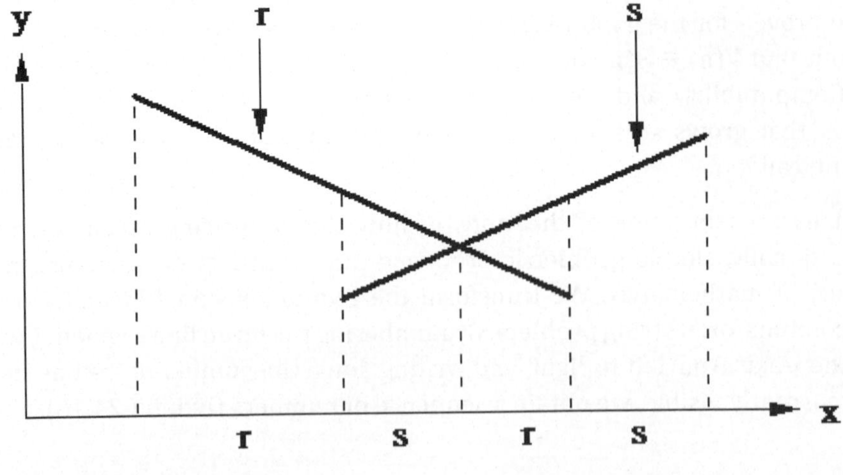

Exhibit 24.17: The subsequence r, s, r, s excludes further occurrences of r.

Example

The sequence for the example above that shows $m \geq 5n - 8$ is

$1, 2, 1, 3, 1, \dots, 1, n-1, 1, n-1, n-2, n-3, \dots, 3, 2, n, 2, n, 3, n, \dots, n, n-2, n, n-1, n.$

Sequences with the properties 1 to 3, called *Davenport-Schinzel sequences*, have been studied in the context of linear differential equations. The maximal length of a Davenport-Schinzel sequence is $k \cdot n \cdot \alpha(n)$, where k is a constant and $\alpha(n)$ is the inverse of Ackermann's function (see "Computability and complexity", Exercise 2) [HS 86]. With increasing n, $\alpha(n)$ approaches infinity, albeit very slowly. This dampens the hope for a linear upper bound for the visibility problem, but does not yet disprove the conjecture. For the latter, we need an inverse: For any given Davenport-Schinzel sequence there exists a corresponding geometric configuration which yields this sequence. An explicit construction is given in [WS 88]. This establishes an isomorphism between the two-dimensional visibility problem and the Davenport-Schinzel sequences, and shows that the size of the output of the two-dimensional visibility problem can be superlinear - a result that challenges our geometric intuition.

Exercises

- Given a set of points S, prove that the pair of points farthest from each other must be vertices of the convex hull H(S).
- Assume a model of computation in which the operations addition, multiplication, and comparison are available at unit cost. Prove that in such a model $\Omega(n \cdot \log n)$ is a lower bound for computing, in order, the vertices of the convex hull H(S) of a set S of n points. *Hint:* Show that every algorithm which computes the convex hull of n given points can be used to sort n numbers.
- Complete the second algorithm for the point-in-polygon test in chapter 24 in the section "The uses of convexity: basic operations on polygons" which computes the crossing number of the polygon G around point P by addressing the special cases that arise when the semi-infinite ray R emanating from P intersects a vertex of G or overlaps an edge of G.
- Consider an arbitrary (not necessarily simple) polygon G (Exhibit 24.18). Provide an interpretation for the winding number w(G, P) of G around an arbitrary point P not on G, and prove that $w(G, P) / 2 \cdot \pi$ of P is always equal to the crossing number of P with respect to any ray R emanating from P.

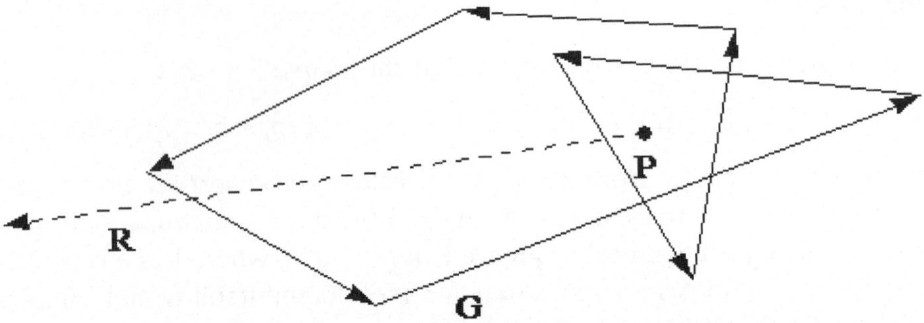

Exhibit 24.18: Winding number and crossing number of a polygon G with respect to P.

- Design an algorithm that computes the area of an n-vertex simple, but not necessarily convex polygon in $\Theta(n)$ time.
- We consider the problem of computing the intersection of two convex polygons which are given by their lists of vertices in cyclic order.
- Show that the intersection is again a convex polygon.
- Design an algorithm that computes the intersection. What is the time complexity of your algorithm?
- *Intersection test for line L and [convex] polygon Q* If an (infinitely extended) line L intersects a polygon Q, it must intersect one of Q's edges. Thus a test for intersection of a given line L with a polygon can be reduced to repeated test of L for intersection with [some of] Q's edges.
- Prove that, in general, a test for line-polygon intersection must check at least n – 2 of Q's edges. *Hint:* Use an adversary argument. If two edges remain unchecked, they could be moved so as to invalidate the answer.
- Design a test that works in time O(log n) for decoding whether a line L intersects a convex polygon Q.
- Divide-and-conquer algorithms may divide the space in which the data is embedded, rather than the set of data (the set of lines). Describe an algorithm for computing the sequence of visible segments that partitions the space recursively into vertical stripes, until each stripe is "simple enough"; describe how you choose the boundaries of the stripes; state advantages and disadvantages of this algorithm as compared to the one described in chapter 24 in the section "Visibility in the plane: a simple algorithm whose analysis is not". Analyze the asymptotic time complexity of this algorithm.

25

Plane-sweep: A General-Purpose Algorithm for two-Dimensional Problems Illustrated Using Line Segment Intersection

Plane-sweep is an algorithm schema for two-dimensional geometry of great generality and effectiveness, and algorithm designers are well advised to try it first. It works for a surprisingly large set of problems, and when it works, tends to be very efficient. Plane-sweep is easiest to understand under the assumption of nondegenerate configurations. After explaining plane-sweep under this assumption, we remark on how degenerate cases can be handled with plane-sweep.

The line segment intersection test

We present a plane-sweep algorithm [SH 76] for the *line segment intersection test*:

```
  Given n line segments in the plane, determine whether any two
intersect;
  and if so, compute a witness (i.e. a pair of segments that
intersect).
```

Bounds on the complexity of this problem are easily obtained. The literature on computational geometry (e.g. [PS 85]) proves a lower bound $\Omega(n \cdot \log n)$. The obvious brute force approach of testing all $n \cdot (n - 1) / 2$ pairs of line segments requires $\Theta(n2)$ time. This wide gap between $n \cdot \log n$ and $n2$ is a challenge to the algorithm designer, who strives for an optimal algorithm whose asymptotic running time $O(n \cdot \log n)$ matches the lower bound.

Divide-and-conquer is often the first attempt to design an algorithm, and it comes in two variants illustrated in Fig. 25.1: (1) Divide the data, in this case the set of line segments, into two subsets of approximately equal size (i.e. n / 2 line segments), or (2) divide the embedding space, which is easily cut in exact halves.

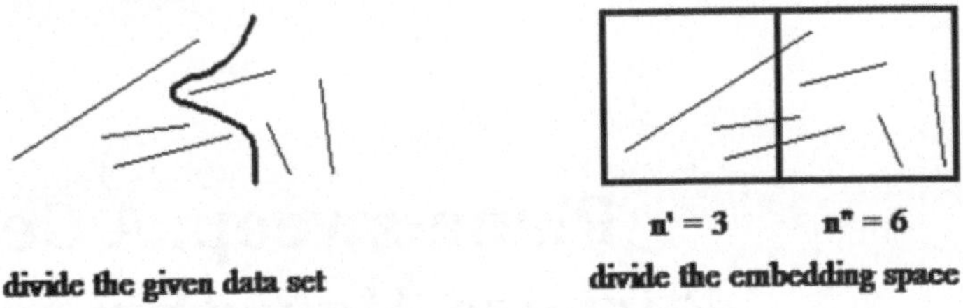

Exhibit 25.1: Two ways of applying divide-and-conquer to a set of objects embedded in the plane.

In the first case, we hope for a separation into subsets S1 and S2 that permits an efficient test whether any line segment in S1 intersects some line segment in S2. Exhibit 25.1 shows the ideal case where S1 and S2 do not interact, but of course this cannot always be achieved in a nontrivial way; and even if S can be separated as the figure suggests, finding such a separating line looks like a more formidable problem than the original intersection problem. Thus, in general, we have to test each line segment in S1 against every line segment in S2, a test that may take Θ(n2) time.

The second approach of dividing the embedding space has the unfortunate consequence of effectively increasing our data set. Every segment that straddles the dividing line gets "cut" (i.e. processed twice, once for each half space). The two resulting subproblems will be of size n' and n", respectively, with n' + n" > n, in the worst case n' + n" = 2 · n. At recursion depth d we may have 2d · n subsegments to process. No optimal algorithm is known that uses this technique.

The key idea in designing an optimal algorithm is the observation that those line segments that intersect a vertical line L at abscissa x are totally ordered: A segment s lies below segment t, written s <L t, if both intersect L at the current position x and the intersection of s with L lies below the intersection of t with L. With respect to this order a line segment may have an upper and a lower neighbor, and Exhibit 25.2 shows that s and t are neighbors at x.

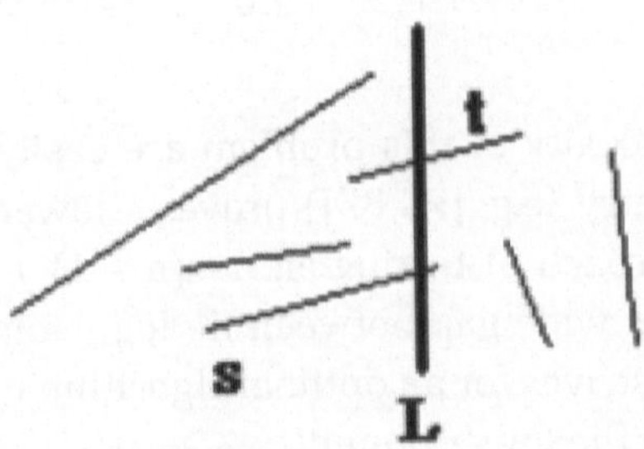

Exhibit 25.2: The sweep line L totally orders the segments that intersect L.

We describe the intersection test algorithm under the assumption that the configuration is nondegenerate (i.e. no three segments intersect in the same point). For simplicity's sake we also assume that no segment is vertical, so every segment has a left endpoint and a right endpoint. The latter assumption entails no loss of generality: For a vertical segment, we can arbitrarily define the lower endpoint to be the "left endpoint", thus imposing a lexicographic (x, y)-order to refine the x-order. With the important assumption of non-degeneracy, two line segments s and t can intersect at x0 only if there exists an abscissa x < x0 where s and t are neighbors. Thus it suffices to test all segment pairs that become neighbors at some time during a left-to-right sweep of L - a number that is usually significantly smaller than n · (n – 1) / 2.

As the sweep line L moves from left to right across the configuration, the order <L among the line segments intersecting L changes only at endpoints of a segment or at intersections of segments. As we intend to stop the sweep as soon as we discover an intersection, we need to perform the intersection test only at the left and right endpoints of segments. A segment t is tested at its left endpoint for intersection with its lower and upper neighbors. At the right endpoint of t we test its lower and upper neighbor for intersection (Exhibit 25.3).

The algorithm terminates as soon as we discover an intersecting pair of segments. Given n segments, each of

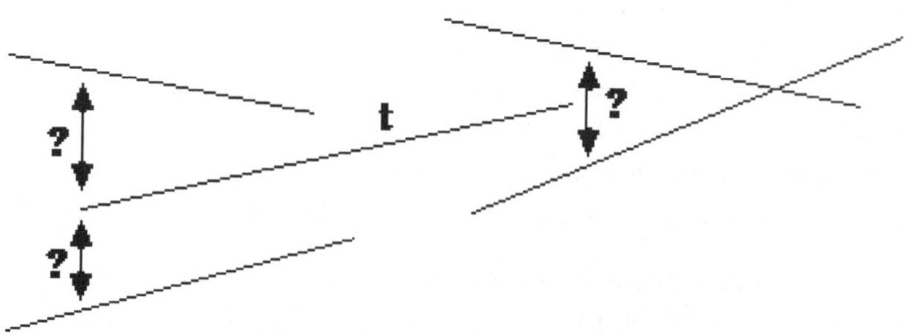

Exhibit 25.3: Three pairwise intersection tests charged to segment t.

which may generate three intersection tests as shown in Exhibit 25.3 (two at its left, one at its right endpoint), we perform the O(1) pairwise segment intersection test at most 3 · n times. This linear bound on the number of pairs tested for intersection might raise the hope of finding a linear-time algorithm, but so far we have counted only the geometric primitive: "Does a pair of segments intersect - yes or no?" Hiding in the background we find bookkeeping operations such as "Find the upper and lower neighbor of a given segment", and these turn out to be costlier than the geometric ones. We will find neighbors efficiently by maintaining the order <L in a data structure called a y- table during the entire sweep.

The skeleton: Turning a space dimension into a time dimension

The name *plane-sweep* is derived from the image of sweeping the plane from left to right with a vertical line (front, or cross section), stopping at every transition point (event) of a geometric configuration to update the cross section. All processing is done at this moving front, without any backtracking, with a look-ahead of only one point. The events are stored in the x-queue, and the current cross section is maintained by the y-table. The skeleton of a plane-sweep algorithm is as follows:

```
initX;  initY;
while  not emptyX  do  { e := nextX;  transition(e) }
```

The procedures 'initX' and 'initY' initialize the x-queue and the y-table. 'nextX' returns the next event in the x- queue, 'emptyX' tells us whether the x-queue is empty. The procedure 'transition', the advancing mechanism of the sweep, embodies all the work to be done when a new event is encountered; it moves the front from the slice to the left of an event e to the slice immediately to the right of e.

Data structures

For the line segment intersection test, the x-queue stores the left and right endpoints of the given line segments, ordered by their x-coordinate, as events to be processed when updating the vertical cross section. Each endpoint stores a reference to the corresponding line segment. We compare points by their x-coordinates when building the x-queue. For simplicity of presentation we assume that no two endpoints of line segments have equal x- or y- coordinates. The only operation to be performed on the x-queue is 'nextX': it returns the next event (i.e. the next left or right endpoint of a line segment to be processed). The cost for initializing the x-queue is O(n · log n), the cost for performing the 'nextX' operation is O(1).

The y-table contains those line segments that are currently intersected by the sweep line, ordered according to <L. In the slice between two events, this order does not change, and the y-table needs no updating (Exhibit 25.4). The y-table is a dictionary that supports the operations 'insertY', 'deleteY', 'succY', and 'predY'. When entering the left endpoint of a line segment s we find the place where s is to be inserted in the ordering of the y-table by comparing s to other line segments t already stored in the y-table. We can determine whether s <L t or t <L s by determining on which side of t the left endpoint of s lies. As we have seen in chapter 14 in the section "Intersection", this tends to be more efficient than computing and comparing the intersection points of s and t with the sweep line. If we implement the dictionary as a balanced tree (e.g. an AVL tree), the operations 'insertY' and 'deleteY' are performed in O(log n) time, and 'succY' and 'predY' are performed in O(1) time if additional pointers in each node of the tree point to the successor and predecessor of the line segment stored in this node. Since there are 2 · n events in

the x-queue and at most n line segments in the y-table the space complexity of this plane-sweep algorithm is O(n).

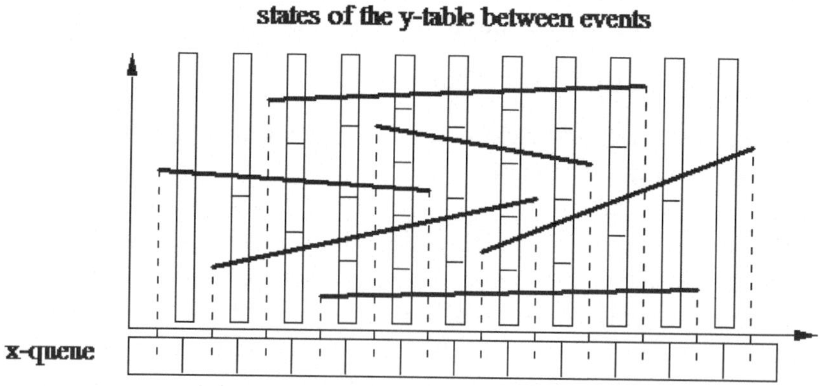

Exhibit 25.4: The y-table records the varying state of the sweep line L.

Updating the y-table and detecting an intersection

The procedure 'transition' maintains the order <L of the line segments intersecting the sweep line and performs intersection tests. At a left endpoint of a segment t, t is inserted into the y-table and tested for intersection with its lower and upper neighbors. At the right endpoint of t, t is deleted from the y-table and its two former neighbors are tested. The algorithm terminates when an intersection has been found or all events in the x-queue have been processed without finding an intersection:

```
procedure transition(e: event);
begin
   s := segment(e);
   if  leftPoint(e)  then  begin
      insertY(s);
   if  intersect(predY(s), s) or intersect (s, succY(s))  then
      terminate('intersection found')
   end
   else  { e is right endpoint of s }  begin
      if  intersect(predY(s), succY(s))  then
         terminate('intersection found');
      deleteY(s)
   end
end;
```

With at most 2 · n events, and a call of 'transition' costing time O(log n), this plane-sweep algorithm needs O(n · log n) time to perform the line segment intersection test.

Sweeping across intersections

The plane-sweep algorithm for the line segment intersection test is easily adapted to the following more general problem [BO 79]:

Given n line segments, report all intersections.

In addition to the left and right endpoints, the x-queue now stores intersection points as events—any intersection detected is inserted into the x-queue as an event to be processed. When the sweep line reaches an intersection event the two participating line segments are swapped in the y-table (Exhibit 25.5). The major increase in complexity as compared to the segment intersection test is that now we must process not only 2 · n events, but 2 · n + k events, where k is the number of intersections discovered as we sweep the plane. A configuration with n / 2 segments vertical and n / 2 horizontal shows that, in the worst case, k ∈ Θ(n2), which leads to an O(n2 · log n) algorithm, certainly no improvement over the brute-force comparison of all pairs. In most realistic configurations, say engineering drawings, the number of intersections is much less than O(n2), and thus it is informative to introduce the parameter k in order to get an output-sensitive bound on the complexity of this algorithm (i.e. a bound that adapts to the amount of data needed to report the result of the computation).

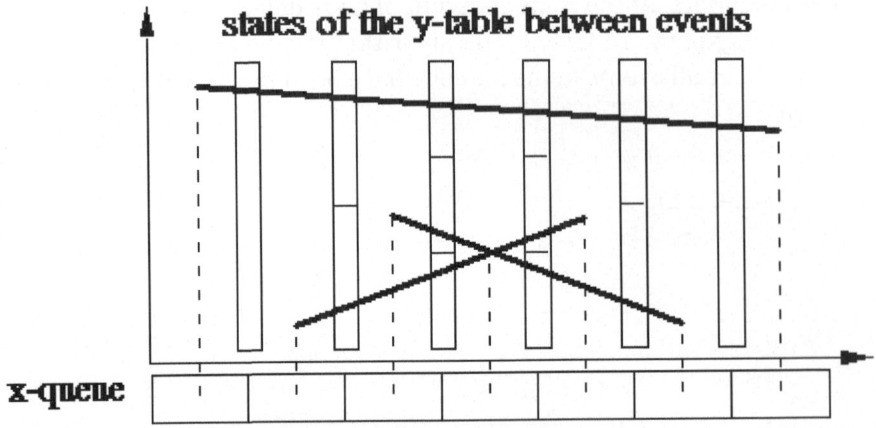

Exhibit 25.5: Sweeping across an intersection.

Other changes are comparatively minor. The x-queue must be a priority queue that supports the operation 'insertX'; it can be implemented as a heap. The cost for initializing the x-queue remains O(n · log n). Without further analysis one might presume that the storage requirement of the x-queue is O(n + k), which implies that the cost for calling 'insertX' and 'nextX' remains O(log n), since k ∈ O(n2). A more detailed analysis [PS 91], however, shows that the size of the x-queue never exceeds O(n · (log n)2). With a slight modification of the algorithm [Bro 81] it can even be guaranteed that the size of the x-queue never exceeds O(n). The cost for exchanging two intersecting line segments in the y-table is O(log n), the costs for the other operations on the y-table remain the same. Since there are 2 · n left and right endpoints and k intersection events, the total cost for this algorithm is O((n + k) · log n). As most realistic applications are characterized by k ∈ O(n), reporting all intersections often remains an O(n · log n) algorithm in practice. A time-optimal algorithm that finds all int ersecting pairs of line segments in O(n · log n + k) time using O(n + k) storage space is described in [CE 92].

Degenerate configurations, numerical errors, robustness

The discussion above is based on several assumptions of nondegeneracy, some of minor and some of major importance. Let us examine one of each type.

Whenever we access the x-queue ('nextX'), we used an implicit assumption that no two events (endpoints or intersections) have equal x-coordinates. The order of processing events of equal x-coordinate is irrelevant. Assuming that no two events coincide at the same point in the plane, lexicographic (x, y)-ordering is a convenient systematic way to define 'nextX'.

More serious forms of degeneracy arise when events coincide in the plane, such as more than two segments intersecting in the same point. This type of degeneracy is particularly difficult to handle in the presence of numerical errors, such as rounding errors. In the configuration shown in Exhibit 25.6 an endpoint of u lies exactly or nearly on segment s. We may not care whether the intersection routine answers 'yes' or 'no' to the question "Do s and u intersect?" but we certainly expect a 'yes' when asking "Do t and u intersect?" This example shows that the slightest numerical inaccuracy can cause a serious error: The algorithm may fail to report the intersection of t and u, which it would clearly see if it bothered to look - but the algorithm looks the other way and never asks the question "Do t and u intersect?"

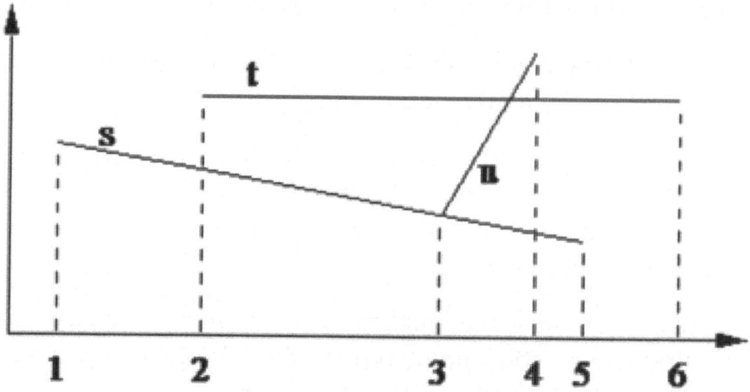

Exhibit 25.6: A degenerate configuration may lead to inconsistent results.

The trace of the plane-sweep for reporting intersections may look as follows:

- s is inserted into the y-table
- t is inserted above s into the y-table, and s and t are tested for intersection: No intersection is found
- u is inserted below s in the y-table (since the evaluation of the function s(x) may conclude that the left endpoint of u lies below s); s and u are tested for intersection, but the intersection routine may conclude that s and u do not intersect: u remains below s
- Delete u from the y-table
- Delete s from the y-table
- Delete t from the y-table

Notice the calamity that struck at the critical step 3. The evaluation of a linear expression s(x) and the intersection routine for two segments both arrived at a result that, in isolation, is reasonable within the tolerance of the underlying arithmetic. The two results together are inconsistent! If the evaluation of s(x) concludes that the left endpoint of u lies below s, the intersection routine *must* conclude that s and u intersect! If these two geometric primitives fail to coordinate their answers, catastrophe may strike. In our example, u and t never become neighbors in the y-table, so their intersection gets lost.

Exercises

- Show that there may be Θ(n2) intersections in a set of n line segments.
- Design a plane-sweep algorithm that determines in O(n · log n) time whether two simple polygons with a total of n vertices intersect.
- Design a plane-sweep algorithm that determines in O(n · log n) time whether any two disks in a set of n disks intersect.
- Design a plane-sweep algorithm that solves the line visibility problem discussed in chapter 24 in the section "Visibility in the plane: a simple algorithm whose analysis is not" in time O((n + k) · log n), where k ▯ O(n2) is the number of intersections of the line segments.
- Give a configuration with the smallest possible number of line segments for which the first intersection point reported by the plane-sweep algorithm in chapter 25 in the section "Sweeping across intersections" is not the leftmost intersection point.
- Adapt the plane-sweep algorithm presented in chapter 25 in the section "Sweeping across intersections" to detect all intersections among a given set of

n horizontal or vertical line segments. You may assume that the line segments do not overlap. What is the time complexity of this algorithm if the horizontal and vertical line segments intersect in k points?

- Design a plane-sweep algorithm that finds all intersections among a given set of n rectangles all of whose sides are parallel to the coordinate axes. What is the time complexity of your algo rithm?

26

The Closest Pair

The problem

We consider the two-dimensional *closest pair problem*: Given a set S of n points in the plane find a pair of points whose distance δ is smallest (Exhibit 26.1). We measure distance using the metric dk, for any k ≥ 1, or d∞, defined as:

$$d_k((x',y'),(x'',y'')) = \sqrt[k]{|x'-x''|^k + |y'-y''|^k}$$
$$d_\infty((x',y'),(x'',y'')) = \max(|x'-x''|, |y'-y''|)$$

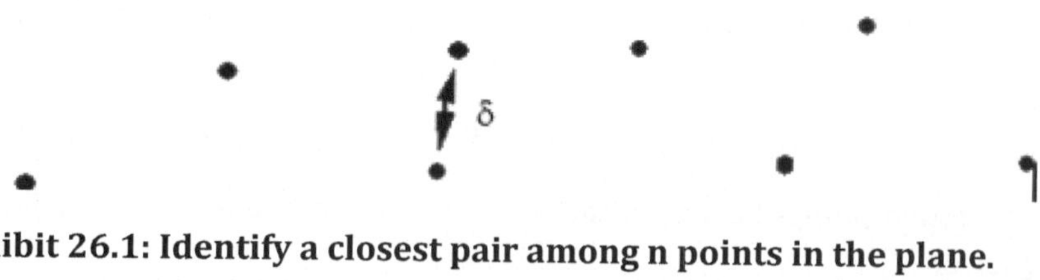

Exhibit 26.1: Identify a closest pair among n points in the plane.

Special cases of interest include the "Manhattan metric" d1, the "Euclidean metric" d2, and the "maximum metric" d∞. Exhibit 26.2 shows the "circles" of radius 1 centered at a point p for some of these metrics.

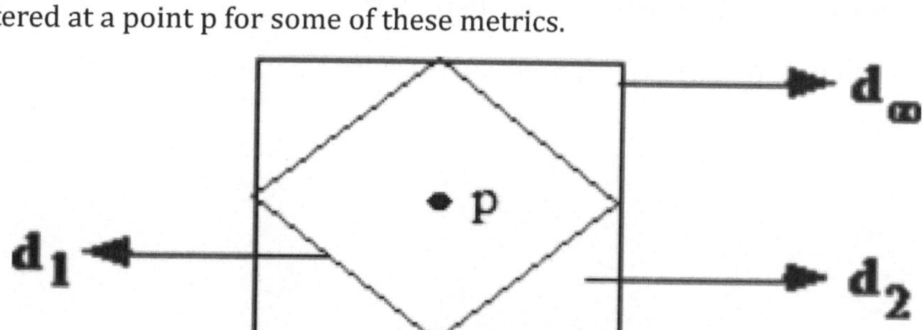

Exhibit 26.2: The results of this chapter remain valid when distances are measured in various metrics.

The closest pair problem has a lower bound $\Omega(n \cdot \log n)$ in the algebraic decision tree model of computation [PS 85]. Its solution can be obtained in asymptotically optimal time $O(n \cdot \log n)$ as a special case of more general problems, such as 'all-nearest-neighbors' [HNS 92] (for each point, find a nearest neighbor), or constructing the Voronoi diagram [SH 75]. These general approaches call on powerful techniques that make the resulting algorithms harder to understand than one would expect for a simply stated problem such as "find a closest pair". The divide- and-conquer algorithm presented in [BS 76] solves the closest pair problem directly in optimal worst-case time complexity $\Theta(n \cdot \log n)$ using the Euclidean metric d2. Whereas the recursive divide-and-conquer algorithm involves an intricate argument for combining the solutions of two equally sized subsets, the iterative plane-sweep algorithm [HNS 88] uses a simple incremental update: Starting with the empty set of points, keep adding a single point until the final solution for the entire set is obtained. A similar plane-sweep algorithm solves the closest pair problem for a set of convex objects [BH 92].

Plane-sweep applied to the closest pair problem

The skeleton of the general sweep algorithm presented in chapter 25 in the section "The skeleton: turning a space dimension into a time dimension", with the data structures x-queue and y-table, is adapted to the closest pair problem as shown in Exhibit 26.3. The *x-queue* stores the points of the set S, ordered by their x-coordinate, as events to be processed when updating the vertical cross section. Two pointers into the x-queue, 'tail' and 'current', partition S into four disjoint subsets:

- The discarded points to the left of 'tail' are not accessed any longer
- The active points between 'tail' (inclusive) and 'current' (exclusive) are being queried
- The current transition point, p, is being processed
- The future points have not yet been looked at
- The y-table stores the active points only, ordered by their y-coordinate.

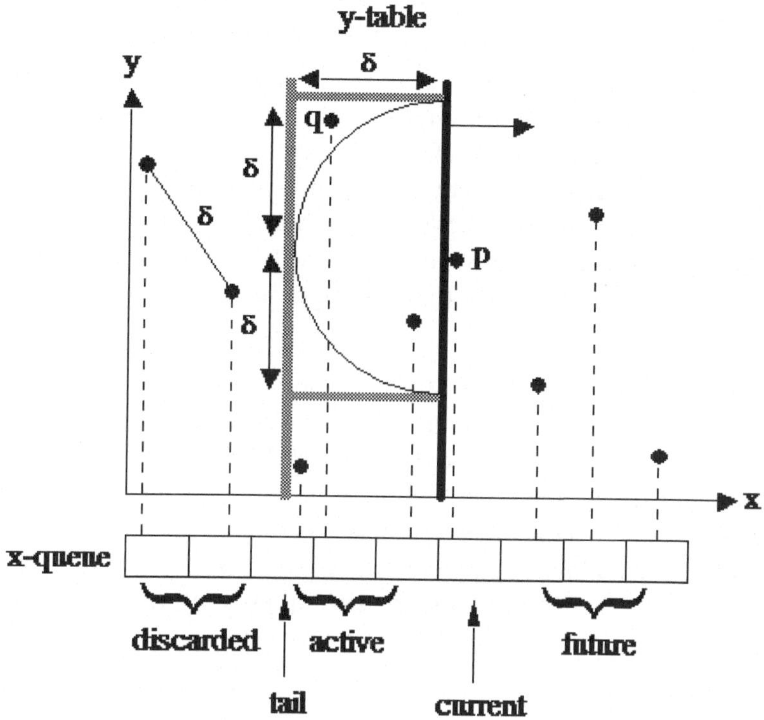

Exhibit 26.3: Updating the invariant as the next point p is processed.

We need to compare points by their x-coordinates when building the x-queue, and by their y-coordinates while sweeping. For simplicity of presentation we assume that no two points have equal x- or y-coordinates. Points with equal x- or y-coordinates are handled by imposing an arbitrary, but consistent, total order on the set of points. We achieve this by defining two lexicographic orders: <x to be used for the x-queue, <y for the y-table:

$$(x', y') <_x :\Leftrightarrow (x' < x'') \vee ((x' = x'') \wedge (y' < y''))$$
$$(x', y') <_y :\Leftrightarrow (y' < y'') \vee ((y' = y'') \wedge ('x < x''))$$

The program of the following section initializes the x-queue and y-table with the two leftmost points being active, with δ equal to their distance, and starts the sweep with the third point.

The distinction between discarded and active points is motivated by the following argument. When a new point p is encountered we wish to answer the question whether this point forms a closest pair with one of the points to its left. We keep a pair of closest points seen so far, along with the corresponding minimal distance δ. Therefore, all candidates that may form a new closest pair with the point p on the sweep line lie in a half circle centered at p, with radius δ.

The key question to be answered in striving for efficiency is how to retrieve quickly all the points seen so far that lie inside this half circle to the left of p, in order to compare their distance to p against the minimal distance δ seen so far. We may use any helpful data structure that organizes the points seen so far, as long as we can update this data structure efficiently across a transition. A circle (or half-circle) query is complex, at least when embedded in a plane-sweep algorithm that organizes data according to an orthogonal coordinate system. A rectangle query can be answered more efficiently. Thus we replace the half-circle query with a bounding rectangle query, accepting the fact that we might include some extraneous points, such as q.

The rectangle query in Exhibit 26.3 is implemented in two steps. First, we cut off all the points to the left at distance ≥ δ from the sweep line. These points lie between 'tail' and 'current' in the x-queue and can be discarded easily by advancing 'tail' and removing them from the y-table. Second, we consider only those points q in the δ-slice whose vertical distance from p is less than δ: $|qy - py| < \delta$. These points can be found in the y-table by looking at successors and predecessors starting at the y-coordinate of p. In other words, we maintain the following invariant across a transition:

- δ is the minimal distance between a pair of points seen so far (discarded or active).
- The active points (found in the x-queue between 'tail' and 'current', and stored in the y-table ordered by y- coordinates) are exactly those that lie in the interior of a δ-slice to the left of the sweep line.
- Therefore, processing the transition point p involves three steps:
- Delete all points q with $qx \leq px - \delta$from the y-table. They are found by advancing 'tail' to the right.
- Insert p into the y-table.
- Find all points q in the y-table with $|qy - py| < \delta$ by looking at the successors and predecessors of p. If such a point q is found and its distance from p is smaller than δ, update δ and the closest pair found so far.

Implementation

In the following implementation the x-queue is realized by an array that contains all the points sorted by their x- coordinate. 'closestLeft' and 'closestRight' describe the pair of closest points found so far, n is the number of points under consideration, and t and c determine the positions of 'tail' and 'current':

```
xQueue: array[1 .. maxN] of point;
closestLeft, closestRight: point;
t, c, n: 1 .. maxN;
```

```
The x-queue is initialized by
  procedure initX;
'initX' stores all the points into the x-queue, ordered by their x-
coordinates.

The empty y-table is created by
  procedure initY;

A new point is inserted into the y-table by
  procedure insertY(p: point);

A point is deleted from the y-table by
  procedure deleteY(p: point);

The successor of a point in the y-table is returned by
  function succY(p: point): point;

The predecessor of a point in the y-table is returned by
  function predY(p: point): point;
```

The initialization part of the plane-sweep is as follows:

```
initX;  initY;
closestLeft := xQueue[1];  closestRight := xQueue[2];
delta := distance(closestLeft, closestRight);
insertY(closestLeft);  insertY(closestRight);
c := 3;
```

The events are processed by the loop:

```
  while c ≤ n do  begin  transition;  c := c + 1;  { next event }
end;
```

The procedure 'transition' encompasses all the work to be done for a new point:

```
procedure transition;
begin

  { step 1: remove points outside the δ-slice from the y-table }
  current := xQueue[c];
  while  current.x - xQueue[t].x ≥ delta  do  begin
    deleteY(xQueue[t]);  t := t + 1
  end;

  { step 2: insert the new point into the y-table }
  insertY(current);

  { step 3a: check the successors of the new point in the y-table }
  check := current;
  repeat
    check := succY(check);
    newDelta := distance(current, check);
    if  newDelta < delta  then  begin
      delta := newDelta;
      closestLeft := check;  closestRight := current;
    end;
   until  check.y - current.y > delta;

   { step 3b: check the predecessors of the new point in the y-
table }
   check := current;
   repeat
     check := predY(check);
     newDelta := distance(current, check);
```

```
        if  newDelta < delta  then  begin
          delta := newDelta;
          closestLeft := check;  closestRight := current;
        end;
      until  current.y - check.y > delta;
    end;  { transition }
```

Analysis

We show that the algorithm described can be implemented so as to run in worst-case time O(n · log n) and space O(n).

If the y-table is implemented by a balanced binary tree (e.g. an AVL-tree or a 2-3-tree) the operations 'insertY', 'deleteY', 'succY', and 'predY' can be performed in time O(log n). The space required is O(n).

'initX' builds the sorted x-queue in time O(n · log n) using space O(n). The procedure 'deleteY' is called at most once for each point and thus accumulates to O(n · log n). Every point is inserted once into the y-table, thus the calls of 'insertY' accumulate to O(n · log n).

There remains the problem of analyzing step 3. The loop in step 3a calls 'succY' once more than the number of points in the upper half of the bounding box. Similarly, the loop in step 3b calls 'predY' once more than the number of points in the lower half of the bounding box. A standard counting technique shows that the bounding box is sparsely populated: For any metric dk, the box contains no more than a small, constant number ck of points, and for any k, ck ≤ 8. Thus 'succY' and 'predY' are called no more than 10 times, and step 3 costs time O(log n).

The key to this counting is the fact that no two points in the y-table can be closer than δ, and thus not many of them can be packed into the bounding box with sides δ and 2 · δ. We partition this box into the eight pairwise disjoint, mutually exhaustive regions shown in Exhibit 26.4. These regions are half circles of diameter δ in the Manhattan metric d1, and we first argue our case only when distances are measured in this metric. None of these half-circles can contain more than one point. If a half-circle contained two points at distance δ, they would have to be at opposite ends of the unique diameter of this half-circle. These endpoints lie on the left or the right boundary of the bounding box, and these two boundary lines cannot contain any points, for the following reasons:

- No active point can be located on the left boundary of the bounding box; such a point would have been thrown out when the δ-slice was last updated.
- No active point can exist on the right boundary, as that x-coordinate is preempted by the transition point p being processed (remember our assumption of unequal x-coordinates).

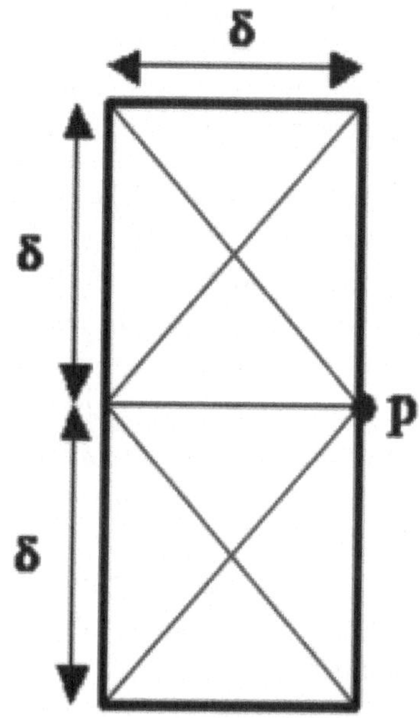

Exhibit 26.4: Only few points at pairwise distance ≥ δ can populate a box of size 2 · δ by δ.

We have shown that the bounding box can hold no more than eight points at pairwise distance ≥ δ when using the Manhattan metric d1. It is well known that for any points p, q, and for any k > 1: d1(p,q) > dk(p,q) > d∞(p,q). Thus the bounding box can hold no more than eight points at pairwise distance ≥ δ when using any distance dk or d∞.

Therefore, the calculation of the predecessors and successors of a transition point costs time O(log n) and accumulates to a total of O(n · log n) for all transitions. Summing up all costs results in O(n · log n) time and O(n) space complexity for this algorithm. Since Ω(n · log n) is a lower bound for the closest pair problem, we know that this algorithm is optimal.

Sweeping in three or more dimensions

To gain insight into the power and limitation of sweep algorithms, let us explore whether the algorithm presented generalizes to higher-dimensional spaces. We illustrate our reasoning for three-dimensional space, but the same conclusion holds for any number of dimensions > 2. All of the following steps generalize easily.

Sort all the points according to their x-coordinate into the x-queue. Sweep space with a y-z plane, and in processing the current transition point p, assume that we know the closest pair among all the points to the left of p, and their distance δ. Then to determine whether p forms a new closest pair, look at all the points inside a half- sphere of radius δ centered at p, extending to the left of p. In the hope of implementing this sphere query efficiently, we enclose this half sphere in a bounding box of side length 2 · δ in the y- and z-dimension, and δ in the x- dimension. Inside this box there can be at most a small, constant number ck of points at pairwise distance ≥ δ when using any distance dk or d∞.

We implement this box query in two steps: (1) by cutting off all the points farther to the left of p than δ, which is done by advancing 'tail' in the x-queue, and (2) by performing a square query among the points currently in the y-z- table (which all lie in the δ-slice to the left of the sweep plane), as shown in Exhibit 26.5. Now we have reached the only place where the three-dimensional algorithm differs substantially. In the two-dimensional case, the corresponding one-dimensional interval query can be implemented efficiently in time O(log n) using find, predecessor, and successor operations on a balanced tree, and using the knowledge that the size of the answer set is bounded by a constant. In the three-dimensional case, the corresponding two-dimensional orthogonal range query cannot in general be answered in time O(log n) (per retrieved point) using any of the known data structures. Straightforward search requires time O(n), resulting in an overall time O(n2) for the space sweep. This is not an interesting result for a problem that admits the trivial O(n2) algorithm of comparing every pair.

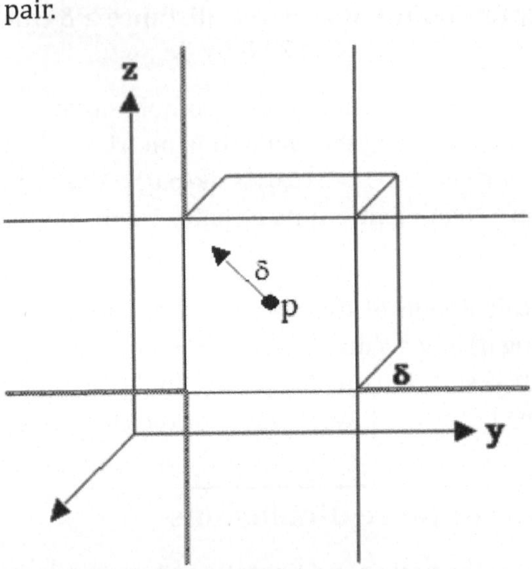

Exhibit 26.5: Sweeping a plane across three-dimensional space. Ideas generalize, but efficiency does not.

Sweeping reduces the dimensionality of a geometric problem by one, by replacing one space dimension by a "time dimension". Reducing a two-dimensional problem to a sequence of one-dimensional problems is often efficient because the total order defined in one dimension allows logarithmic search times. In contrast, reducing a three-dimensional problem to a sequence of two-dimensional problems rarely results in a gain in efficiency.

Exercises

- Consider the following modification of the plane-sweep algorithm for solving the closest pair problem [BH 92]. When encountering a transition point p do not process all points q in the y-table with $|qy - py| < \delta$, but test only whether the distance of p to its successor or predecessor in the y-table is smaller than δ. When deleting a point q with $qx \leq px - \delta$ from the y-table test whether the successor and predecessor of q in the y- table are closer than δ. If a pair of points with a smaller distance than the current δ is found update δ and the closest pair found so far. Prove that this modified algorithm finds a closest pair What is the time complexity of this algorithm?

- Design a divide-and-conquer algorithm which solves the closest pair problem. What is the time complexity of your algorithm? *Hint:* Partition the set of n points by a vertical line into two subsets of approximately n / 2 points. Solve the closest pair problem recursively for both subsets. In the conquer step you should use the fact that δ is the smallest distance between any pair of points both belonging to the same subset. A point from the left subset can only have a distance smaller than δ to a point in the right subset if both points lie in a $2 \cdot \delta$-slice to the left and to the right of the partitioning line. Therefore, you only have to match points lying in the left δ-slice against points lying in the right δ-slice.

Index

9 789360 842727

Printed at Repro India Ltd.